OCEANOGRAPHY

Upper photograph: A cliff face on the northwest coast of the island of Gotland in the Baltic Sea, showing a fossil reef from the Silurian Period; now partly eroded, it forms a headland through differential erosion, because the reef is a cemented mass compared with the soft muds which surrounded it. The reef core (centre) does not exhibit bedding because it is composed of masses of fossilised corals, sponges and algae; the flank beds (top) dip to the right, composed of fossil debris draped over the reef. This reef grew when the Baltic region was in a tropical climate *c.* 20° S of the equator, 430 million years ago. During the last ice age, this region was buried by an ice sheet, but its current topography is partly due to isostatic rebound following deglaciation. Gotland is a natural laboratory for the study of fossil reefs and has some of the best examples in the world.

Lower photograph: An eroded sea stack (the 'mushroom rock') of Pleistocene limestone, at the Perachora Peninsula, near Corinth, Greece. The stack has been undermined by modern marine dissolution processes because sea level has been relatively stable in this region in recent millennia. The result, a marine notch, is one example of a range of features used to estimate relative sea-level change in the region.

In these two pictures, how many aspects of modern and ancient ocean-related processes can you identify?

OCEANOGRAPHY

An Earth Science Perspective

Steve Kershaw

Department of Geography and Earth Sciences
Brunel University

with contributions from

Andy Cundy

Southampton Oceanography Centre
University of Southampton

First published in 2000 by:
Stanley Thornes (Publishers) Ltd
Delta Place
27 Bath Road
Cheltenham
Glos.
GL53 7TH
United Kingdom

00 01 02 03 / 10 9 8 7 6 5 4 3 2 1

A catalogue record for this book is available from the British Library
ISBN 0 7487 5442 3

Figures 3.8, 3.9, 3.11, 4.6a, 4.6b, 5.9, 6.2, 10.11 by Barking Dog Art.

References to this book should be made as follows:
Kershaw, S. (2000) *Oceanography: An Earth Science Perspective*. Stanley Thornes, Cheltenham.

Typeset by Florence Production Ltd, Stoodleigh, Devon
Printed and bound in Great Britain by Redwood Books, Trowbridge, Wiltshire

CONTENTS

PREFACE

Oceanography (the study of oceans and seas) is traditionally seen as the domain of marine scientists, and although a great deal of knowledge exists about the modern oceans, much of this has not been in the gaze of Earth scientists. On the other hand, oceanography in Earth history (palaeoceanography) is essentially geological and relies on the interpretation of geological material. Much palaeoceanography is published in a wide range of journals, mostly geological, which are usually not read by geographers and marine scientists. So there is a huge body of information about the Earth's oceans and their processes scattered about the literature, with little integration of the range of data on modern and ancient oceans in a form readily accessible to students.

This book aims to redress the imbalance, by drawing together all the key strands of ocean study into a holistic view of ocean processes, ancient and modern. The book therefore is a starting place for an Earth science or geography student to gain a good grasp of ocean processes, and facilitates an understanding of ancient ocean processes for marine scientists. It is aimed at all undergraduates, as well as postgraduates, academics and others in a variety of related and wider disciplines who wish to get a balanced understanding of the issues in oceanography. Furthermore, by demonstrating the array of interlinks between the various aspects of ocean study against the backdrop of change in Earth systems, we hope that an understanding of ancient, recent and modern ocean systems will be valuable to students wishing to appreciate current global environmental issues. For example, changes in global heat distribution are of key concern in the much-publicised global warming debate; these changes appear to be principally mediated by thermohaline circulation within oceans. Also, the fact that the oceans can store 30× more heat than the atmosphere means that small changes in ocean heat distribution have huge atmospheric implications. Ocean systems have been shown to undergo rapid change, with catastrophic effects. We see this with the ENSO changes highlighted in recent years, but there are more sinister features which have emanated from the study of the last glacial phase; for example, current ideas are showing that ice sheets can grow and wane much faster than was previously realised. Change in these systems can take place on decade scales, so that climate changes, naturally involving ocean change, occur within the human lifetime scale.

The arguments presented throughout the book are intended to challenge the reader to inquire about ocean processes, and to read beyond the confines of this text. For that reason, in appropriate parts of the book, sources of information are quoted as much as possible, so that readers can investigate the arguments more deeply, if desired. Also, the text is deliberately controversial in places, by questioning the roots of many aspects of oceanography. Only by questioning will the science of oceanography be pushed forward, so it is hoped that student readers will take note. Although the book includes as many important references as possible, apologies are offered to those authors whose work is under-represented. Therefore we would be pleased to receive comments on any part of this book by interested parties.

ORGANISATION OF THIS BOOK

The book has three parts. Part A considers essential features of the oceans, and is written in such a way that people who have little previous knowledge of physics and chemistry will be able to grasp the key processes, for later application in the book and elsewhere. Part B shows how oceans have developed through the geological past, and demonstrates long- and short-term processes, of great relevance to understanding current and recent ocean change. Part C addresses anthropogenic aspects of the oceans, and draws on the knowledge given in Parts A and B, to aid our understanding of ocean management. Ocean process and change is the key theme throughout the book. Throughout the book, most maps are drawn onto a Mercator projection base, and reaches to both poles, thereby enormously enlarging higher latitude regions, but critically allowing all land and water regions to be displayed. The large overlap of left and right margins of the base map permits an easy appreciation of relationships between all neighbouring land and ocean areas.

ACKNOWLEDGEMENTS

This book evolved from two sources: (a) geology and geography undergraduate courses at Brunel University, and previously at the West London Institute of Higher Education, which merged with Brunel in 1995; and (b) research work at both Brunel and Southampton Universities. The knowledge gained from these sources includes information from a wide range of field sites around the world, studied as part of teaching and research programmes, and we are grateful to the universities for the financial support and encouragement received in course and research development.

Our research has also been supported by a range of funding bodies, including NERC, the Royal Society, the Geological Society of London, the Geologists' Association, the European Community, the Robertson Group, the American Chemical Society and the China National Petroleum Corporation.

We thank colleagues in the Department of Geography and Earth Sciences at Brunel, and also acknowledge the direct and indirect assistance we have received from departments and colleagues in other establishments, as follows.

For Steve Kershaw: Trevor Laverock (North Cestrian Grammar School, Altrincham, Cheshire) introduced geology, and epitomises the adage that a teacher never knows the limit of his influence!; Fred Broadhurst (ex-Manchester) and Robert Riding (Cardiff) kindled an interest in sedimentary geology and palaeontology; Ron West (Kansas), Al Fagerstrom (Oregon) and Frank Brunton (Sudbury, Ontario) discussed many aspects of marine sedimentology; the administrators of the famous Allekvia field station on Gotland, Sweden, who facilitate research there (which has much oceanographic significance), provided a forum for much late-night discussion.

For Andy Cundy: Geoff Millward, Keith Dyer, Mike Rhead and others at Plymouth University for inspiring an interest in all things oceanographic, and Ian Croudace and John Thomson at the Southampton Oceanography Centre for support and advice.

We also thank Ian Francis, formerly at Chapman & Hall Publishers, for his assistance at the beginning of this project, and Catherine Shaw and Chris Wortley at Stanley Thornes, for overseeing the final production. This book is written for students, so we thank innumerable students over the years, whose enthusiasm for Earth Sciences helped us to develop courses which gave the impetus to write this book. Finally we thank our wives, Li and Tamsin, and our parents, Reg and Ruth, and Brian and Mary, for the support and encouragement over the years in doing such strange things as collecting odd fossils and getting up to our knees in mud; we wouldn't have missed it for the world! In particular, Li Guo is thanked for her patience throughout the writing of this book, and without her support it would probably not have surfaced.

Steve Kershaw
Andy Cundy

PART A

THE ESSENTIALS OF MODERN OCEANOGRAPHY

INTRODUCTION: BASICS OF OCEANOGRAPHY FOR EARTH SCIENTISTS

1

1.1 THE LIMITS OF OCEANOGRAPHY

Oceanography is the study of processes in oceans and seas, which, together with lakes and rivers, represent the Earth's hydrosphere; the surface marine water is irregularly distributed over the globe (Figure 1.1). Oceanography involves four interlinked aspects: physical, chemical, biological and geological, all operating within the hydrosphere. However, the hydrosphere is only one of several overlapping entities which combine to make the Earth's surface systems work; altogether these are hydrosphere, cryosphere, atmosphere, biosphere and geosphere. Oceans and seas may be thought of as lying on the geosphere and under the atmosphere, while much of the biosphere is immersed in the oceans. Of course, this view is too simple, because there is so much overlap and interchange between these components, that the Earth's surface is really an ocean–atmosphere–land linked system (Figure 1.2). To illustrate these relationships more fully, note the following points:

1. Water is continuously cycled through the upper layers of the solid Earth (crust and upper mantle), exchanging dissolved and particulate matter, and it has been estimated that about every 10 million years, the volume of water cycled is equivalent to the entire ocean mass. Because there is reasonable evidence that the first oceans were present by 3900 million years ago (Ma), and maybe earlier, that means a large number of water molecules have been through the crust and upper mantle 390 times. Because the hydrological cycle brings water onto land, some of those water molecules might be in your body as you read this!
2. Ocean basin geometry (i.e. three-dimensional shape) and circulation are greatly controlled by plate tectonics, and tectonic change over time has a profound influence on oceans and climate.
3. The oceans receive huge amounts of material from terrestrial erosion, and exchange gases and particles with the atmosphere.
4. The oceans have been substantially modified by organic activity through geological time.
5. The oceans store 97.2% of Earth's water, compared with the atmosphere (0.001%), ice caps and glaciers (2.15%), subsurface water (0.625%) and surface water (0.017%) (Murray, 1992). Oceans couple with the atmosphere to form a system responsible for distributing heat around the Earth, and thus play a major role in the Earth's climate system. Ocean water has about 30 times more heat capacity than the atmosphere, so small changes in ocean circulation lead to profound climatic modification, as illustrated by the ENSO events of the Pacific, and (possibly) the Younger Dryas event *c.*11 000 years ago.
6. The oceans generate about one-third of global primary production, and are a valuable biological and food resource. Over

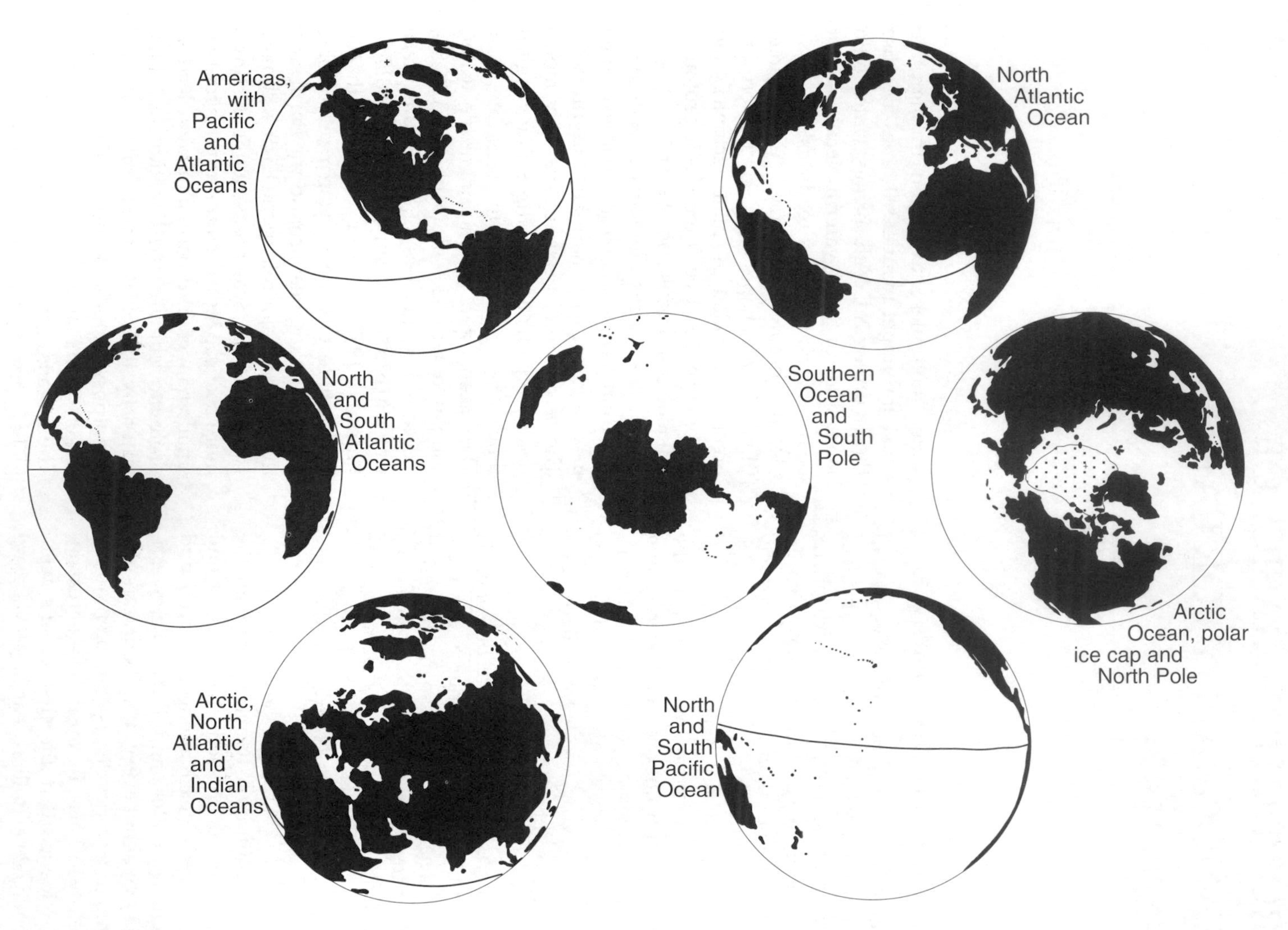

Figure 1.1 The globe from several different viewpoints, to illustrate the large area covered by oceans, and the uneven distribution of both land and oceans. Remember that the Earth's continents and oceans are wrapped around its surface, all too easy to forget when studying features drawn using the Mercator projection used in most of the book.

recent decades the potential of oceans for mineral and energy resources has become increasingly recognised.

So, it is clear that to define oceanography simply as the study of processes in oceans and seas masks the complexity of interactions beyond their physical margins. Such interactions also involve variations in orbital characteristics of the Earth, causing long-term changes in the amount of sunlight reaching any spot on the Earth's surface, so that the limits of what oceanography can tell us lie well outside the Earth's perimeter. Time, too, has a claim on oceanography; appreciation of both present and past ocean processes is an essential (though controversial) prerequisite in the field of prediction and modelling of future potential climatic change, and may aid such critical economic considerations as coastal management.

The geological aspects of oceanography have also been called marine geology. That term is not used in this book because of the changing trend in recent years of the perception of what constitutes the Earth Sciences. Thus, the emphasis on interactions between hydrosphere, atmosphere and geosphere go beyond the traditional limits of marine geology, and, in this book, we also include human impacts on oceans (Part C). Thus, the Earth sciences now justifiably have a claim on geological, physical–geographical and human–geographical aspects. Because oceanography impacts on all of these, this book is presented as a view of oceanography against the backdrop of Earth sciences.

1.2 FOUR OR FIVE OCEANS AND MANY SEAS

Oceans cover approximately 361 million km^2 (about 71%) of the Earth's surface. They have an average depth of 3730 m, compared with the land's average height of 840 m; the difference is attributable to the enormous variation of ocean floor topography due to mid-ocean ridges, which are the highest mountains on Earth. The Earth's land masses have an areal extent of about 149 million km^2, and are unevenly distributed around the Earth (Figure 1.1), the major landmasses being concentrated in temperate latitudes in the Northern Hemisphere (although it was different in the geological past). The Southern Hemisphere contains only about 30% of the Earth's land, mainly in tropical and polar latitudes. Between latitude 55° and 65° in the Southern Hemisphere, a complete ring of water encircles the Antarctic continent, and permits a circumpolar current. The present configuration of continents divides the modern ocean system into four: the Pacific, Atlantic, Indian and Arctic Oceans (Figure 1.3). A fifth ocean, the Antarctic (or Southern) Ocean, is included by some authors, justified because the Antarctic Circumpolar Current acts as a physical barrier to interchange of water currents and biota with the three oceans surrounding it. However, because the Antarctic Ocean is not well defined by surrounding land, its waters are usually taken to comprise the southern extension of the Atlantic, Indian and Pacific oceans. Smaller areas enclosed (at least partly) by islands or continents are seas, although the term sea can encompass distinct portions within some oceans (e.g. Sargasso Sea in the North Atlantic Ocean). Some seas are landlocked (e.g. Caspian Sea) and others nearly (the North Sea) or almost so (e.g. Black Sea and even the Mediterranean). Below is a brief description of the four oceans.

The Pacific Ocean (meaning tranquil) is the largest, approximately circular with a diameter of *c.*13 000 km and average depth 4.2 km, and contains the deepest recorded ocean depth of 11.5 km (Mindanao Trench near the Philippines). Active volcanism along the eastern, northern and western edges of the Pacific led to the term 'ring of fire', accompanied by frequent earthquake activity. Islands are common in the Pacific, occurring as arcs, chains or isolated landforms.

The Atlantic Ocean (named by the Romans, who derived the word from the Atlas

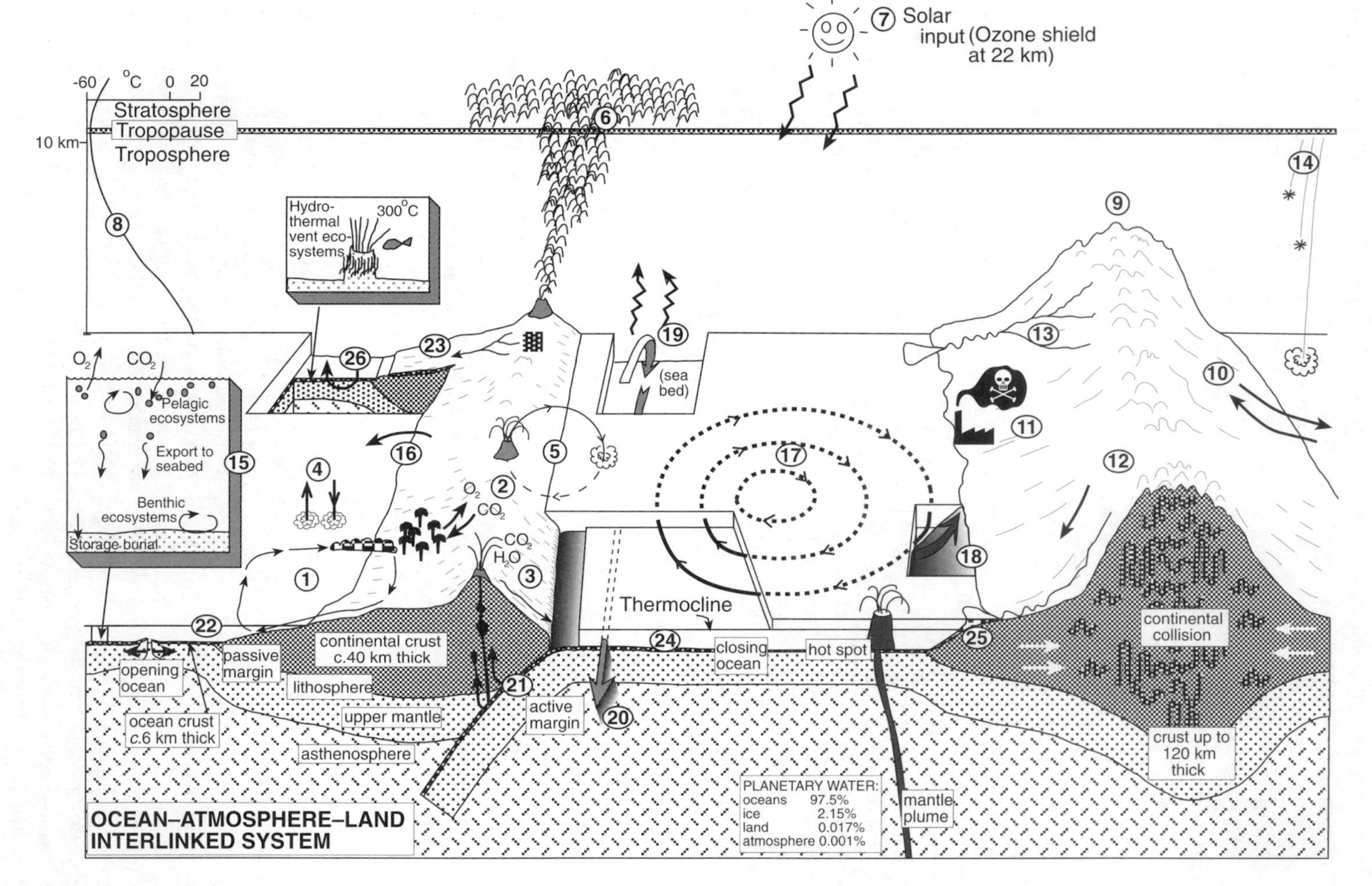

Figure 1.2 A simple cartoon illustrating the interlinked nature of oceans, atmosphere and land. The front face of the diagram is a vertical cross-section through a continental interior to an ocean basin. The transfer of material between the three components is shown by arrows. Features are numbered as below, and appropriate cross-references to other chapters are given.

1. Hydrological cycle. Water is a liquid at surface temperatures due to hydrogen bonding (Appendix 1).
2. Land ecosystems interchange gases with atmosphere and enhance terrestrial weathering
3. Atmospheric volcanic eruptions add gases to atmosphere, especially water and CO_2. Ash is carried to sea, and lavas and pyroclastic flows may reach the sea; rivers transport volcanic products to the sea (Chapter 5).
4. Ocean and atmosphere interchange gases and aerosols; the generation of dimethyl sulphide (DMS) by algae in the ocean is passed to the atmosphere to affect cloud formation (Chapter 12).
5. Carbon and some nutrient cycles are partly represented here (Chapters 5, 6).
6. Explosive eruptions from acid volcanoes penetrate the tropopause; the products (ash, gases and aerosols) therefore spread globally (e.g. Mt Pinatubo, Phillipines, 1991).
7. Solar energy drives the Earth's climate and ocean systems.
8. Atmospheric temperature curve falls throughout the troposphere to the tropopause; above this, temperature inversion blocks most interchange between troposphere and stratosphere, so weather systems are trapped within the troposphere.
9. Mountain barrier created by continental collision gives regional and global climate change because it interrupts global atmospheric circulation, and influences weathering (cf. 12) (e.g. Himalayas).
10. Monsoon winds are enhanced by large coastal mountain ranges, and drive seasonal currents (e.g. Indian Ocean) (Chapter 3).
11. Anthropogenic pollution into atmosphere, thence to ocean; also to ocean directly, and indirectly via ground water (Chapters 11, 12).
12. Terrestrial weathering and erosion draws CO_2 from the atmosphere, and aids climatic cooling (Chapters 8, 9, 10).
13. Sediment and nutrient input to oceans (Chapter 5).
14. Cosmic particles arriving on the surface add *c.*30 000 tonnes/a to the Earth's mass. An uncertain amount of water also arrives this way. The possibility of extraterrestrial organic matter has not been disproved.
15. Oceanic productivity is greatest near land, but calcareous and siliceous oozes form from pelagic production. Overall, oceans generate *c.*1/3 of global productivity (Chapter 6).
16. Aeolian dust from arid regions carries iron to the ocean, possibly an important nutrient source.
17. Shallow water ocean gyres are formed by friction of the moving air (not shown) over the ocean surface (Chapter 3).
18. Coastal upwelling is driven by shallow-water ocean gyres, drawing nutrient-rich waters to the surface (Chapter 3).
19. In polar regions, warm surface waters travelling from the tropics lose their heat to the atmosphere, and generate cold oxic water which sinks to the ocean floor (Chapter 3).
20. Cold deep water travels the world's bottom currents of the thermohaline circulation; this system gives long-term planetary heat distribution (Chapter 3).
21. Not only do subduction zones recycle oceanic crust, but ocean floor sediments are also recycled. Calcareous and silicate sediments are destroyed, and the CO_2 locked up in limestone sediments (24) is returned to the atmosphere. Also, ocean water lowers the temperature at which hot rock will melt, and plays an important part in assisting the plate tectonic machine (Chapter 4).
22. Mineral deposits on ocean floor include Fe-rich and Mn-rich nodules (Chapter 5).
23. Land-derived shelf sediments load the continental margin and assist isostatic depression, as does glacial ice (Chapter 4).

24, 25. Sediments deposited on ocean floor (24), and on continental margins (25), provide much of the geological record for palaeoceanography. Calcareous sediments (limestones) and organic matter products (oil and gas) contain CO_2 that was once in the atmosphere. Approximately, for each CO_2 molecule locked in the seabed and sedimentary rocks, there is one molecule of O_2 in the atmosphere.

26. The crustal–ocean factory: the system of interchange between the ocean, crust and upper mantle.

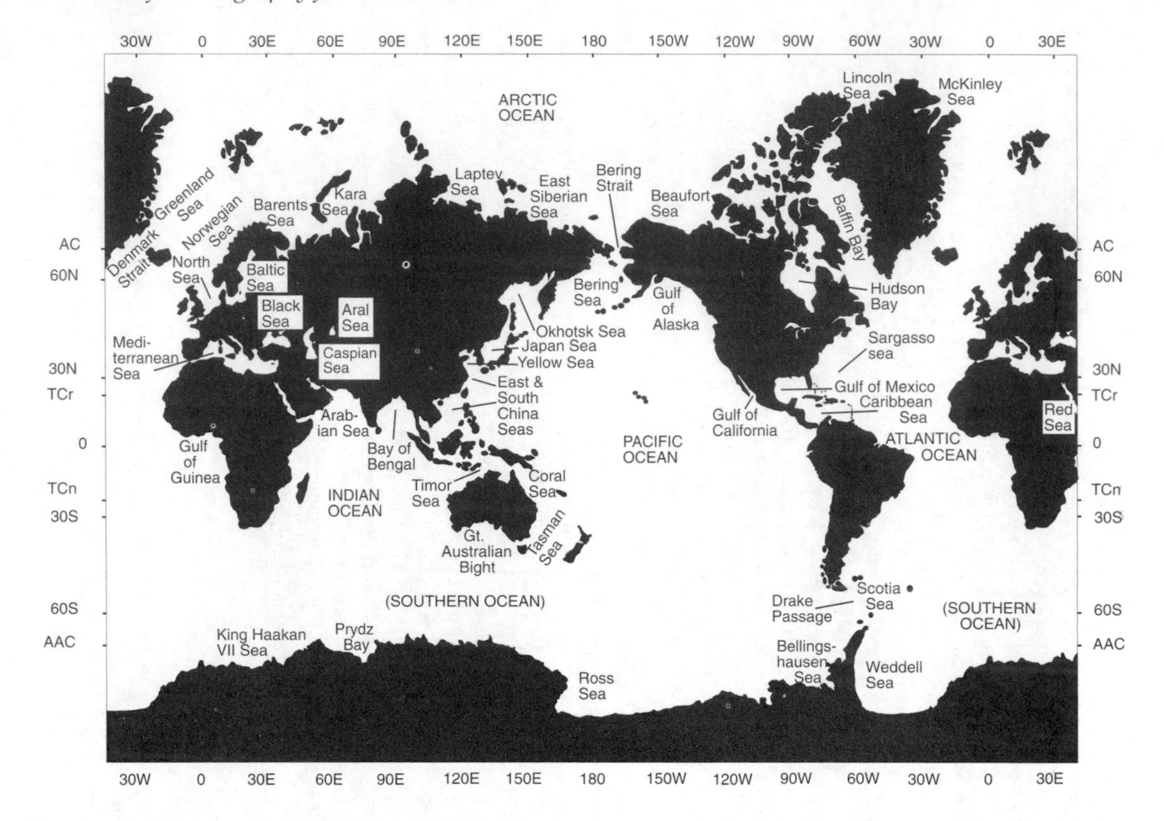

Figure 1.3 A map of the world showing location and names of oceans and major seas. Note the Southern Ocean is labelled in brackets because it is not recognised by many workers.

Mountains which lie at the eastern edge of the then unexplored ocean) is narrower and S-shaped, with an average width of *c*.6000 km and average depth of 3.6 km. The Atlantic is dominated by a central submarine mountain chain, the Mid-Atlantic ridge, which has a powerful influence on the path of deep water cold currents (see Chapter 3).

The Indian Ocean is largely in the Southern Hemisphere and forms a large triangle between the Indian subcontinent in the north and Antarctica in the south. Its average width is *c*.7000 km, with an average depth of 3.7 km.

The Arctic Ocean is shallow and small compared with the other oceans, with an average width of *c*.4400 km and average depth of *c*.1.1 km. Largely landlocked, much of its surface is covered by relatively permanent sea-ice (the polar ice pack), which extends during winter months. The Arctic Ocean is separated from (a) the Atlantic by a large, discontinuous submarine high, east and west of Iceland, of critical importance in controlling deep ocean circulation in the region; and (b) the Pacific by the narrow Bering Strait through which current flow is limited.

1.3 GEOMORPHOLOGY OF THE OCEAN FLOOR

Motion of the Earth's tectonic plates (Figure 1.4) is achieved by spreading, subduction and transform faulting, and consequently the ocean floor contains a variety of large physiographic features. These can be divided broadly into 3 zones: the continental margins, deep-ocean basins and midocean ridges, together with relatively minor features of ocean

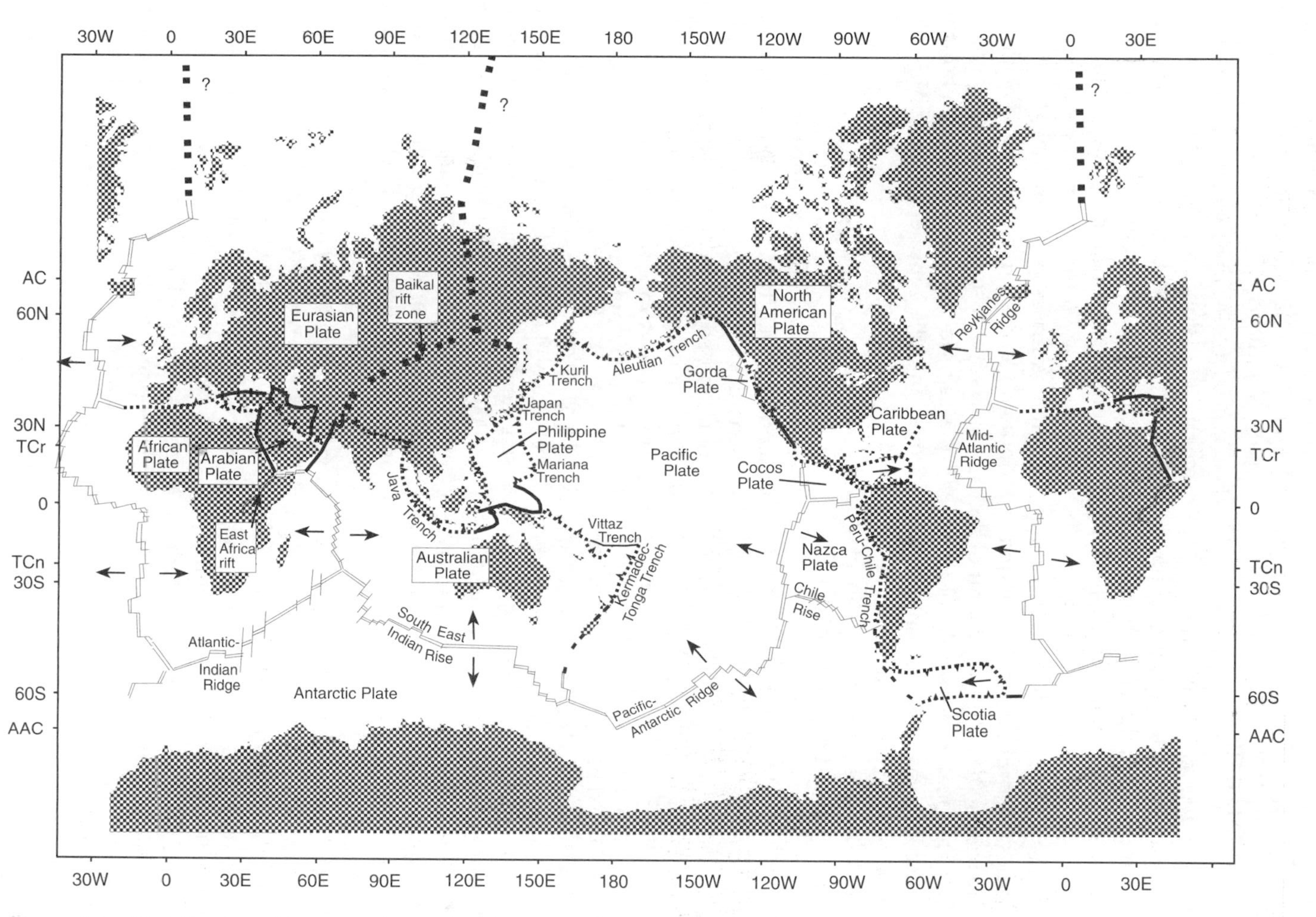

Figure 1.4 Distribution of tectonic plates, showing their direction of movement. Note that this configuration is only the current setting, and in a few tens of million years' time it will differ a little, therefore changing the shape, size and interactions between the oceans and seas.

trenches (Figures 1.5 and 1.6). Their main characteristics are as follows.

1.3.1 CONTINENTAL MARGINS

These are submerged edges of continental crust covered, to a large extent, by terrigenous debris eroded from the adjacent continents. Lesser amounts of carbonates (limestones) are present on today's margins, but in the geological past limestones have been dominant at certain times. Morphology of modern continental margins has been profoundly influenced by Late Quaternary sea-level fluctuations, because large parts of the continental shelves were exposed by lowered sea levels as recently as 18 000 years ago. The continental margins generally consist of 3 components:

1. *Continental shelf*: a nearly flat plain or terrace bordering the continent. The continental shelf slopes at about 1° towards the ocean basin (Figure 1.6). Its width varies: from a few km along the Pacific coast of North and South America, up to more than 1000 km in the Arctic Ocean. Many continental shelves have an irregular topography of banks, valleys and basins, while others are relatively smooth, related to tectonic history. The seaward limit of the continental shelf is defined by a distinct change in gradient, termed the shelf break, at water depths ranging from 35–250 m.
2. *Continental slope*: occurs seaward of the continental shelf, where the slope angle averages 4°. Angles vary significantly according to the supply of terrigenous sediment, and the tectonic setting of the margin, and extends to depths of 2–3 km. The continental slope is often dissected by submarine canyons, formed by erosion during low sea-level stands and subsequently enlarged by submarine sediment flows. These act as chutes to transport sediment from continental margins to the deep ocean. The sloping angle of continental margins therefore facilitates turbidity currents, of dense, sediment-laden water flowing downslope at speeds of up to *c*.20 m/s (about 45 mph) in the case of the Grand Banks earthquake, 1929 (Pinet, 1992, p. 110).
3. *Continental rise*: this lies at the toe of the continental slope, and consists of a wedge of terrigenous sediment several km thick, which has an average slope of 1° and has accumulated through transportation of sediment down the continental slope. Wide continental rises occur in the Atlantic and Indian Oceans; those in the Pacific are narrower or in some areas absent (replaced by deep-sea trenches where subduction zones abut the continental margin).

1.3.2 DEEP OCEAN BASINS

These lie seaward of the continental margins, at depths greater than 3 km. Their morphological character is largely governed by plate tectonic and sedimentation processes, or the degree to which sediments cover the irregular volcanic topography of the seafloor (Figure 1.6). The following major features occur:

1. *Abyssal plains and hills*: abyssal plains are extremely flat, broad areas of land-derived sediments, with gradients not exceeding 1:1000. They are the Earth's flattest areas, and have formed through burial of the irregular volcanic topography of the original ocean floor by thick deposits of sedimentary material. Abyssal hills are elongated hills, less than 900 m high and up to 100 km wide, formed by volcanism or from folded sediments.
2. *Seamounts*: these are mostly extinct volcanoes greater than 900 m high, occurring in groups or as isolated features. Guyots are flat-topped seamounts which at one time reached the sea surface, and have had their tops truncated by wave action.
3. *Deep-sea trenches*: these are steep sided, long, often narrow basins, with water depth between 5 and 12 km (3–5 km deeper than the adjacent ocean floor). Trenches are

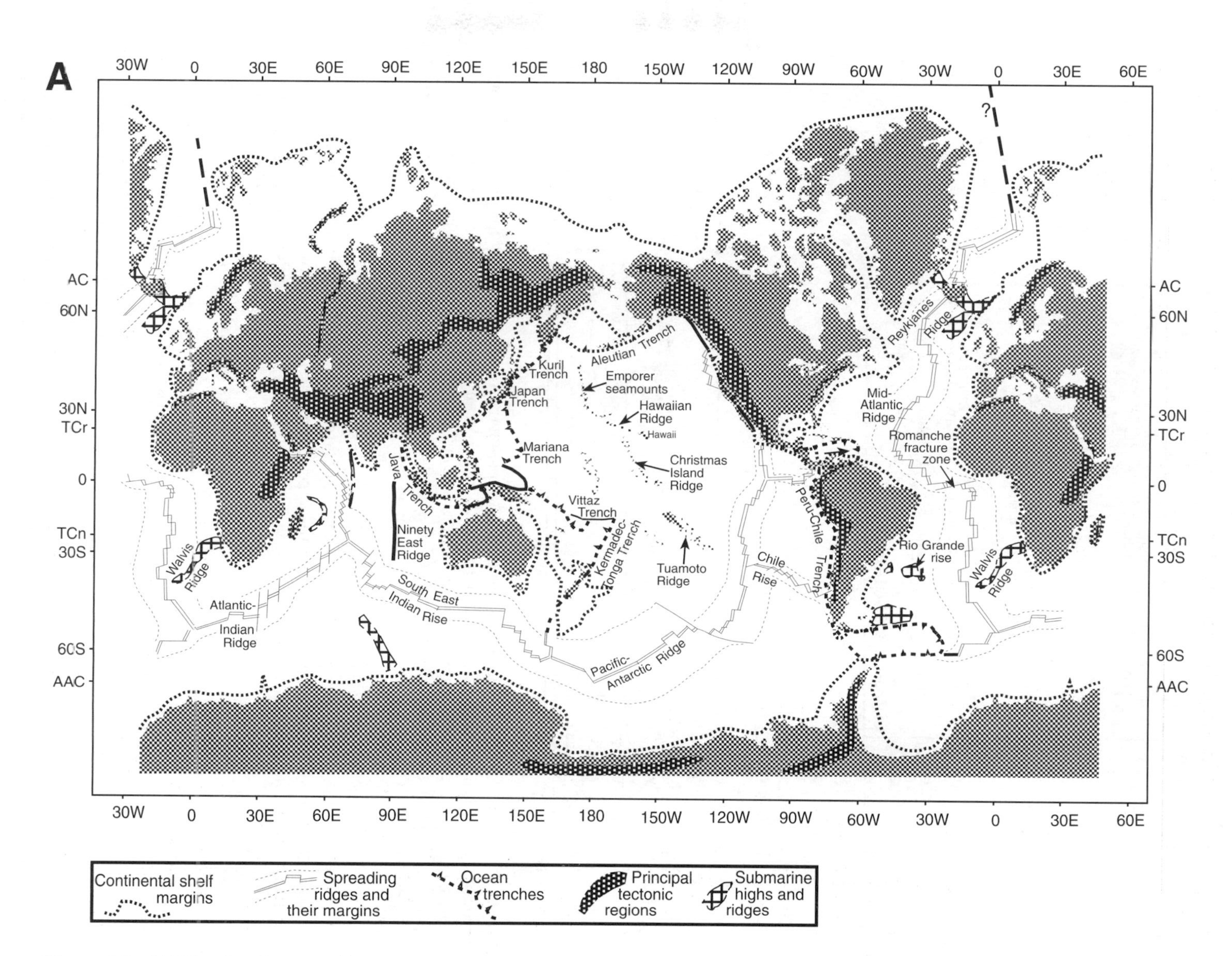

Figure 1.5 (A) The distribution of major topographic features of oceans, showing a close relation with the tectonic plates (Figure 1.4). Key features are labelled (figure continued overleaf).

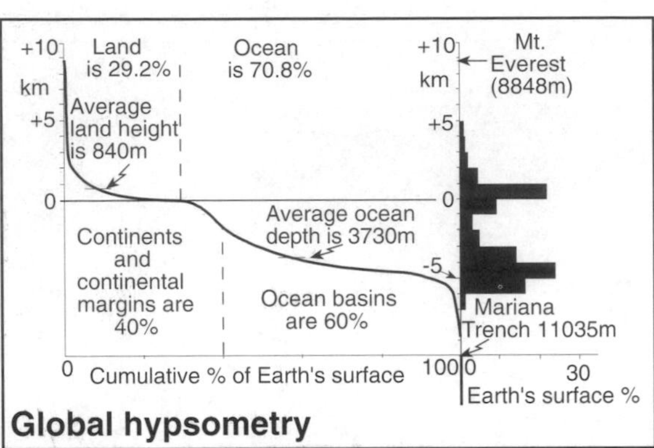

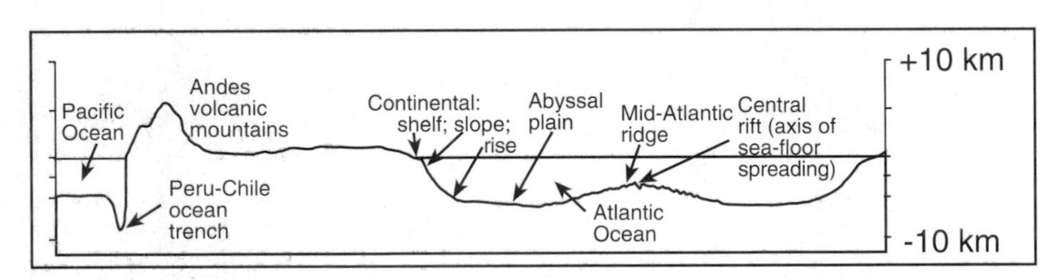

Figure 1.5 (*continued*) (B) shows height distribution (hypsometry) of world surface topography; (C) shows the topographic profile from eastern Pacific to west Africa. Ocean floor topographic variation also affects height variations on the sea surface, clearly seen by radar imaging (Seasat satellite); locations of ocean ridges and trenches can be detected at sea level. ((C) from Bearman, 1989a; reproduced with permission of the Open University.)

formed at plate boundaries, where ocean crust subducts beneath the less dense continental crust or beneath ocean crusts.

1.3.3 MID-OCEAN RIDGES

These are the world's highest mountains and form an almost continuous chain through the ocean basins (Figure 1.5), with crests 2–3 km below the ocean surface, up to 4 km above the surrounding ocean floor (in some places, e.g. at Iceland, the ridge actually reaches the ocean surface). Ridges occur at plate boundaries and are areas of seismic activity and ocean crust production. Their summits are either broadly convex or cut by rift valleys, and are offset by fracture zones arranged perpendicularly to the ridge crest (Figure 1.6B).

1.4 CHANGE IN THE OCEAN SYSTEM

The present ocean features outlined above are only the current situation of a long period of change, which continues. Such change is a major theme of this book, and a detailed treatment of the relationships between oceans and

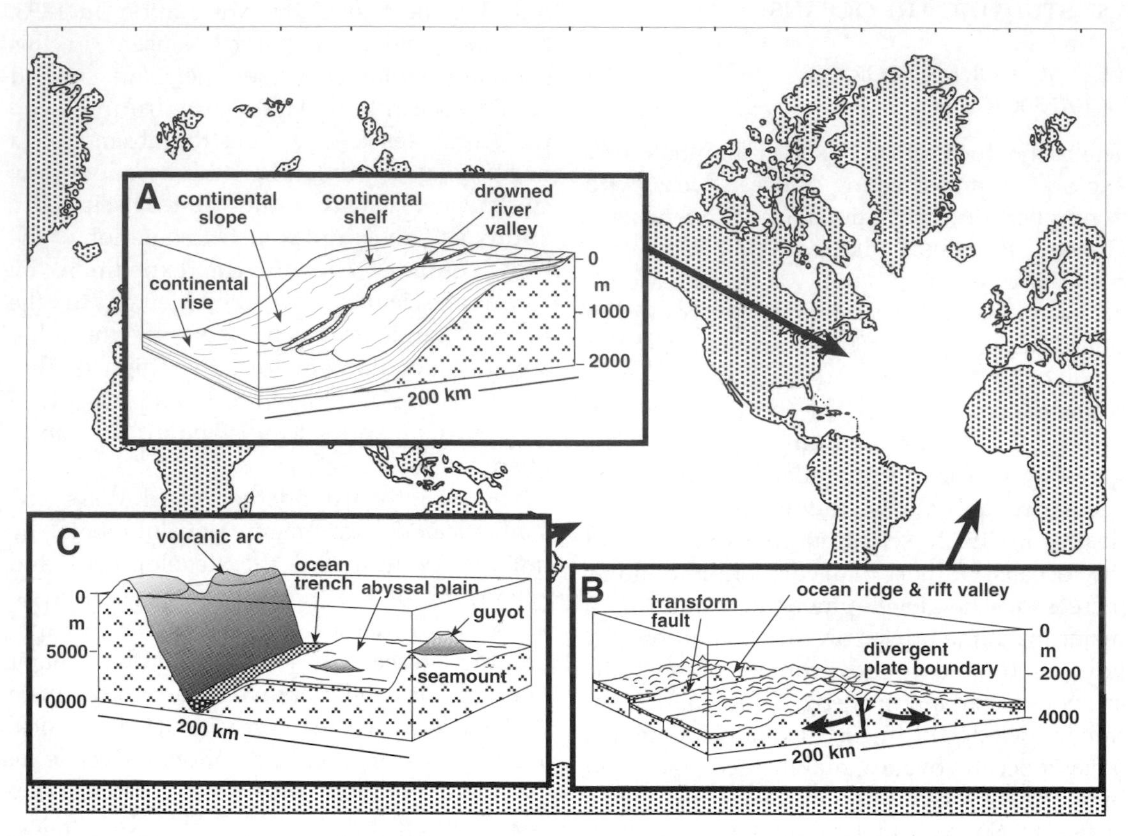

Figure 1.6 Details of ocean floor topography. Note the close relationship between topography and tectonic setting. (A) A passive continental margin (no subduction zone) in Eastern USA; the submarine valley (based loosely on the Hudson Canyon) was formed during sea-level lowstand in the last glacial phase, and is now drowned; (B) mid-ocean ridge system in the central Atlantic, located at a spreading ridge (constructive plate boundary); (C) ocean trench and associated island arc in Western Pacific, based on the Tonga Trench; the trench lies above a subduction zone.

plate tectonic processes is given in Part B. The critical point to emphasise at this stage is that ocean evolution and plate tectonics are integrally linked because of the interchange between ocean water and the crust and mantle, and the changing configuration of the continents and ocean basins with time.

Ocean water circulation throughout the depth and breadth of the oceans acquired its existing pattern only about 30 million years ago, when the opening of the Austral and Drake passages in the southern oceans allowed the unrestricted flow of the Antarctic circumpolar current. Furthermore, over the past 125 000 years, the blink of an eye in geological terms, sea level has fluctuated by more than 100 m in response to the waxing and waning of ice sheets, and ocean circulation has undergone rapid changes, with severe effects on climate. The short-term future of the ocean system, however, is arguably more dependent on the actions of human society. As human populations have grown, and technology has increased, humans have become important agents of change in the ocean system. This anthropogenic contribution to ocean change is discussed in Part C.

1.5 STUDYING THE OCEANS

1.5.1 A SHORT HISTORY OF OCEANOGRAPHY

The early study of oceans was dominated by exploration and charting of ocean areas, with a number of civilisations (the Phoenicians, Greeks, Romans, Vikings) expanding the boundaries of the 'known world' (the Mediterranean Sea and its limits). Lunar control of tides was recognised by Greek and English scholars, and current and wave observations recorded by da Vinci. In the centuries following the 3-year circumnavigation of the globe by Ferdinand Magellan (completed by Sebastian del Cano following Magellan's death) in 1522, scientific investigations of the oceans, rather than exploration alone, increased. The development of Britain as a major maritime power prompted a number of voyages funded by government or scientific societies to expand the geographical and scientific knowledge of the oceans. James Cook led 3 major ocean voyages, making geological and biological observations, Fitzroy commanded a cruise to South America on which Darwin collected data for his theory of evolution and on coral reef development, and John and James Ross recovered and analysed samples from the bottom of the high-latitude oceans. This culminated in 1876, when the British ship HMS Challenger completed a 4-year voyage around the Atlantic, Pacific and Southern Indian Oceans, describing seafloor sediments and ocean currents, and determining the chemical composition of seawater and the depth distributions of marine organisms.

The Challenger expedition is considered by many oceanographers to represent the birth of oceanography as a true science. Major developments were by then occurring elsewhere in Europe and in North America. Maury published his successful book *The Physical Geography of the Sea* in the USA in 1855, and recognised the need for international cooperation in the ocean sciences, and Agassiz founded the first US marine station. In 1893, the Norwegian explorer Nansen in his specially reinforced vessel, the Fram, studied Arctic Ocean circulation by drifting with the Arctic sea ice, proving the absence of a north-polar continent. Modern oceanography commenced with the interdisciplinary approach and complex instrumentation of the 20th century. A two-year expedition by German scientists on the Meteor to the South Atlantic used echo sounding and other advanced equipment to gather high quality depth, salinity, temperature and dissolved oxygen data, and set the standard for subsequent surveys.

The periods around the World Wars saw sophisticated equipment developments by military bodies, and the development and expansion of oceanographic research institutes, with dedicated vessels for oceanographic research. More recently, major international collaborative efforts have been initiated, such as the International Geophysical Year (1957–58), the International Decade of Ocean Exploration (the 1970s) and the Deep Sea Drilling Project (1968–75). The palaeoceanographic section of this book uses information partly derived from ocean floor drilling. Reorganised as the International Program of Ocean Drilling in 1975, this project continues, using a newer vessel, the Joides Resolution, along with other international projects to examine the ocean's role in the global carbon cycle (Joint Global Ocean Flux Study – JGOFS), ocean circulation (World Ocean Circulation Experiment – WOCE), ocean biology (Global Ocean Ecosystem Dynamics – GLOBEC) and others. The advent of satellite oceanography has added a new dimension to oceanography, allowing data to be collected rapidly over vast areas, rather than the point measurements usually taken from research vessels. Satellites are being increasingly used to assess both large- and small-scale ocean processes, with a large increase in the applications of radar.

1.5.2 PROBLEMS AND TECHNIQUES

Two major problems exist in getting information from the oceans: their great size, and their opacity to light. The large size of oceans, and the various scales of the processes that occur in them, mean that observations taken from a single location at one point in time are of limited use in determining ocean processes. Ocean research cruises are extremely expensive, and information gained is in the form of a series of point measurements over a limited time period. It is difficult, if not impossible, to extrapolate these measurements to gain a thorough understanding of complex ocean processes. Due to the turbulent nature of the oceans, measurements from a ship cannot be taken as quickly as the ocean changes. A recent development in remote sensing techniques, however, has considerably improved this situation. Satellite measurements allow oceanographers to survey large areas of ocean rapidly and inexpensively, providing valuable data on sea surface temperature and productivity. Recent developments allow the entire depth of the oceans to be examined via altimetry, by measuring the shape of the sea surface using radar pulses bounced off the sea surface and measuring their travel time. If the oceans were totally at rest, then each point of the ocean surface would be perpendicular to the local force of gravity. The shape of this (equipotential) surface depends on the shape of the ocean floor, and recently released altimetry data allow the ocean floors to be mapped to an unprecedented degree of accuracy. Also, since the ocean obviously is not at rest, detailed altimetry measurements over time allow variations in the tilt of the oceans' surface to be determined. This allows ocean circulation (current speed and direction) to be computed. Putting this theory into practice is extremely difficult, although the recent initiation of the French/American TOPEX/POSEIDON project will allow global ocean circulation, and its variability over as little as weekly intervals, to be determined (Wunsch, 1992).

The oceans are opaque to light waves (and other electromagnetic radiation), which will penetrate only the surface few metres in even the clearest ocean waters. This meant that until recently much of the deep ocean remained unseen, and studies relied on samples dredged up to the surface. Developments in the use of submersibles has allowed parts of the deep ocean to be explored directly. These allow deep sea organisms to be examined *in situ*, sampling areas to be examined before collection of sediment, rock or water samples, and direct measurements of deep sea physical, chemical and biological parameters to be made, rather than the hit or miss approach of dropping a dredge or grab from a ship up to 4 km above. Submersibles have been important tools in deep sea exploration, and an early example in 1977 was the use of the submersible Alvin in a study of hydrothermal vents in the 2500 m deep rift valley of the Galapagos ridge. During this expedition hydrothermal vent communities, and their chemosynthesis-based ecosystems, were observed for the first time using cameras and sensors mounted on Alvin. As technology and submersible design improves, the use of submersibles is likely to significantly increase in the future. Nevertheless, despite the plethora of techniques, there are major gaps in understanding. This reflects the complexity of the ocean system and the fact that oceanography is a comparatively young science. Despite the major advances made over the last century or so, much of the ocean, particularly the deep sea areas and their biology and water circulation, remains relatively poorly understood.

1.6 THE AMAZING PROPERTIES OF WATER

To complete this introduction, we cannot overemphasise the importance of the nature of water, discussed in Chapter 2. If the world was a fair place, H_2O should be a gas at Earth surface temperatures and pressures. The elements near oxygen in the Periodic Table of elements make simple compounds with hydrogen, which are all familiar: H_2S, HCl, HF, NH_3, CH_4; all are gases. That water is the only one forming a liquid under surface conditions is due to its peculiar arrangement of the H and O atoms forming a type of chemical bonding called hydrogen bonding. Not only does hydrogen bonding permit water to be a liquid, it also imparts a critical ability of water to store heat, and therefore importantly to affect climate. If it lacked these odd properties, we would have no rivers, no oceans, no water-mediated erosion, no aquatic-controlled deposition and, of course, no life! Earth would also probably lack a plate tectonic system because water carried down subduction zones assists the melting of crustal rocks, essential to keep the system moving.

1.7 CONCLUSION

This simple introduction shows how important it is to have a holistic view of oceans, drawing on evidence from a wide range of sources. Perhaps it is rather fanciful to claim that oceanography is the ultimate Earth science, but given its importance to the modern world, this is probably not far wrong. Examination of the oceans begins in Chapter 2, with a description of the properties of water, necessary for a useful understanding of how oceans work. Also, in the Appendix there is a beginner's guide to the terminology of chemical processes, to assist those students who lack a background in chemistry.

1.8 SUMMARY

1. The study of oceanography is limited principally by the imagination, because the interwoven relationship between oceans, atmosphere and the Earth's interior makes the ocean margins difficult to define; this results in an ocean–atmosphere–land interaction system. Also, ocean sediments record fluctuations in orbital processes, and therefore reveal aspects of astronomic change affecting the Earth.
2. Seawater covers 71% of the Earth's surface, as four (or five) oceans and many seas, drawing attention to its prime importance.
3. The geomorphology of the ocean floor is much more spectacular than terrestrial landscapes, and reflects the evolution of the plate tectonic system.
4. Oceans are difficult to study because of their inaccessibility and their opacity to monitoring systems. Access via submersibles and sea-floor drilling, accompanied by satellite technology, is gradually improving knowledge.
5. The oceans would not exist but for the peculiar properties of water, which make it a liquid at surface temperatures; otherwise it would be a gas.

THE NATURE OF SEAWATER AND ITS IMPORTANCE IN OCEAN PROCESSES 2

2.1 INTRODUCTION

Earth is unique amongst the planets of the solar system in that its surface has abundant free water, in a constant state of movement termed the hydrological cycle (Figure 1.2). In 1998 water was found on the now-frozen surfaces of the Moon and Mars, leading to speculation about potential for life outside the Earth. On a global scale, the Earth's surface may be viewed as a giant distillation unit, powered by the Sun, where water is continuously evaporated from the oceans and the land, and returned as precipitation. The oceans form the largest reservoir, holding around 97% of the Earth's surface water.

Water plays a major role in controlling the distribution of heat around the Earth's surface, and in moderating temperature and climatic extremes. Water is extremely well suited to these roles, largely due to its unique physical and chemical properties. Liquid water has also been instrumental in the history of life on Earth, and life almost certainly developed in oceans before it migrated to land. All organisms contain liquid water to carry dissolved substances around and between cells, and moderate temperature fluctuations. Therefore you need to appreciate the properties of water in relation to ocean processes, considered in this chapter.

2.2 PROPERTIES OF WATER

The following is a brief survey of the chief properties of water, and their implications in oceanography. The first, hydrogen bonding, influences many of the others, and serves to show just what an unusual substance water really is. The Appendix gives a simple introduction to chemistry for those lacking sufficient background, and you should read the Appendix now before continuing.

2.2.1 HYDROGEN BONDING

Hydrogen bonding occurs because water molecules do not have an equal distribution of electric charge across the molecule, resulting in a degree of attraction between oppositely charged ends of neighbouring water molecules. This critical feature means that water is a liquid with high heat capacity (p.18), while similar chemicals that lack hydrogen bonding are gases at Earth surface temperatures.

The presence of hydrogen bonds also causes water to have several other properties, in particular the following.

1. High viscosity and low compressibility, important in the control of suspended and mobile objects in oceans, especially the biota and fine-grained inorganic particles and organic detritus. Water might not seem particularly viscous, but if you compare the ease of running on a road with the difficulty of running in the shallow end of a swimming pool, its significance is impressive.
2. High surface tension. Water molecules in the surface layer in contact with the atmosphere have a greater attraction for each other than for air molecules, and this unequal attraction gives the water surface its familiar 'skin' used by lightweight animals. Its importance is greater in still water such as lakes, but less

critical under wave influence in the ocean; nevertheless, the ocean surface is an important interface between the masses of ocean and atmosphere, across which interchange of gases and solids occurs.

2.2.2 THE THREE PHASES OF WATER

Water is one of the few substances on Earth to occur naturally in 3 phases: solid (ice), liquid (water) and 'gas', i.e. water vapour, not a true gas (Figure 2.1A). The phase of water present in any situation depends on how ordered its structure is. The vapour phase is characterised by a relatively random and loosely attracted arrangement of molecules, whereas liquid and solid phases have molecules that are increasingly strongly bonded. As with other substances, the density of water is greater than water vapour, because the molecules are more ordered and closer together. The solid phase possesses the most orderly molecular arrangement, and is the densest in most substances; however, water expands when it freezes, shown by the fact that ice is less dense than water and floats. Water's anomalous behaviour is due to its chemical structure and hydrogen bonding.

2.2.3 HEAT CAPACITY

Heat capacity is the ability to store heat, defined as the amount of heat required to change the temperature of 1 gram of a substance by 1°C. Increasing temperature in a substance leads to an increase in the motion of its atoms, but some of the heat absorbed is used up in breaking the attractive bonding forces between atoms and molecules, and does not contribute to its temperature; this is latent

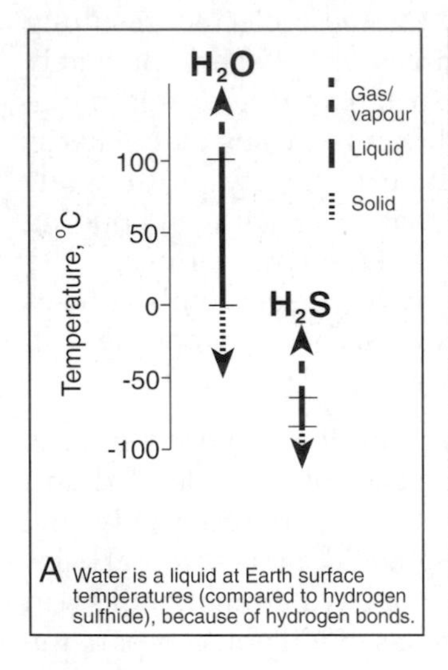

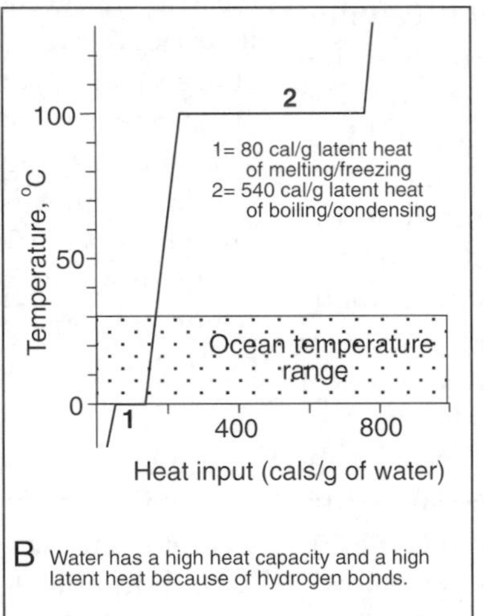

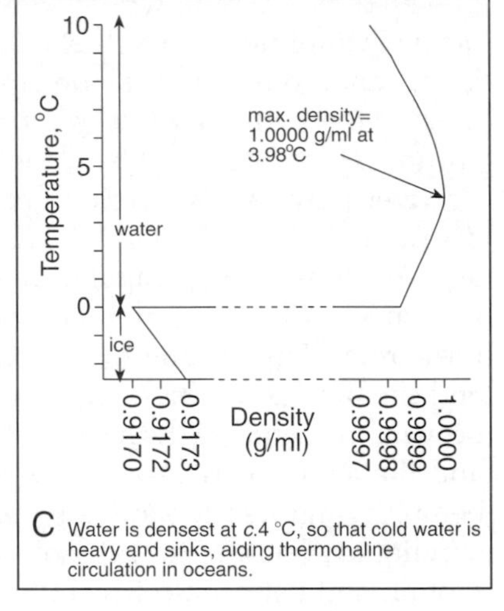

Figure 2.1 Physical properties of water and their oceanographic importance. (A) At the Earth's surface temperature, water is a liquid, but a closely related compound, hydrogen sulphide (H_2S) is a gas at those temperatures, which is more typical of the chemical properties of such compounds. Water's liquid nature is due to hydrogen bonding (see Appendix). (B) Water has a high heat capacity and a high latent heat; large amounts of heat are required to change water between liquid and solid, and between liquid and vapour states; oceans are an important heat store and exert a long-term control on climate. (C) Water density exhibits an odd trend in relation to temperature: its greatest density is at *c.*4°C; above and below that it expands. Density variations are an important aspect of the formation and transport of ocean water bodies, especially involved in deep-ocean circulation of the global conveyor, and its climatic implications. Adapted from P. Pinet, *Oceanography: An Introduction to the Planet Oceanus* (1992), Jones and Bartlett Publishers, Sudbury, MA, USA.

(*hidden*) heat, which has no effect on the temperature of the material. Substances with hydrogen bonds absorb more latent heat than substances lacking hydrogen bonds, because there is great strength of attraction between neighbouring molecules in materials with hydrogen bonds, and some of the heat is used to break a number of those bonds. So water has a higher heat capacity than other materials because of its greater latent heat properties.

Latent heat also is involved whenever substances change their phase (i.e. from solid to liquid and liquid to gas/vapour, and back again). The graph in Figure 2.1B shows how much heat is required to change water between the solid, liquid and vapour states. The only thing that makes something solid is the attractive bonding forces between atoms and molecules; if bonding is sufficiently broken, the material will change phase, and the energy required is latent heat. Thus latent heat of melting turns ice at 0°C into water at 0°C by breaking a large number of hydrogen bonds, but there are still a lot of bonds that are *not* broken (giving clusters of water molecules, see Appendix); thus it is liquid. The rest of the bonds break when latent heat (of vaporisation) is absorbed to turn water at 100°C into a vapour (steam) at 100°C. Therefore, when compared with its nearest chemical neighbour, H_2S, ice has a higher melting temperature, and water has a higher vaporisation temperature, so H_2S is a gas at normal temperatures (Figure 2.1A).

For the oceans, the consequences of water's high heat capacity are of truly monumental proportions. Heat is transferred to and from water with difficulty, and thus prevents extreme ranges in temperature. So the oceans heat up and cool far more slowly than land, moderating coastal climates by absorbing heat during the summer and releasing it in the winter. Also, large amounts of heat are transferred around the globe by ocean currents, and have a great influence on climate. The deep ocean circulation, which probably controls much of the long-term global climate, involves a great deal of latent heat.

2.2.4 DENSITY

Pure water also has odd density patterns. Most liquids become denser as they freeze, because lower temperatures mean less molecular motion. Because of hydrogen bonding, water, in contrast, has its greatest density (1 g cm^{-3}) at *c*.4°C (actually 3.98°C). At lower and higher temperatures it expands (Figure 2.1C); ice has a more open structure than water and floats, having expanded by about 10% in volume.

When seawater freezes, something else odd happens. Seawater is a solution of great complexity, and the most abundant dissolved material is salt (NaCl). 70% of the dissolved salt is excluded from the ice (van Andel, 1994, p. 201), so the water surrounding the ice is enriched in salt. The implications of this are examined in Chapter 3, in relation to the thermohaline circulation mechanism.

2.2.5 THERMAL EXPANSION

Liquids expand and contract with changes of temperature. The oceans contain a lot of water; cooling and warming causes volume changes (Figure 2.1C). Calculations show that warming the ocean volume from 0°C to 12°C leads to a sea level rise of 3 m (Jeppsson, 1990), and therefore this is an added dimension to global warming.

2.2.6 WATER AS A SOLVENT

Water is an extremely effective solvent due to its ability to dissolve at least a small amount of virtually every known substance; this is a consequence of the dipole structure of H_2O molecules, which allows water to dissolve most substances extremely well (see Appendix); notable exceptions are the non-aqueous phase liquids (NAPLs), such as oils, which can cause severe pollution problems in the sea and also in aquifers. The fact of dissolution makes chemical reactions easier, therefore indicating that water is a key mediator in chemical processes in oceans. In fact the role of water is so critical that the uptake of essential nutrients in ionic form by organisms depends on the dipolar water molecules to keep ions separate; therefore it is likely that life on Earth critically depends on this mechanism.

2.3 SEAWATER CHEMICAL COMPOSITION

2.3.1 CATEGORIES OF COMPOSITION

Because water is such an effective solvent, a wide range of substances are found dissolved in ocean waters (Figure 2.2), and concentration varies with depth (Figure 2.3). The amount of a substance dissolved in seawater depends on its crustal abundance, and how reactive it is. Once dissolved in seawater, a substance may simply stay in solution, or may react with and attach to sediments or sinking particulates, or may be taken up by organisms (discussed in Chapter 6). The substances dissolved in seawater are called solutes, and show a vast range of concentrations; they can be divided into 5 main groups:

1. major constituents
2. minor (or trace) constituents
3. nutrients
4. gases
5. organic compounds.

It is important to note the difference between *dissolution* and *suspension*. Dissolution is determined by the particle size. Particles are said to be dissolved when they will pass through a filter with a pore size of 0.4 μm; anything larger is simply suspended in the water. This division between dissolved and particulate species is made for convenience and is somewhat arbitrary. The five main groups are introduced here, and more information is provided in Chapter 4.

2.3.2 MAJOR CONSTITUENTS

The major constituents of seawater are those which occur at concentrations of greater than 1 part per million (ppm), and account for over 99.9% of the total dissolved salts in seawater. The most common solutes are Na and Cl (Figure 2.2), followed by magnesium (Mg), sulfate (SO_4^{2-}), calcium (Ca), potassium (K), bromine (Br), bicarbonate (HCO_3^-), boric acid (H_3BO_4), strontium (Sr) and fluorine (F). While these elements may be rapidly removed from solution by chemical and biological processes, removal generally has little overall effect on the elements' concentration. As the concentrations of major constituents are typically stable over time, they are said to exhibit conservative behaviour. The rate at which water mixes through the ocean is much faster than the rate at which these constituents are supplied or removed, so variations in their concentrations from place to place are usually affected only by physical processes such as evaporation, ice melting and mixing. So, while the total amount of dissolved salts (or salinity) may vary across the oceans, the ratio of these elements to each other, and to total salinity, is almost constant under most conditions (important exceptions are discussed in Chapter 4). This is termed the principle of constant proportions, and has important implications for the determination of salinity.

2.3.3 MINOR (OR TRACE) CONSTITUENTS

These are inorganic consituents that are present at very low concentrations, usually less than 1 ppm, and many are found at the parts per billion (ppb) or parts per trillion level. Trace constituents are involved in many chemical and biological reactions, and this causes their concentrations to vary significantly from area to area, and over time. They are said to exhibit non-conservative behaviour. Concentrations for some trace constituents are shown in Figure 2.2.

2.3.4 NUTRIENTS

These are components which are essential to organic growth, and four are given prominence. Most important is phosphorus (P), and to a lesser extent nitrogen (N); silica is important for siliceous plankton, which are

Figure 2.2 Features of dissolved material in the oceans, and relevant comparisons with the atmosphere (A) Elements, compiled from Bearman (1989b) (*=nutrients)

Element		*Concentration (ppm)*	*Dissolved compounds and ions*	*Total content in oceans (tonnes)*
Major constituents (>1 ppm)				
Chlorine	Cl	1.95×10^4	Cl^-	2.57×10^{16}
Sodium	Na	1.077×10^4	Ca^+	1.42×10^{16}
Magnesium	Mg	1.290×10^3	Mg^{2+}	1.71×10^{15}
Sulphur	S	9.05×10^2	SO_4^{2-}, $NaSO_4^-$	1.2×10^{15}
Calcium	Ca	4.12×10^2	Ca^{2+}	5.45×10^{14}
Potassium	K	3.80×10^2	K^+	5.02×10^{14}
Bromine	Br	67	Br^-	8.86×10^{13}
Carbon	C	28	HCO_3^-, CO_2, CO_3^{2-}	3.7×10^{13}
* Nitrogen	N	11.5	N_2 gas, NO_3^-, NH_4^+	1.5×10^{13}
Strontium	Sr	8	Sr^{2+}	1.06×10^{13}
Oxygen	O	6	O_2 gas	7.93×10^{12}
Boron	B	4.4	$B(OH)_3^-$, $B(OH)^{4-}$, $H_2BO_3^-$, H_3BO_4	5.82×10^{12}
Silicon	Si	2	$Si(OH)_4$	2.64×10^{12}
Fluorine	F	1.3	F^-, MgF^+	1.72×10^{12}
Minor constituents (<1 ppm)				
Argon	Ar	0.43	Ar gas	5.86×10^{11}
Lithium	Li	0.18	Li^+	2.38×10^{11}
Rubidium	Rb	0.12	Rb^+	1.59×10^{11}
*Phosphorus	P	6×10^{-2}	HPO_4^{2-}, PO_4^{3-}, $H_2PO_4^-$	7.93×10^{10}
Iodine	I	6×10^{-2}	IO_3^-, I^-	7.93×10^{10}
Barium	Ba	2×10^{-2}	Ba^{2+}	2.64×10^{10}
Molybdenum	Mo	1×10^{-2}	MoO_4^{2-}	1.32×10^{10}
Arsenic	As	3.7×10^{-3}	$HasO_4^{2-}$, $H_2AsO_4^-$	4.89×10^9
Uranium	U	3.2×10^{-3}	$UO_2(CO_3)_2^{4-}$	4.23×10^9
Vanadium	V	2.5×10^{-3}	$H_2VO_4^-$, HVO_4^{2-}	3.31×10^9
Aluminium	Al	2×10^{-3}	$Al(OH)_4^-$	2.64×10^9
*Iron	Fe	2×10^{-3}	$Fe(OH)_2^+$, $Fe(OH)_4^-$	2.64×10^9
Nickel	Ni	1.7×10^{-3}	Ni^{2+}	2.25×10^9
Titanium	Ti	1×10^{-3}	$Ti(OH)_4$	1.32×10^9
Zinc	Zn	5×10^{-4}	$ZnOH^+$, Zn^{2+}, $ZnCO_3$	6.61×10^8
Caesium	Cs	4×10^{-4}	Cs^+	5.29×10^8
Chromium	Cr	3×10^{-4}	$Cr(OH)_3$, CrO_4^{2-}	3.97×10^8
Antimony	Sb	2.4×10^{-4}	$Sb(OH)_6^-$	3.17×10^8
Manganese	Mn	2×10^{-4}	Mn^{2+}, $MnCl^+$	2.64×10^8
Krypton	Kr	2×10^{-4}	Kr gas	2.64×10^8
Selenium	Se	2×10^{-4}	SeO_3^{2-}	2.64×10^8
Neon	Ne	1.2×10^{-4}	Ne gas	1.59×10^8
Cadmium	Cd	1×10^{-4}	$CdCl_2$	1.32×10^8
Copper	Cu	1×10^{-4}	$CuCO_3$, $CuOH^+$	1.32×10^8
Tungsten	W	1×10^{-4}	WO_4^{2-}	1.32×10^8
Germanium	Ge	5×10^{-5}	$Ge(OH)_4$	6.61×10^7
Xenon	Xe	5×10^{-5}	Xe gas	6.61×10^7
Mercury	Hg	3×10^{-5}	$HgCl_4^{2-}$, $HgCl_2$	3.97×10^7
Zirconium	Zr	3×10^{-5}	$Zr(OH)_4$	3.97×10^7
Bismuth	Bi	2×10^{-5}	BiO^+, $Bi(OH)_2^+$	2.64×10^7
Niobium	Nb	1×10^{-5}	not known	1.32×10^7
Tin	Sn	1×10^{-5}	$SnO(OH)_3^-$	1.32×10^7
Thallium	Tl	1×10^{-5}	Tl^+	1.32×10^7

Figure 2.2 *(continued)*

Element		*Concentration (ppm)*	*Dissolved compounds and ions*	*Total content in oceans (tonnes)*
Thorium	Th	1×10^{-5}	$Th(OH)_4$	1.32×10^{7}
Hafnium	Hf	7×10^{-6}	not known	9.25×10^{6}
Helium	He	6.8×10^{-6}	He gas	8.99×10^{6}
Beryllium	Be	5.6×10^{-6}	$BeOH^{+}$	7.40×10^{6}
Gold	Au	4×10^{-6}	$AuCl_2$	5.29×10^{6}
Rhenium	Re	4×10^{-6}	ReO_4^{-}	5.29×10^{6}
Cobalt	Co	3×10^{-6}	Co^{2+}	3.97×10^{6}
Lanthanum	La	3×10^{-6}	$La(OH)_3$	3.97×10^{6}
Neodymium	Nd	3×10^{-6}	$Nd(OH)_3$	3.97×10^{6}
Silver	Ag	2×10^{-6}	$AgCl_2^{-}$	2.64×10^{6}
Tantalum	Ta	2×10^{-6}	not known	2.64×10^{6}
Gallium	Ga	2×10^{-6}	$Ga(OH)_4^{-}$	2.64×10^{6}
Yttrium	Y	1.3×10^{-6}	$Y(OH)_3$	1.73×10^{6}
Cerium	Ce	1×10^{-6}	$Ce(OH)_3$	1.32×10^{6}
Dysprosium	Dy	9×10^{-7}	$Dy(OH)_3$	1.19×10^{6}
Erbium	Er	8×10^{-7}	$Er(OH)_3$	1.06×10^{6}
Ytterbium	Yb	8×10^{-7}	$Yb(OH)_3$	1.06×10^{6}
Gadolinium	Gd	7×10^{-7}	$Gd(OH)_3$	9.25×10^{5}
Praseodymium	Pr	6×10^{-7}	$Pr(OH)_3$	7.93×10^{5}
Scandium	Sc	6×10^{-7}	$Sc(OH)_3$	7.93×10^{5}
Lead	Pb	5×10^{-7}	$PbCO_3$, $Pb(CO_3)_2^{-}$	6.61×10^{5}
Holmium	Ho	2×10^{-7}	$Ho(OH)_3$	2.64×10^{5}
Lutetium	Lu	2×10^{-7}	$Lu(OH)$	2.64×10^{5}
Thulium	Tm	2×10^{-7}	$Tm(OH)_3$	2.64×10^{-5}
Indium	In	1×10^{-7}	$In(OH)_2^{+}$	1.32×10^{5}
Terbium	Tb	1×10^{-7}	$Tb(OH)_3$	1.32×10^{5}
Tellurium	Te	1×10^{-7}	$Te(OH)_6$	1.32×10^{5}
Samarium	Sm	5×10^{-8}	$Sm(OH)_3$	6.61×10^{4}
Europium	Eu	1×10^{-8}	$Eu(OH)_3$	1.32×10^{4}
Radium	Ra	7×10^{-11}	Ra^{2+}	92.5
Protactinium	Pa	5×10^{-11}	not known	66.1
Radon	Rn	6×10^{-16}	Rn gas	7.93×10^{-4}
Polonium	Po	–	PoO_3^{2-}, $Po(OH)_2$?	–

(B) Gases in ocean and atmosphere, compiled from Segar (1998)

Gas	*% in atmosphere*	*% in surface oceans*	*% total for oceans*
Nitrogen (N_2)	78	48	11
Oxygen (O_2)	21	36	6
Carbon dioxide (CO_2 in atmosphere; $CO_2 + HCO_3^{-} + CO_3^{2-}$ in oceans)	0.04	15	83

(C) Gaseous flow rates between ocean and atmosphere, compiled from Segar (1998)

	Gas	*Direction of Flow*	*Transfer g/a*
Sulphur dioxide	SO_2	to ocean	1.5×10^{14}
Nitrous oxide	N_2O	to atmosphere	1.2×10^{14}
Carbon monoxide	CO	to atmosphere	4.3×10^{13}
Methane	CH_4	to atmosphere	3.2×10^{12}
Methyl iodide	CH_3I	to atmosphere	2.7×10^{11}
Dimethyl sulphide	$(CH_3)_2S$	to atmosphere	4.0×10^{13}

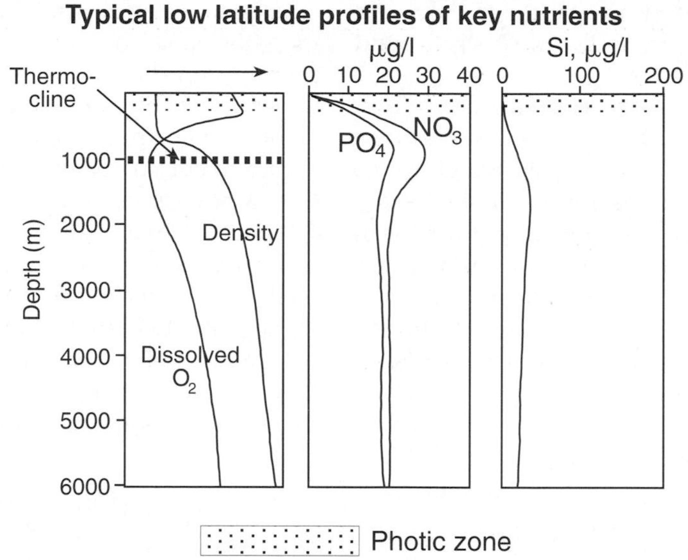

Typical low latitude profiles of key nutrients

Figure 2.3 Vertical distribution of some dissolved constituents through various locations in the oceans.

important in the modern oceans, and iron (Fe) plays a part in plant and bacterial activity. Many trace constituents show nutrient-type behaviour due to involvement in biological processes. Again, these exhibit non-conservative behaviour due to uptake and release by organisms, but typical concentrations are shown in Figures 2.2 and 2.3.

2.3.5 DISSOLVED GASES

The main gases dissolved in seawater are nitrogen (N_2), oxygen (O_2), carbon dioxide (CO_2), hydrogen (H_2) and the noble gases argon (Ar), neon (Ne) and helium (He) (called noble because they are very unreactive and do not form compounds easily). N_2 and the noble gases are conservative, their distribution being controlled by physical processes such as the mixing of water bodies and the effect of temperature and salinity on their solubility. In contrast, O_2 and CO_2 are non-conservative and their concentrations are strongly dependent on photosynthesis and respiration processes (the former consumes CO_2 and releases O_2, the latter consumes O_2 and releases CO_2).

2.3.6 ORGANIC MATERIAL

This grouping includes a variety of organic (carbon-containing) molecules, which typically occur at low concentration and are produced by both metabolic and decay processes in plants, microorganisms and animals. Various carbohydrates, lipids (fats) and proteins have been identified.

2.4 THE CHEMICAL AND PHYSICAL STRUCTURE OF THE OCEANS

So far in this chapter, we have discussed the chemical features of seawater, its behaviour and its ability as a solvent. These processes can now be applied to understanding the behaviour of ocean systems.

2.4.1 SALINITY

The total content of dissolved materials in seawater is expressed in terms of salinity. Since most of the dissolved material in seawater is anions and cations, salinity can be thought of as being the total mass of dissolved salts. The units used are parts per thousand (‰). Average salinity in the oceans is around 35‰ salts by weight.

Unfortunately for oceanographers, salinity cannot be determined simply by evaporating a known volume of seawater and weighing its salt content, because some components are volatile, and even an apparently dry salt residue will retain some H_2O molecules. Therefore, salinity is determined indirectly, either through titration, or more commonly now through conductivity measurements. In the former method a titration against chloride with silver nitrate is carried out, and, based on the principle of constant proportions, salinity is calculated using the following equation:

$$\text{Salinity}\ (‰) = 1.80655 \times \text{chlorinity}\ (‰)$$

where chlorinity is the total amount of halogens (i.e. F, Cl, Br, I and At) expressed as g/kg. The more common method of determining salinity today uses the principle that the electrical conductivity of seawater is directly proportional to its ion content. Electrical conductivity, and hence salinity, is measured using a salinometer, an electrical probe which has been calibrated against a water sample of known salinity (usually an internationally accepted salinity standard such as Standard Sea Water).

Variations in salinity occur in the oceans as a result of variations in the rate of fresh water input from precipitation, ice-melting and rivers, and in the rate of water removal by evaporation and ice formation. Surface seawater salinity has a close direct relationship with the overall balance of evaporation and precipitation (Figure 2.4). High precipitation in the tropics and mid-latitudes depresses surface water salinity in those locations, while clear skies, dry air and high evaporation in the subtropical ocean increase surface water salinity, and show a relatively clear latitudinal dependence (Figure 2.4). Open ocean salinity values range from 33–37‰, although higher values occur in regions of high evaporation (e.g. Mediterranean, 39‰, and the Red Sea, 41‰) and lower values near the mouths of major rivers (e.g. the Amazon) or areas of ice melt. Importantly, the Atlantic Ocean has higher salinity than the Pacific, which appears to be a function of atmospheric processes, and is suspected to be a key feature of the thermohaline circulation, discussed in Chapter 3.

2.4.2 TEMPERATURE

2.4.2.1 Temperature and latitude

The surface temperature of the oceans varies seasonally, and latitudinally as a function of solar insolation (radiation from the Sun that reaches the Earth's surface) (Figure 2.5). In the spring and autumn equinoxes, the Sun is at its highest angle over the equator and its lowest angle over the polar regions, so that, at low latitudes, a beam of solar radiation is spread over a smaller area, and passes through a smaller thickness of atmosphere, than at higher latitudes. Reflection of solar radiation is also increased at high latitudes, because of the low angle of incidence. The result of these features is that equatorial regions receive four times as much heat as the poles (Murray, 1992). This imbalance is responsible for driving the large-scale atmospheric and ocean circulation, which acts to re-distribute heat from the warm equatorial regions to the cool polar regions. Globally, isotherms (lines of constant sea surface temperature) run east–west, parallel to

Figure 2.4 Seawater salinity and its variations. Lines of equal salinity are haloclines. (A) Global ocean surface salinity; note it is higher at the equator and highest in enclosed seas. (B) Vertical and lateral salinity variations in the Atlantic Ocean. (C) Salinity profile of the sea surface; note that equatorial salinity can be lower because of increased rainfall in tropics. (D) Corollary of (C), showing evaporation and precipitation characteristics with latitude. (A) Reproduced with permission from Tait, *Elements of Marine Ecology* (1968), Butterworth Heinemann Publishers, a division of Reed Educational and Professional Publishing, Oxford. (B) Reproduced with permission from Bearman (1989b).

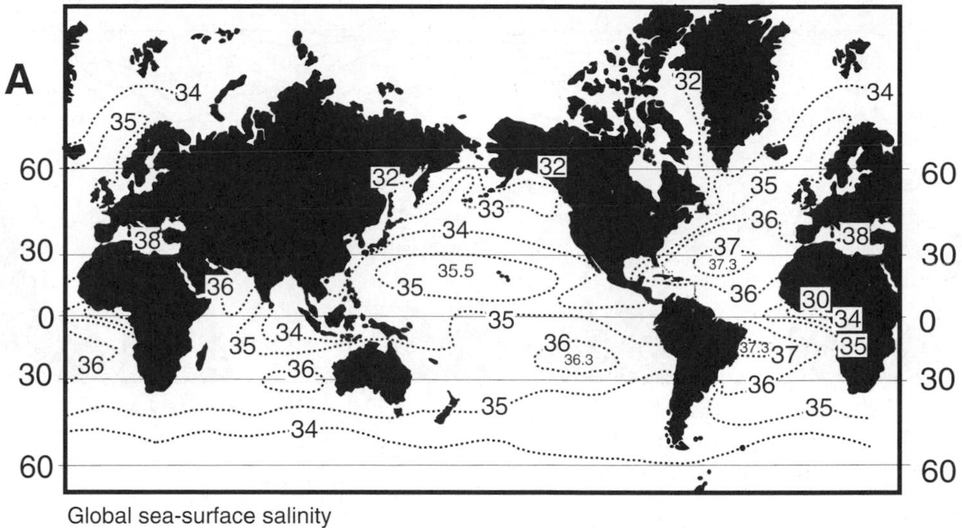

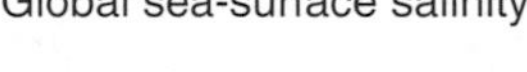
Global sea-surface salinity

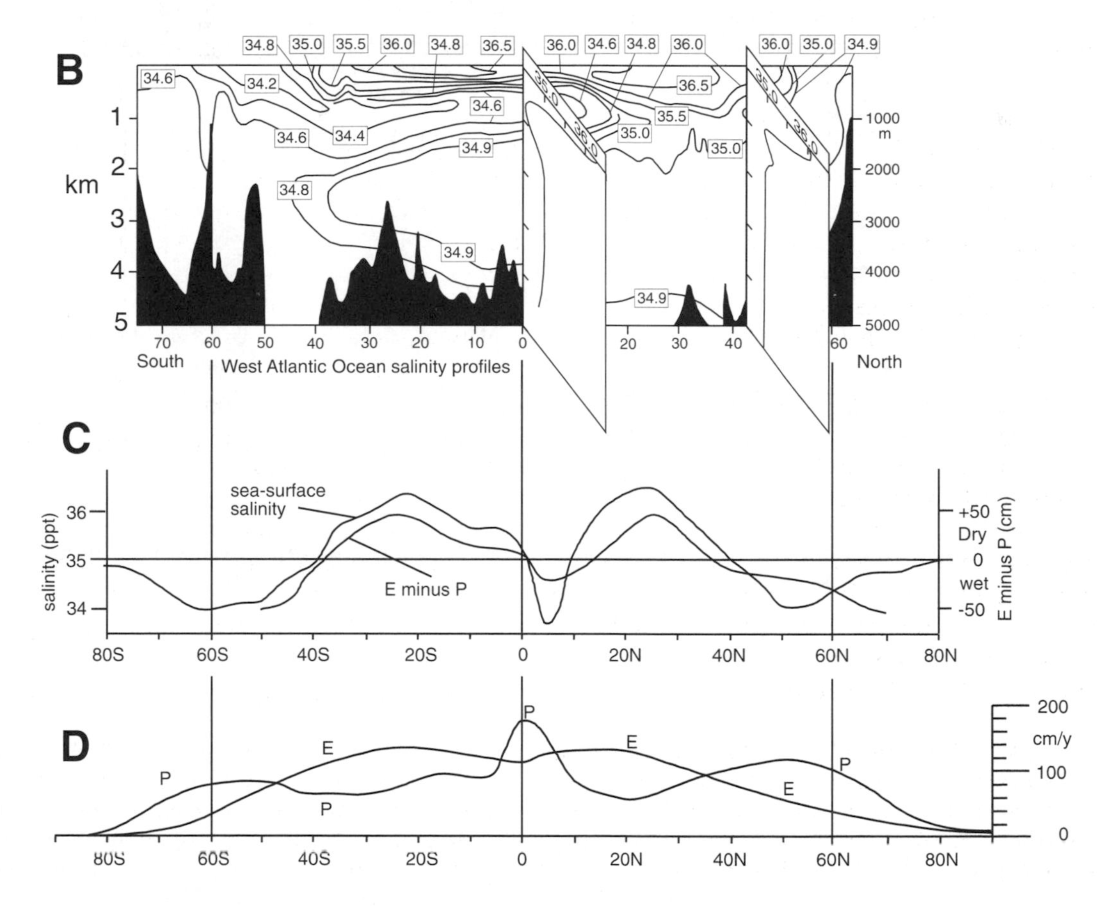

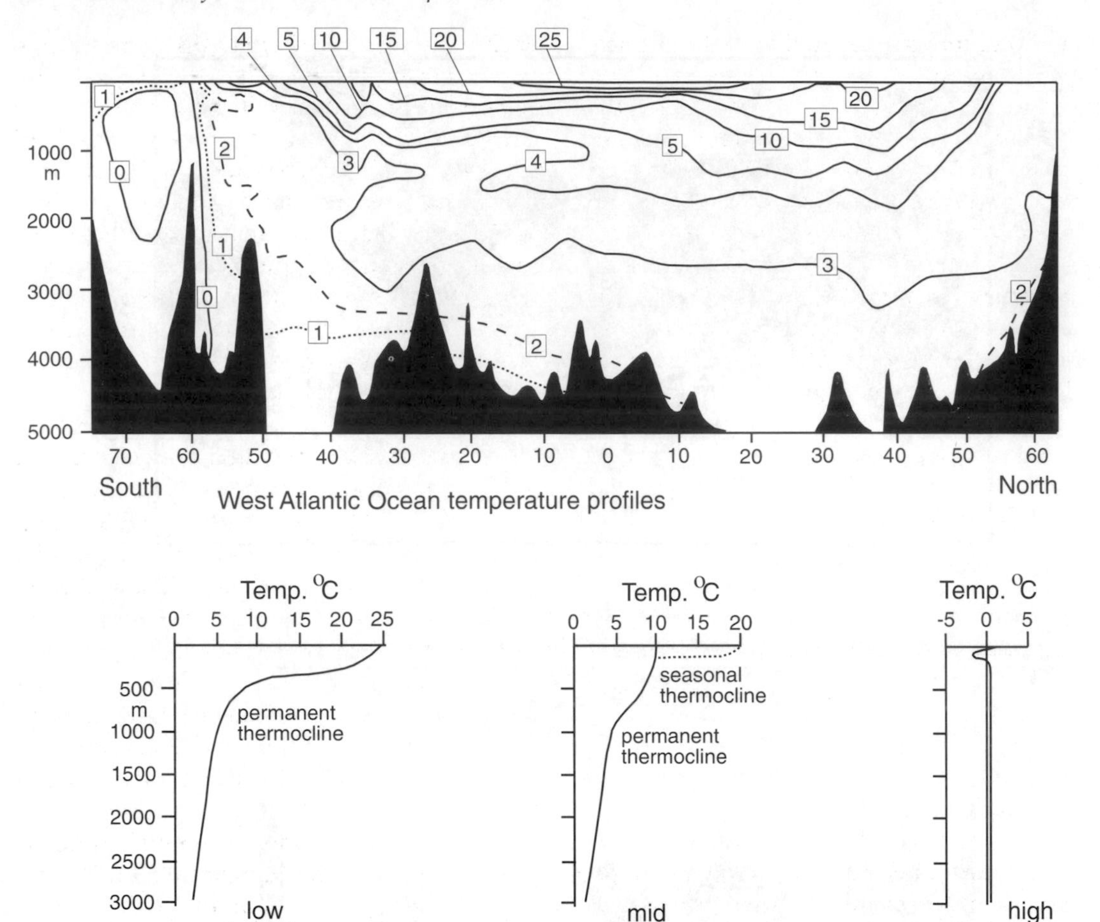

Figure 2.5 Cross-section of the Atlantic Ocean showing thermal structure, together with generalised vertical temperature distribution in the low, mid and high latitudes. Reproduced from Bearman, 1989b, with permission from the Open University. Lower section reproduced from Pickard and Emery, Descriptive Physical Oceanography – an introduction, 4th edn, Pergamon Press.

lines of latitude, and sea surface temperature decreases away from the tropics (Figures 2.5, 3.15). Variations in the east–west trends of the isotherms are caused by surface ocean currents, which transport warm water poleward at the western edges of the oceans, and cool water equatorward at the eastern edges. The actual position of the isotherms varies seasonally, as solar heating is greatest during the summer. Generally, sea surface temperature ranges from −2°C (near sea ice at high latitudes) to 30°C (near the equator), although locally temperatures may be even higher (e.g. 32°C in the Persian Gulf, and much larger in areas of hydrothermal activity).

2.4.2.2 Vertical distribution of temperature

The oceans have a layered thermal structure, illustrated in Figure 2.5. The vertical water column in low and mid-latitudes can be divided into 3 general zones in terms of temperature structure. An upper, wind-mixed zone exists from the surface to 50–200 m depth, with temperature similar to that at the surface. Below this, temperature rapidly decreases in a zone between 200 and 1000 m deep, called the thermocline zone. At greater depths water temperatures are generally much cooler, and temperature changes only slowly with depth (deep water zone). Hence warm surface water floats over colder, denser water, separated by the thermocline zone. This pattern (Figure 2.5) is subject to seasonal variations, as summer heating causes a seasonal thermocline to develop in mid (and more weakly in high) latitudes, which is destroyed by surface cooling and wind mixing during winter. The thermocline is strongest in the low-latitude tropical oceans and weakest at polar latitudes, due to lower solar input in polar regions. Intense cooling at the sea surface and strong wind mixing may produce an almost isothermal water column in polar regions.

2.4.3 DENSITY

Temperature and salinity act together (with pressure) to determine the density of a particular body of water. Relative density determines the depth at which a distinct body of water will be found: denser waters sink while less dense waters tend to lie at shallower depths. Density differences between water bodies are responsible for deep ocean circulation (Chapter 3). Typical ocean water has densities between 1020 and 1070 kg m^{-3}, while fresh water has a density of *c.*1000 kg m^{-3}. For convenience, oceanographers express density using the symbol σ_t, and is normally expressed to 2 decimal places; extremely small density differences can cause sinking of water bodies. Since density cannot be measured directly this precisely, it is usually calculated from accurate temperature and salinity values using a complex formula or tables. Warm water is less dense than cold; high salinity water is more dense than low salinity water.

Temperature exerts a stronger control on density than does salinity in the open ocean, so it is by no means uncommon to find a layer of warm, saline water overlying a layer of cold, fresher water. Vertical density profiles for the ocean are shown in Figure 2.6, the rapid change of temperature with depth in the thermocline producing a strong density gradient, or pycnocline. This divides the water column into 3 layers, a warm surface layer of low density, overlying a zone of change in density (the pycnocline), in turn overlying a cool, denser deep layer. This creates a strongly stratified structure, which is very stable, whereby less dense water sits on top of more dense water. It takes a large amount of energy to mix water across the pycnocline, and so large-scale vertical movement and mixing are minimised. In high latitudes of the polar regions, the weak thermal profile gives rise to a water column that is only

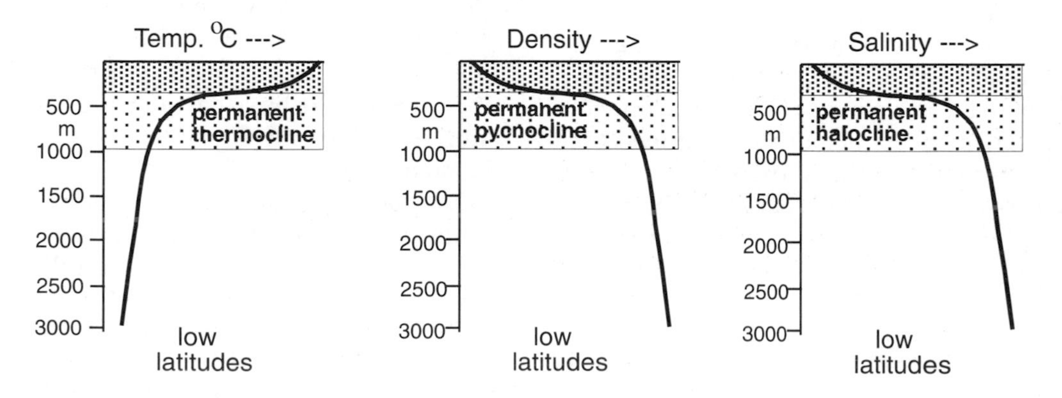

Figure 2.6 Generalised vertical profiles in oceans at low latitudes.

weakly stratified, the density shows little variation with depth and so vertical mixing or overturning can occur (Figure 2.6). Surface water cooling during winter produces a layer of dense surface water and an unstable situation where denser water sits on top of less dense water, shown by a slight decrease in density with depth. The polar surface water can then sink (Figure 2.6), with consequences discussed in Chapter 3.

2.5 OCEAN–AIR INTERFACE

This operates as a complex barrier zone between the two fluids of air and water. Gases, vapours and particles cross the ocean–atmosphere interface in both directions, and there is a considerable body of research on the nature of the interface. Much ocean produce stays in the top layer of the ocean, recognised as a surface microfilm (Chester, 1990; Liss and Duce, 1997) *c.*1 mm thick. Above the sea surface is a *c.*0.5 mm boundary layer of reduced atmospheric turbulence.

The ocean surface collects a range of materials, much organic matter being released by dead plankton (e.g. amino acids, proteins, fatty acids, lipids and phenols), and some is brought to the surface by rising air bubbles. As well as organic breakdown products, toxins of various types accumulate in the surface microfilm. The now-famous dimethyl sulphide (DMS), released by plankton and transferred to the atmosphere as a cloud-seeding agent, has been proposed as a Gaian-type mechanism of climate control, and has attracted much interest in the mitigation of global warming. In this mechanism, higher air temperatures at the ocean surface are partly caused by reduced cloud cover; plankton release DMS to the atmosphere to induce cloud formation, thereby encouraging cooling. This biotically mediated regulatory process is the essence of the Gaia theory.

Surface effects of increased UV-B radiation resulting from ozone destruction by CFCs is well appreciated on a personal basis by members of the human population, but effects on reactions and biota in the surface layer are yet to be evaluated; UV-B can penetrate several metres of surface water. While the interchange of material across the ocean–air interface, and the frictional forces of wind on water, cause a linked system, the surface is also a barrier, and the interrelationship is a complex one.

2.6 CONCLUSION

This chapter provides an overview of the main features of ocean water which control ocean systems. The physical and chemical characteristics of ocean water interact with the solar input to form a complex control system influencing the entire Earth's surface. The key controls are salinity and temperature, because they influence density of water masses, critical in driving ocean circulation. What is clear is that the study of oceans involves processes at different scales. At the molecular scale, the peculiar properties of water are translated to the global scale, because there is such a large amount of water in the oceans.

2.7 SUMMARY

The characteristics of seawater include the following factors, all of which have a critical impact on the operation of ocean systems.

1. Water structure is dominated by hydrogen bonding, which is the attractive force holding water molecules together; without it, water would be a gas, not a liquid, at surface temperatures, and so there would be no oceans.
2. Water has a high heat capacity because of hydrogen bonding; thus water stores much heat and has a moderating effect on climate, preventing extremes of cold and heat.
3. Water expands when it freezes, and sea ice excludes most of its salt on freezing. Thus not only does sea ice float, the remaining water is richer in salt, which makes it

denser, and it sinks, aiding the deep ocean circulation.

4. Because of hydrogen bonding, water is an excellent solvent, and seawater carries a heavy load of dissolved material, critical for both organic and inorganic reactions.
5. Seawater salinity and temperature are important controls on density. Vertical and lateral temperature and salinity differences produce bodies of water with differing relative densities, which may stratify the water column, or alternatively cause it to overturn.

THE MOVEMENT OF OCEAN WATERS 3

3.1 INTRODUCTION – THE PERSPECTIVE OF SEAWATER CIRCULATION

Ocean waters are in a state of constant movement, stirred and mixed by two major forces: *advective* (those which cause displacement of water) and *turbulent* (involving random motion of water molecules). These mix huge quantities of water, together with their load of dissolved and suspended substances and gases, both vertically, between surface and deep waters, and horizontally, around the Earth. Because of the high heat capacity of water highlighted in Chapter 2, large-scale ocean circulation is responsible for approximately one-third of the long-term net transfer of heat energy from the tropics to the poles, and is therefore an important part of the global climate system. Furthermore, circulation in the deep ocean aerates bottom waters and causes mixing between the ocean basins. Thus the concept of ocean water movement is simple, but the details are horrendously complex because of the problems in identifying the shape and distribution of water masses. Complexity of water circulation is also due to the different scales of motion in the ocean, from small eddies up to ocean-scale gyres.

Because of its physical properties, ocean water tends to form masses which maintain their integrity for long periods, due largely to salinity (S) and temperature (T) characteristics acquired at the surface. Once water masses lose contact with the surface, T and S can only be changed by slow mixing with other water bodies. Thus the typical maximum age (= length of time it maintains its integrity) of a deep ocean water mass is around 1500 years (Crowley and North, 1991, p. 28), a short time in a geological context, attesting to the efficiency of ocean mixing and stirring. This chapter aims to identify the critical aspects of modern ocean water movement, which has applications not only for the geological past, but for modelling future changes.

Ocean flow falls into two broad categories (Figure 3.1). Firstly, the ocean surface waters are wind-driven; they flow in the shallow water above the thermocline, and have been well documented for hundreds of years because they constrain the motion of sailing ships. Secondly, deeper water circulation beneath the thermocline is driven by differences in density between water masses. The thermocline therefore operates as a barrier between the two circulation patterns, so that they are effectively decoupled, except in places where deep circulation intersects the surface (especially the North and South Atlantic). The thermocline is not a continuous or uniform feature, becoming less important in polar regions because of the lower temperatures of surface waters (Figure 2.6). The patterns of movement of both surface and deep ocean circulation are considerably modified by the irregular distribution of continental masses. Finally, the topography of the ocean floor has a strong impact on deep circulation, because of the barriers created by ocean-floor ridges.

Overall, four groups of water masses have been identified (Pinet, 1992, p. 183): *central waters* represent all the water from the surface to the base of the thermocline, and therefore contain the surface circulation system described in Section 3.2; *intermediate waters* lie beneath central waters, down to about 2 km, beneath which are *deep waters* and the (coldest,

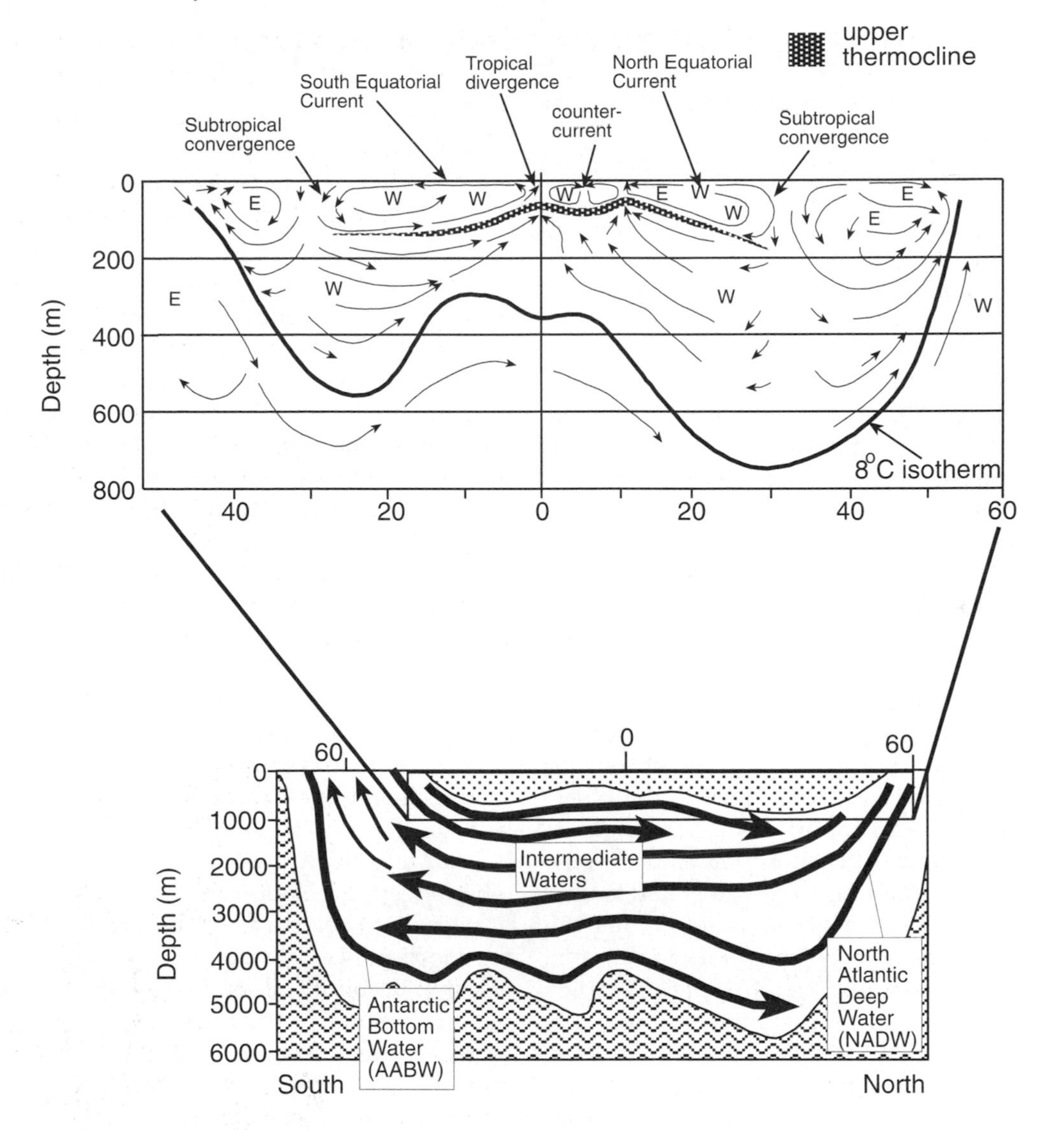

Figure 3.1 Schematic N–S cross-section of the Atlantic Ocean showing the distribution of surface and deep waters. The boundary between the two is indicated by an isotherm at *c.*8°C; thus the thermocline is not only a feature of lower to mid latitudes, but is also a band, rather than a line. The fundamental differences in flow attributes between surface and deep waters is shown: surface waters show limited vertical ellipsoidal motion above the thermocline, while deep waters are volumetrically much more abundant and develop large distance transport; this difference is the reason why deep waters are so important in heat transport around the Earth's surface. Upper diagram redrawn from Wells (1997), *The Atmosphere and Ocean: a Physical Introduction*, © John Wiley and Sons Ltd, reproduced with permission; lower diagram redrawn from Bearman (1989b), reproduced with permission from the Open University.

of polar origin) *bottom waters*. This observation is of critical importance not only in understanding how the modern oceans work, but therefore also in reconstructing ocean changes in the recent past.

Understanding the nature of ocean circula-

tion is of great value to Earth scientists attempting to model flow in ancient oceans, but complexities also lie in defining the exact positions of the continents through time. As a result, the more recent past has been modelled with greater confidence than older times. Change in ocean circulation has been shown to be rapid in some cases, with large implications for regional and global climates, and these aspects are dealt with in Part B of this book.

3.2 SURFACE OCEAN CIRCULATION

3.2.1 GENERAL FEATURES

Surface ocean waters form a pattern of sufficient permanence that the different regions are named (Figure 3.2). This pattern is due to established wind-driven water flow which has a variety of features, as follows (Figure 3.3).

1. Water flow is dominated by five large wind-induced elliptical currents (gyres) which occur in the North and South Atlantic, North and South Pacific and Indian Oceans. Northern hemisphere gyres have clockwise flow, while Southern Hemisphere gyres flow anticlockwise. Eastward-flowing countercurrents move along the equator between the gyres.
2. Gyres transport warm low-latitude waters towards the poles at the western edges of the ocean basins, and cool high-latitude waters towards the equator at the eastern edges of the ocean basins.
3. Current flow in gyres is not uniform, the strongest currents occur on their western

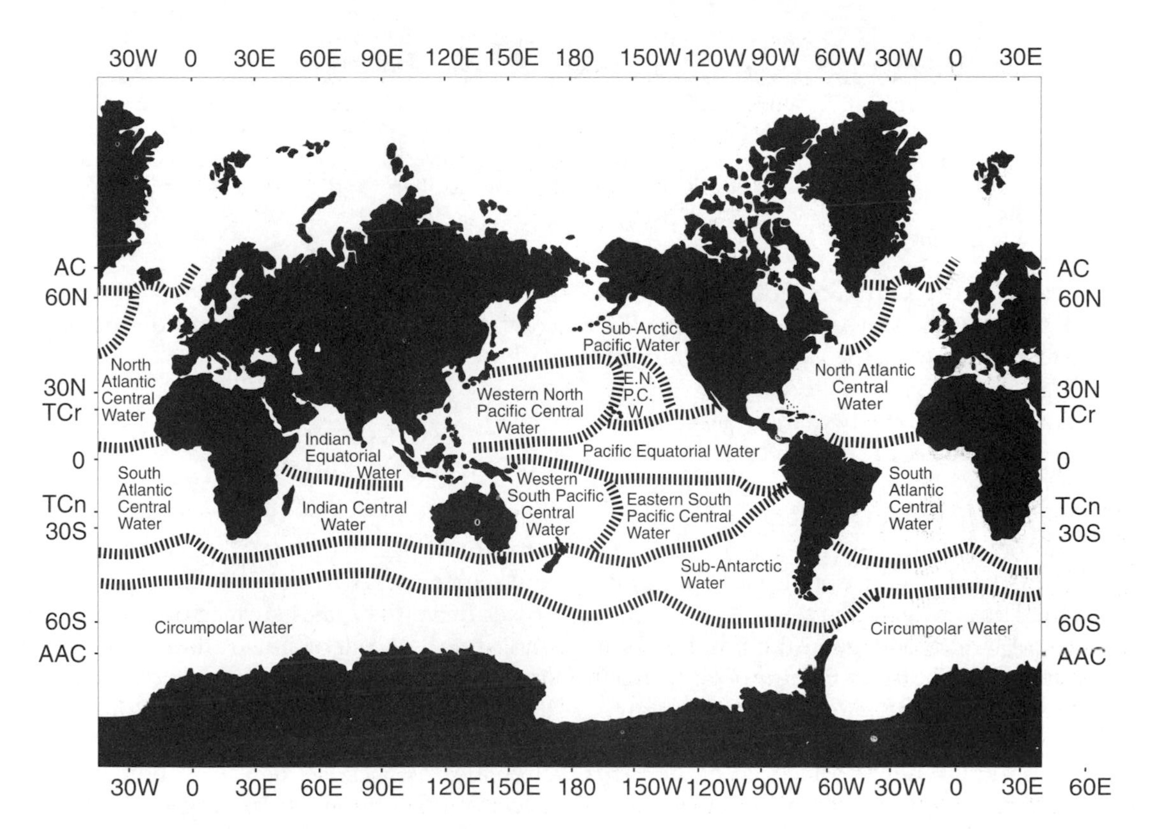

Figure 3.2 Distribution of permanent surface water masses, which are controlled by wind, as illustrated in Figures 3.3 and 3.4. Reproduced from Bearman (1989b), with permission.

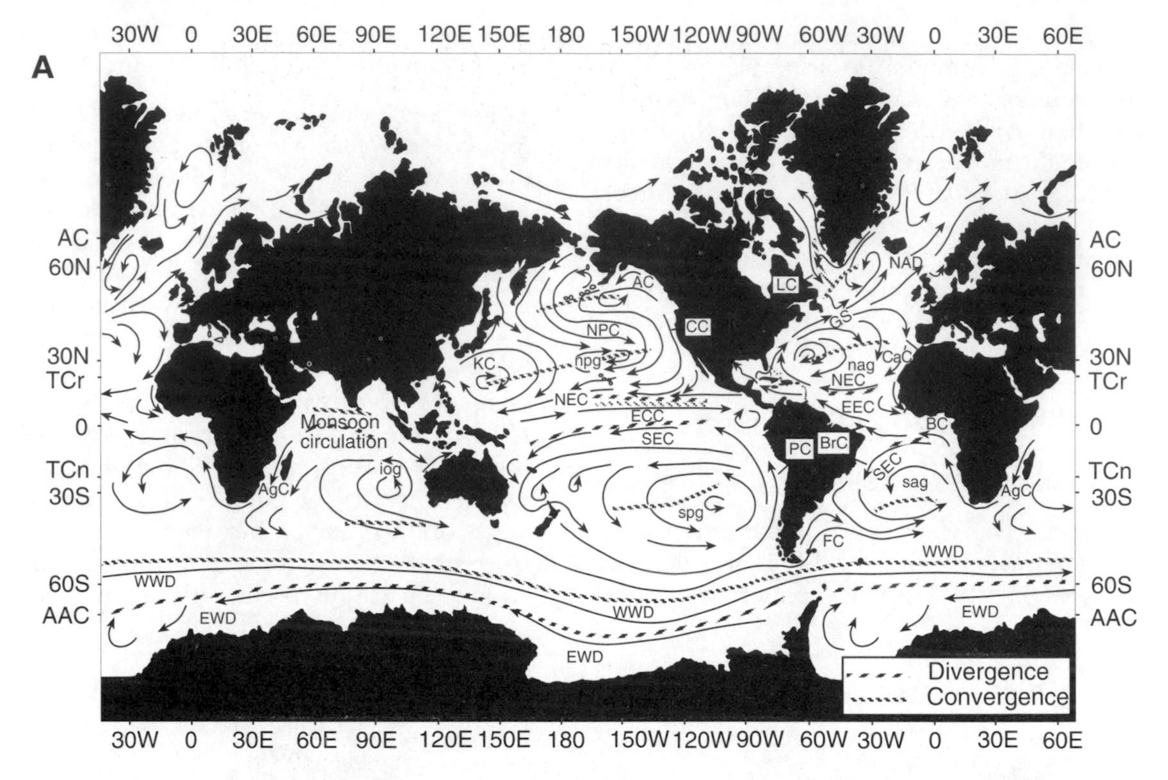

Figure 3.3 Ocean surface currents. (A) Location and flow patterns of global surface ocean currents. Note particularly the clockwise and anticlockwise flow of Northern and Southern Hemisphere gyres respectively.

sides (e.g. the Gulf Stream, the Kuroshio Current) with slower currents on their eastern sides (e.g. the Canary Current, the Benguela Current).

4. Currents in the Northern Indian Ocean vary seasonally due to the reversal of winds (monsoon means *season* in Portuguese, Malay and Arabic). Seasonal reversals are common globally, but are intensified in the Indian Ocean because of the close proximity of the Himalayas to the sea.
5. Lack of land blockage in the Southern seas allows the Antarctic Circumpolar Current (West Wind Drift) to circle the Antarctic continent. This is the only place in the present oceans where surface ocean currents run uninterrupted around the globe. There are at least two other instances of proposed circumglobal currents in geological history: the mid-Palaeozoic and Late Cretaceous, and their implications are discussed in Part B.

3.2.2 EFFECTS OF ATMOSPHERIC CIRCULATION PATTERNS

Winds blowing over ocean surfaces exert a stress (or drag) on the water, generating waves and currents. Atmospheric circulation processes drive the major surface ocean currents, and are an excellent illustration of the interlinked nature of ocean and atmosphere. We therefore need to consider the rudiments of atmospheric motion; the Earth's atmosphere circulates as a consequence of the combined effect of (a) differential heating of the Earth's surface between tropics and poles (Figure 3.4), and (b) the rotation of the Earth.

B

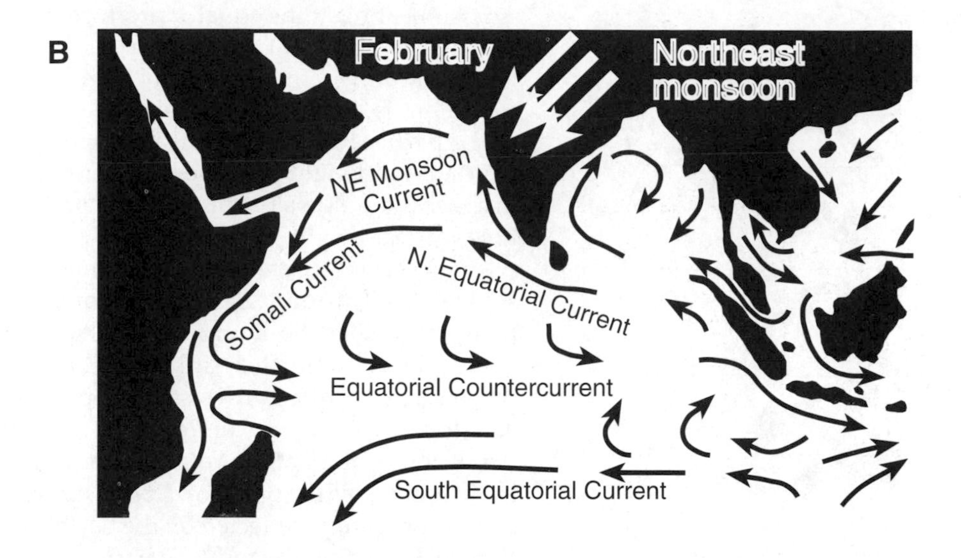

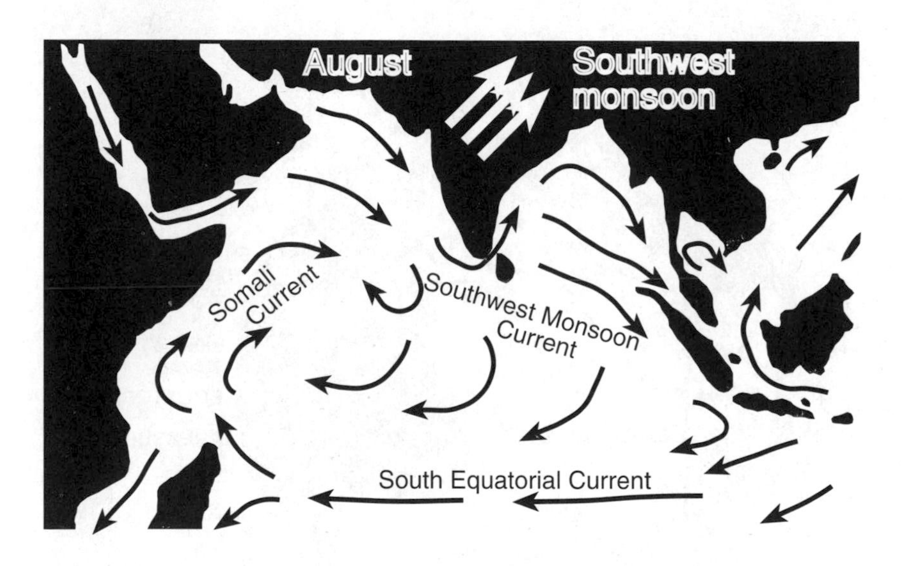

Figure 3.3 (*continued*) (B) All waters bordering continents have seasonal flow because of differences between heat capacity of water and land in winter and summer, but the Indian Ocean is particularly affected by the presence of the Himalaya mountain mass, accentuating monsoon circulation.
AC = Alaska Current; AgC = Agulhas Current; BC = Benguela Current; BrC = Brazil Current; CC = California Current; CaC = Canary Current; ECC = Equatorial Countercurrent; EWD = East Wind Drift; FC = Falklands Current; GS = Gulf Stream; iog = Indian Ocean Gyre; KC = Kuroshio Current; LC = Labrador Current; NAD = North Atlantic Drift; nag = North Atlantic Gyre; NEC = North Equatorial Current; NPC = North Pacific Current; npg = North Pacific Gyre; PC = Peru Current; sag = South Atlantic Gyre; SEC = South Equatorial Current; spg = South Pacific Gyre; WWD = West Wind Drift.

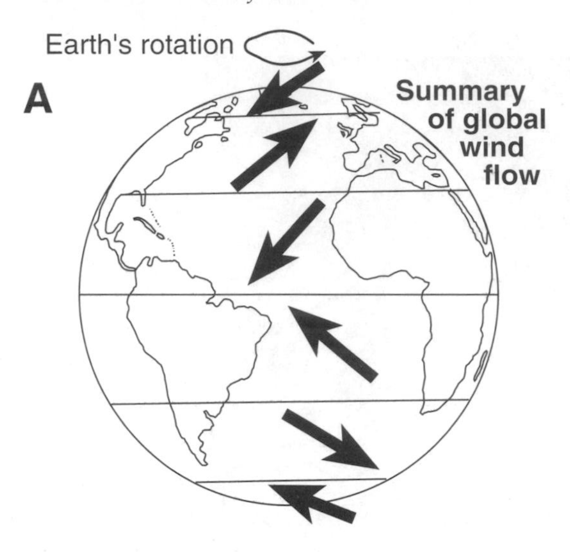

Figure 3.4 (A) Summary of major wind flow patterns, for reference when examining (B) and (C).

Solar heating at the equator produces warm, low density, buoyant air which rises, creating low pressure in the equatorial regions. The rising air is deflected by the tropopause (a major barrier to further vertical air movement at 10–15 km height because of a temperature increase), and flows towards North and South Poles at that height. Because the air cools as it rises, it becomes denser and descends in the subtropics, at around 30° latitude N and S, and therefore creates the subtropical high pressure zone (Figure 3.4). This sinking air diverges when it reaches the Earth's surface: part flows equatorwards, and undergoes apparent deflection by the Earth's rotation (forming the trade winds), while part flows polewards, and is also apparently deflected by the Earth's rotation to form the westerlies, characteristic of

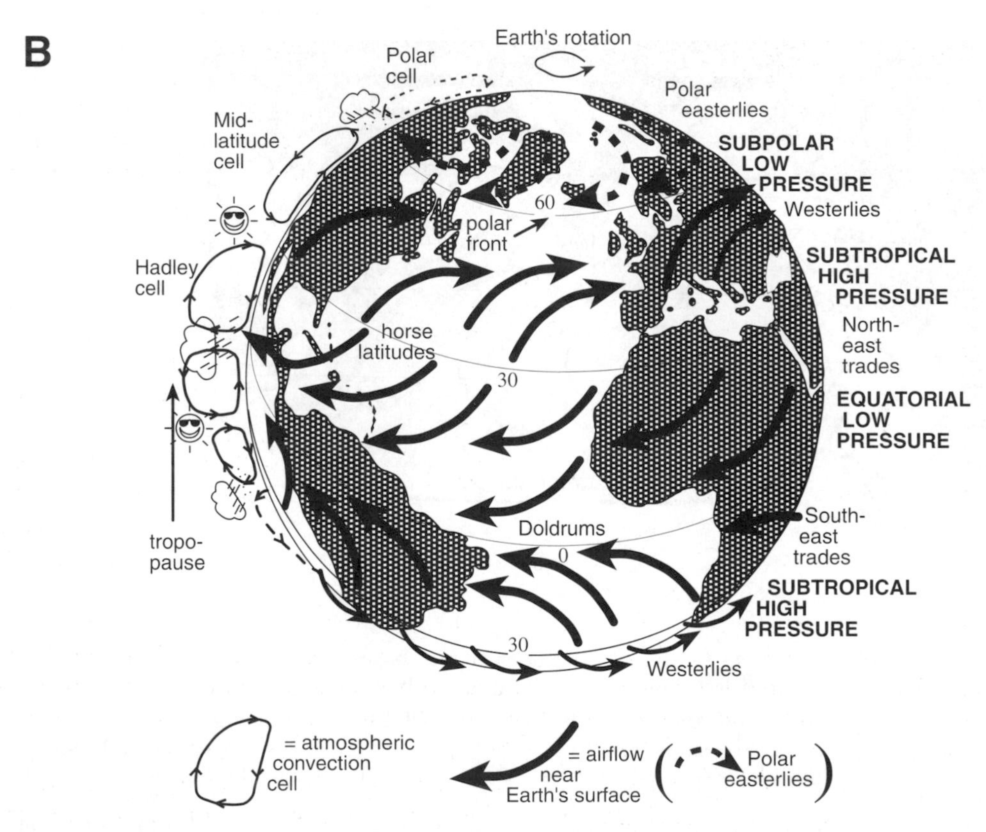

Figure 3.4 (*continued*) (B) Atmospheric circulation cells combined with the Earth's rotation (Coriolis force) to produce global wind patterns; Northern Hemisphere emphasis.

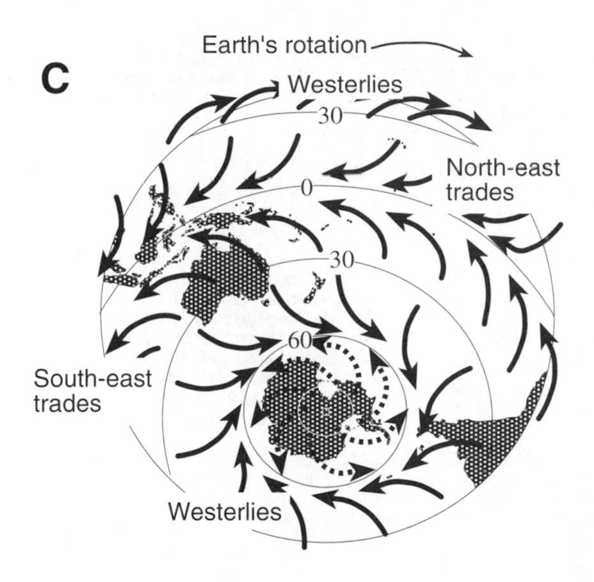

Figure 3.4 (*continued*) (C) Wind patterns; Southern Hemisphere emphasis.

mid-latitude areas. The deflections are due to the Coriolis effect, explained in Section 3.2.3. Cold, dense air sinks over the polar regions and flows equatorwards as the polar easterlies, which meet the westerlies at the Polar Front, at around 60°N and S.

The pattern of atmospheric circulation would be uniform if the Earth's surface was flat and was entirely ocean or entirely land, but because of the combined effect of physical intrusion of the continents into moving air, and differences between the heat capacity of water and land, the circulation is uneven, resulting in a complex flow pattern. This pattern includes a series of irregular ellipses of wind flow over land and ocean, induced by the topographic variation of the Earth's surface; they change with the seasons, and do not correspond to the ocean flow patterns. Surface ocean currents are constrained by the blocking effect of continents; water obviously cannot flow over the continents, only around them, and therefore turns into gyres (Figure 3.3), which in turn influence the water beneath them. A discussion of the controls and implications of clockwise and anticlockwise flow on atmosphere and oceans follows in the next section.

3.2.3 CORIOLIS EFFECT: PRINCIPLES

Moving air and water masses are influenced by the Earth's rotation in a process called Coriolis deflection. This effect causes water and air currents to be apparently deflected to the right of their direction of motion in the Northern Hemisphere, and to the left in the Southern Hemisphere. It is important to realise that Coriolis deflection is not a real force; movement of an air mass over the Earth's surface *appears* to a viewer, on the ground, to have been deflected, when in fact a viewer in a space vehicle, looking down, would see the movement of the air mass in a straight line as the Earth rotates beneath it. Coriolis deflection depends on the fact that the Earth's speed of rotation is fastest at the equator, and gradually declines to zero at the poles (see Figure 3.6). A desk globe of the Earth serves to show how this works. The globe rotates on its spindle, and a speck of dust stuck to central Africa (on the globe's equator) moves very quickly in a circle, and covers a distance equal to the circumference of the globe for each time it goes round. Dust in higher latitudes (e.g. Northern Europe or the northern USA) goes a shorter distance in the same time, and therefore moves slower; dust on the North Pole just rotates (see Figure 3.6).

On the Earth, a solid object, such as an aeroplane, or a less solid but relatively integral body, such as an air or ocean water mass, which travels from the equator northwards, will be moving over ground which is decreasing in speed. Now, because air heated at the equator rises and travels both north and south, northward-moving air passes over ground which is travelling slower the further north it gets, and therefore the air mass runs ahead (i.e. eastwards), in a curve, from the line of longitude it began from; it is *apparently* deflected to the east (Figure 3.5). If this air had coloured smoke injected into it, a man on

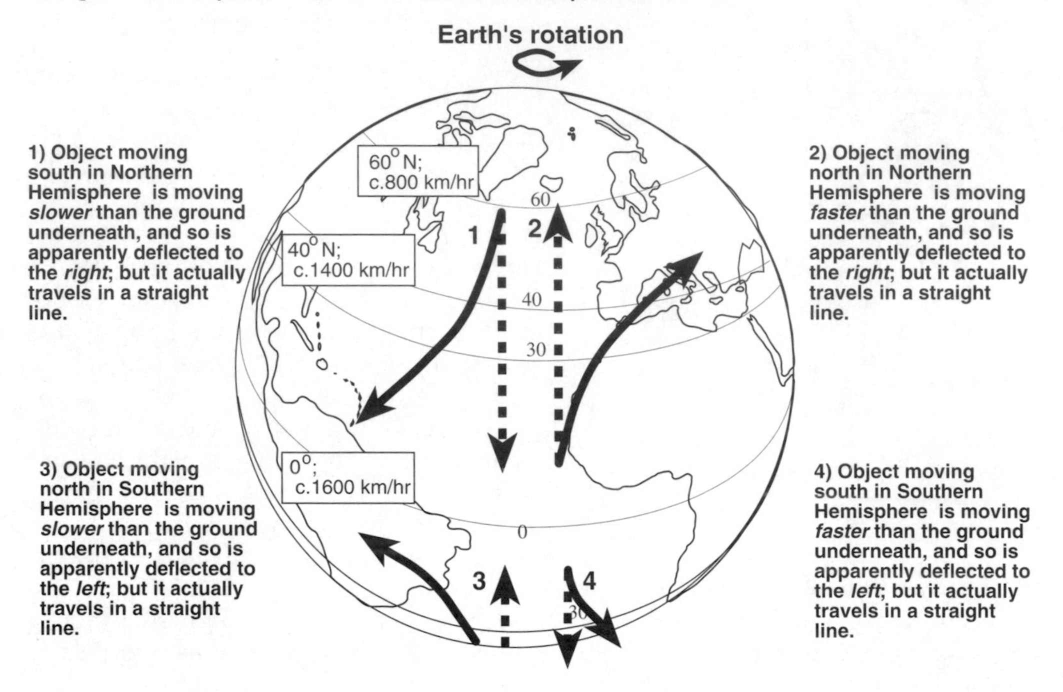

Figure 3.5 Simple illustration of the operation of the Coriolis force.

the ground looking in the same direction as the wind flow (north) would see the smoke deflected to his right. This air cools, as we have seen, and descends in the subtropics; near the ground its increased friction with the Earth's surface slows, but does not stop, the moving mass. The mass splits, part travels further north, and continues to curve ahead (eastward) of the slower Earth rotation to the north, generating westerly winds. The other part travels south, and, having been previously slowed down by friction with the ground, travels *slower* than the rotating Earth beneath it; that air therefore runs behind (westward), in a curve, of the line of longitude from which it started its journey south, forming the north-east trade winds. Note that as this air moves south, it flows over ground which is moving faster the further south it goes, with its fastest at the equator. A woman standing in the subtropics, facing in its direction of flow (south), would see coloured smoke, entrained in the air, also deflected to her right.

Figure 3.6 Speeds of rotation (to nearest 5m) at various latitudes, relative to a viewer on the Moon.

Latitude (deg)	*Speed of rotation ($m\ s^{-1}$)*
0	450
30 (subtropical high)	390
60 (subpolar low)	225
80	80

Therefore, freely moving objects travelling either north or south in the Northern Hemisphere are seen to curve to the right. In contrast, those in the Southern Hemisphere curve to the *left* of the observer who looks in the same direction as the flow, because the direction of viewing is opposite to that in the Northern Hemisphere (Figure 3.5). Furthermore, the amount of Coriolis deflection increases from poles to equator because of the increase in speed of rotation of the Earth towards lower latitudes, as shown in Figure 3.6. In physics, a force which changes direction is said to accelerate. Coriolis acceleration is mathematically equal to $2\Omega v \sin L$, where Ω = Earth's angular velocity of rotation, v = current velocity and L= latitude. This equation means that Coriolis acceleration is proportional to current velocity, and latitude. (Remember that, in physics, acceleration can mean a change in direction, without a change in velocity.)

This set of flow patterns is rather confusing at first, but is logically deduced, and easily observed. Perhaps the most confusing aspect is that Coriolis deflection itself is an illusion; the air/water/aeroplane movements are actually in straight lines, they only *appear* to an observer to curve left or right because of differences in the rotation speed of the Earth from pole to equator. Nevertheless, Coriolis deflection has a real effect on the air and ocean surface water circulation, in the formation of ellipitical masses of moving air and water.

Finally, note that we have only dealt here with the lower levels of atmospheric circulation; we have not yet considered the effects of Coriolis force in the oceans (see the next section). Furthermore, we exclude from this text the effects of the jet stream (higher atmospheric level) and its influence on generating cyclones and anticyclones.

3.2.4 CORIOLIS EFFECT IN OCEANS: OCEAN GYRES, EKMAN TRANSPORT AND UPWELLING

In oceans, the Coriolis effect operates much more slowly than in the air, because of the much higher viscosity and density of water, which therefore responds more slowly to the latitudinal variation of rotational speed. Also, the surface ocean circulation, which is caused by transfer of frictional energy from air to water, is only *c*.3% efficient, so surface currents are only *c*.3% of the velocity of an overlying air mass. Therefore in the short term, the atmosphere controls the distribution of heat over the Earth's surface (approximately 80% of surface heat is carried in the atmosphere and 20% in the ocean). Day-to-day climatic change is governed by the atmosphere, but in the longer term, the oceans have a much more powerful effect. This is because of the higher heat capacity of water, able to transport much larger amounts of heat, but more slowly, largely because of the greater viscosity of water; the differences are of great significance in deep ocean circulation, also because of its larger volume which can carry more heat (Chapter 2), discussed in Section 3.3.

As wind stress initiates water movement at the ocean's surface, the water at the very top of the ocean surface begins to move. Due to Coriolis deflection, the surface current so generated flows at approximately 45° to the right of wind direction in the Northern Hemisphere, and to the left in the Southern Hemisphere. The surface water layer exerts frictional drag on the water underneath it, which therefore begins to move. Drag involves a loss of energy, so the lower water moves slower than the water above it; it is therefore affected more severely by the Coriolis force, falls behind the upper water and so is deflected slightly to the right of the surface water. The drag of the wind is transmitted downwards to deeper levels in this way, and water at successively deeper levels moves slower, and is deflected slightly to the right of the water above, producing a spiralling current called the Ekman spiral (Figure 3.7). This is really a poor name because each level of water is moving *horizontally; this water does not corkscrew downwards, and so is not a spiral actually*. Because of friction, the speed of the current declines with depth. Although in detail the water motion is in many directions, calculations indicate that the bulk of the water (the net transport) moves at a 90° angle

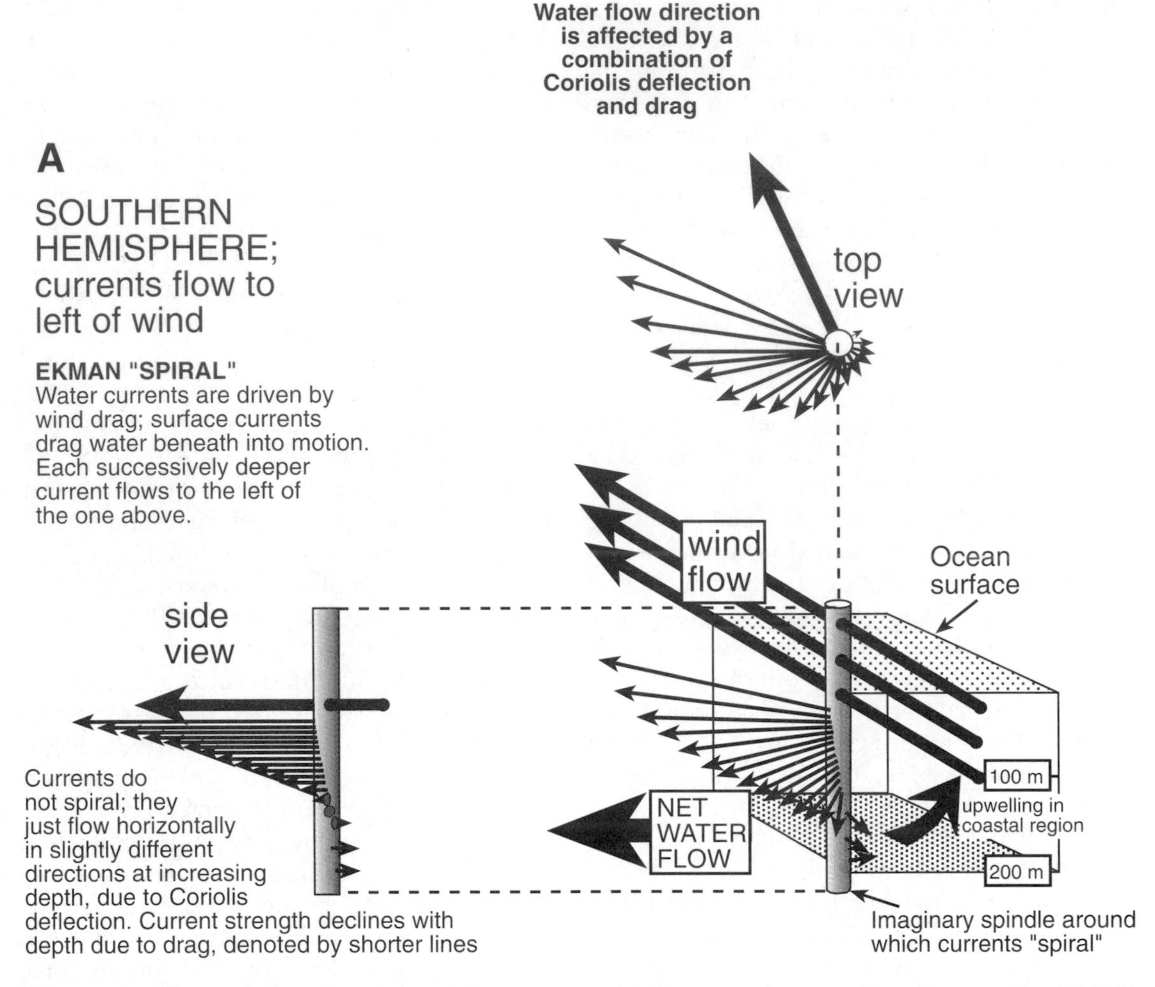

Figure 3.7 Diagram of the Ekman spiral flow. (A) Principal features of current flow, governed by Coriolis forces, illustrated using the Southern Hemisphere as an example.

to the right of the generating wind in the Northern Hemisphere; this is called Ekman transport.

Strong, persistent winds flowing parallel to the coast may induce Ekman transport, with net water flow at right angles to the coast. If Ekman transport draws water away from the coast, that water is replaced by deeper water rising to the surface (upwelling). Conversely, Ekman transport towards coastlines piles water against the coast, to induce down-welling. Upwelling and down-welling can also occur in the open ocean due to divergence and convergence of currents. Upwelling areas are highly productive, bring nutrients from below the thermocline (phosphates, nitrates, silicates, etc.) to surface waters which have been depleted of nutrients by biological demands. Examples of upwelling areas include the Canary current, California current, Peru current, Benguela current (these are all currents on the eastern boundaries of the oceans they flow in), Somali Current (caused by monsoon winds), and Antarctic divergence (produced by polar easterlies and westerlies) (Figure 3.3).

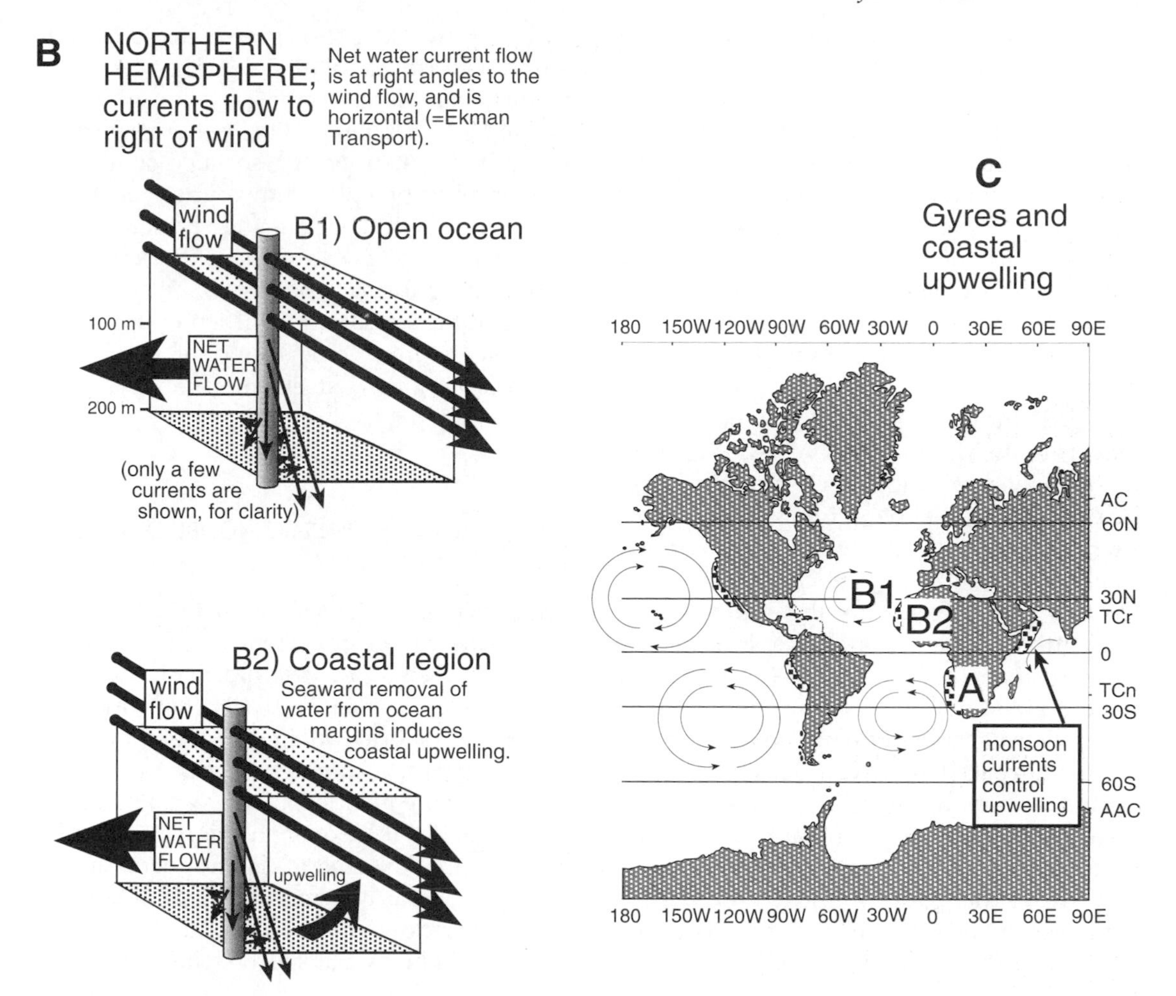

Figure 3.7 (*continued*) (B) Ekman processes near land generate coastal upwelling, illustrated for the Northern Hemisphere. (C) Locations of coastal upwelling in relation to ocean surface gyres, controlled by Ekman processes.

3.2.5 SEASONAL VARIATION: THE ASIAN MONSOON

The difference between the high heat capacity of water and the much lower heat capacity of land can have a considerable impact on ocean circulation. The large circulating gyres of the Pacific and Atlantic Oceans are relatively stable features, because these oceans are wide and are bounded by land of generally low topography, limiting the influence of heating and cooling of surrounding landmasses on their large-scale circulation. Surface circulation patterns in the Indian Ocean, in contrast, show considerable seasonal variation as a result of the climatic effect of the topographically prominent Asian landmass to the north of the ocean (Figure 3.3).

Land surfaces heat up and cool down far more rapidly than does the ocean, so they heat or cool air above them, respectively. In winter, cold, dense (therefore high pressure), dry air develops over land and flows from land to sea. In summer warm, less dense (therefore low pressure), moisture-laden air flows onto land from the sea. The effect of this is intensified in the Indian Ocean because of its relatively small size (confined between Africa to the west, and Indonesia/Australia to the east) and the geological accident that the Himalayas are so close to the ocean. This region is very cold in winter because of its height, so that sinking of the northward-flowing tropical air at tropospheric height is magnified, building up a mass of cold, dense air near the ground. This powerful air mass travels southwestwards into the northern Indian Ocean (remember – air deflects to the right as it travels south because it passes over ground which is moving faster the further south it goes, giving it a southwestward direction). These winds (the Northeast Monsoon) induce an anticlockwise current (the Somali Current) in the Indian Ocean. During the spring, the high pressure weakens as the land surface is warmed (in turn warming the air above it), and by the summer a strong thermal low pressure area (warm, rising air) develops over Asia. This vertical convection draws in warm, moist air travelling northeast from the Indian Ocean (the Southwest Monsoon), causing monsoon rains over the Indian subcontinent and parts of southeast Asia. The airflow and associated Indian Ocean current patterns reverse direction over 4–6 weeks.

The Somali Current flows at *c*.7.2 km/h (comparable with the powerful Florida Current) and, during the summer months, forms the western arm of a strong gyre completed by the Monsoon Current and the South Equatorial Current (Figure 3.3). As the South Equatorial Current approaches Madagascar it diverges, promoting upwelling of nutrients and high rates of productivity. Further north over the Arabian continental shelf, southwest monsoon winds (summer) cause offshore movement of water as a consequence of Ekman transport, again causing upwelling and high biological productivity. During autumn and winter, winds associated with the northeast monsoon cause the Somali current to weaken and reverse, and a cyclonic (anticlockwise) circulation develops.

This example not only serves to show how air and surface circulation patterns form an integrated system with associated continental topography, but also draws attention to the problems of reconstructing ancient ocean currents. It is not sufficient to know the positions of former continents, but also their topographic variations (difficult, if not impossible in most cases), and of course, the climatic regime, itself not always well constrained.

3.2.6 THE GEOSTROPHIC MODEL OF OCEAN FLOW

The previous sections show that surface winds drive surface ocean currents, and the Ekman effect induces net transport of water at right angles to the wind direction, inducing upwelling or downwelling in the appropriate circumstances. Although ocean gyres are created by the interaction between atmosphere and ocean, the full explanation is more complex. Taking the Northern Hemisphere as an example, westerly winds in the mid-latitudes produce a flow of water towards the south due to Ekman transport, whereas Ekman transport associated with the trade winds (blowing towards the west) induces a flow of water to the north. These currents converge in the mid-ocean, and build up a mound of water in the centre of the gyres (Figure 3.8). This produces a sloping sea surface, and consequently a Pressure Gradient Force (PGF), and leads to a more complete model of gyre flow, called the geostrophic model (Greek *geo* = earth, *strophe* = turning). What follows requires careful reading.

Pressure is equal to force divided by the area over which the force acts. Hence, at a given depth z, pressure is equal to the weight of a 1 m × 1 m water column of height z (plus atmospheric pressure), according to the following relationship:

$$P = F/A = (mg)/A = (rvg)/A = (rAzg)/A$$

(where P = pressure, F = force, A = area, m = mass, g = acceleration due to gravity, r = density, v = volume, z = water depth).

This means that horizontal differences in pressure can be caused by:

1. differences in density between 2 points on the same horizontal surface (variations in r); or
2. variations in height of the water surface, or a sloping sea surface (variations in z) (the area, A, and gravitational constant, g, will not change).

The difference in pressure causes water to flow from the area of higher pressure to the area of lower pressure, forming a PGF, which increases as density differences increase, or the sea surface slope becomes steeper. Piling up of water into a mound by the westerlies and trade winds means that PGF will act on the ocean surface, forcing the water to flow downslope away from the surface of the mound. As this water begins to flow it is deflected by the Coriolis effect (to the right of the direction of motion in the Northern Hemisphere), and as it accelerates the Coriolis deflection increases (because Coriolis deflection is proportional to current velocity) until a balance is achieved, the PGF dragging the water downslope is equal to the Coriolis effect pushing it to the right (uphill), and the water flows parallel to lines of latitude (often referred to as zonal transport), a condition known as geostrophic flow. Strictly speaking, this is an approximation; the current actually curves slightly and is called a gradient current.

The continents interrupt the zonal transport pattern and deflect the westerly flow polewards and the easterly flow equatorwards, producing closed ocean gyres. Ocean circulation reconstructions indicate a well-developed North Atlantic Ocean gyre by the early Tertiary, thereby giving some idea of the longevity of that part of the modern surface circulation system.

3.2.7 WESTERN BOUNDARY INTENSIFICATION

Section 3.2.6 explains the formation of gyres, but why are currents stronger on the western sides of the gyres?

The large ocean gyres are asymmetric, having intensified current flows on the western peripheries of the ocean basins (e.g. Gulf Stream, Kuroshio Current, Brazil Current). The Gulf Stream and the Kuroshio Current are narrow streams (typically < 50–75 km wide) with speeds reaching 3–10 km/h. At the eastern side of the gyres currents are broad, relatively sluggish and flow towards the equator (e.g. Canary, California, Peru and Benguela Currents) – several 1000 km wide with current speeds usually < 1 km/h. The difference between eastern and western currents is called western boundary intensification (Figure 3.9), and can be explained, for the purposes of this text, by the combination of two effects: the rotation of the Earth, and the action of trade winds, which both cause seawater to pile up at the western edges of the ocean basins. The trade winds draw water towards the equator from the subtropics of both hemispheres, which causes a steep water slope in equatorial regions, and consequently large pressure gradient, intensifying the surface currents. Breakdown of the large-scale action of amassing water on the western margins of the Pacific Ocean is part of the process leading to El Niño, examined at the end of this chapter.

3.2.8 IMPORTANCE OF EDDIES

The large surface-current systems in the oceans are well-defined features, but are not static; they are dynamic components of a global circulation pattern which operates at a variety of scales and may evolve quite rapidly. Increasingly, the importance of currents and their variations is being recognised, both in surface and deep ocean circulation, generating eddies due to local turbulence. Eddies are

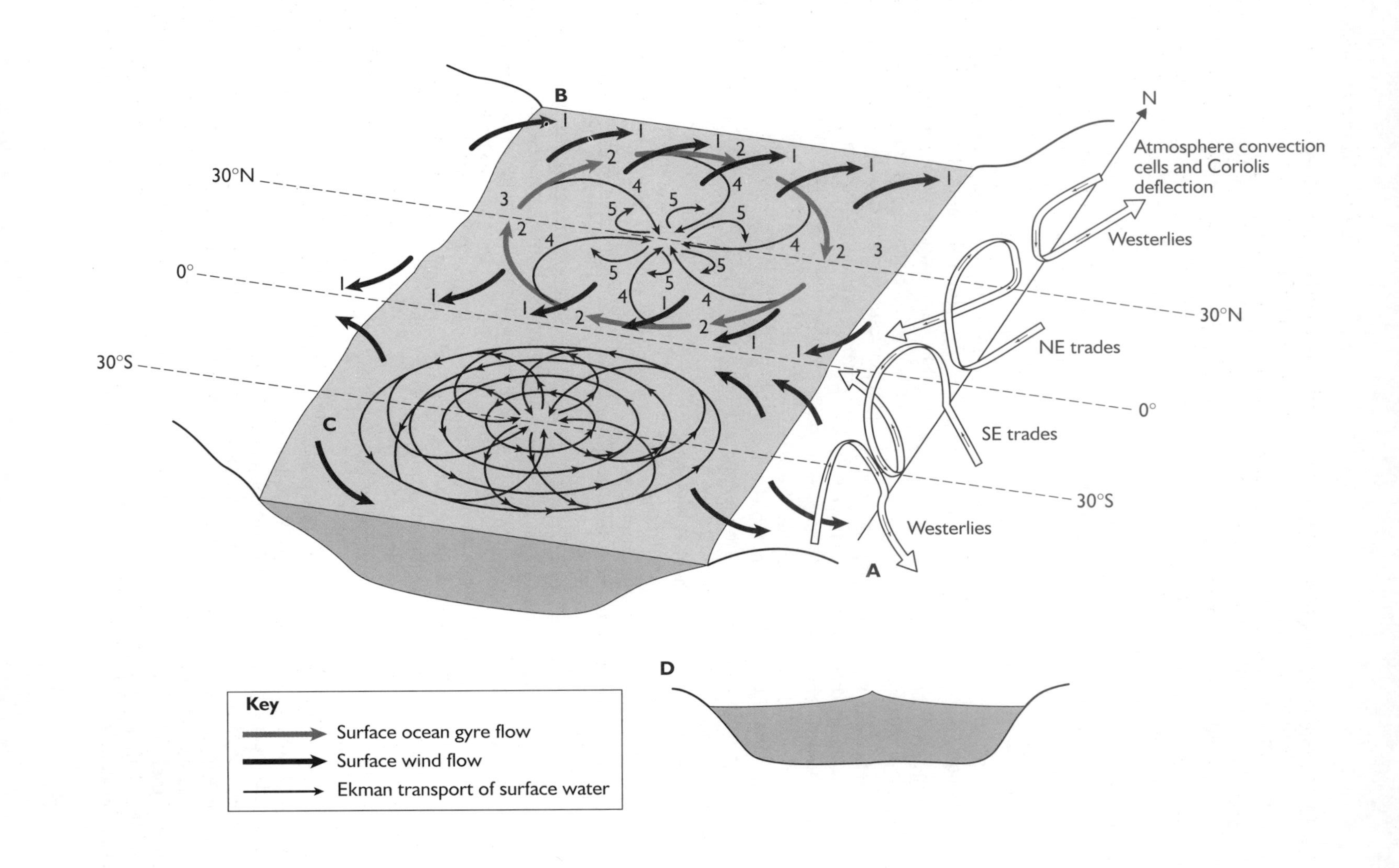

B
N
Atmosphere convection cells and Coriolis deflection
Westerlies
30°N
0°
30°S
NE trades
SE trades
Westerlies
A
C
D
Key
Surface ocean gyre flow
Surface wind flow
Ekman transport of surface water

probably responsible for a major part of the energy transfer in the ocean.

One of the best studied examples are eddies associated with the Gulf Stream. The Gulf Stream is a major current system in the North Atlantic, a powerful western boundary current with current flows reaching 9.7 km/h, transporting up to 150 × 10^6 m^3 seawater per second. It forms a narrow (< 115 km) band of fast-flowing water which separates the cooler coastal waters off North America (termed slope water) from the warm and salty waters of the Sargasso Sea. The Sargasso Sea is a relatively stagnant, 700–800 m deep warm water mass near the centre of the North Atlantic circulation gyre (Figure 3.3), named after the *Sargassum* weed which floats on its surface. The north-western margin of the Gulf Stream is usually well defined, its southeast margin less so. It does not follow a straight path but meanders, much as a river does. Eventually meanders may become so pronounced that a ring of water pinches off, in a similar mechanism to an ox-bow lake on land (Figure 3.10). Rings which form to the south of the Gulf Stream entrain cold slope water and transfer it to the Sargasso Sea (cold-core rings), those which form to the north entrain Sargasso Sea water and transfer it to the cooler water off the North American coast (warm-core rings). Individual rings may persist for months or even years, and are important in mixing cold slope water with Sargasso Sea water, and in carrying the chemical characteristics and biological populations of slope water into the Sargasso Sea (and vice versa).

3.2.9 SPECIAL CONDITIONS IN THE ANTARCTIC REGION

A complex junction of waters develops around Antarctica, called the Antarctic Convergence (Pinet, 1992, p. 388), created by the coincidental location of the subpolar low pressure zone at 60°S and the circum-Antarctic ring of unconstrained water (Figure 3.3). In this zone are two currents, the West Wind Drift north of 60°S and the East Wind Drift, south of that line. The main circumpolar current is the West Wind Drift, because the East Wind Drift is interrupted by the irregularity of the Antarctic coastline, forming subsidiary gyres (Figures 3.3, 3.11).

3.2.10 WAVES AND TIDES

Mention of waves and tides completes this

Figure 3.8 Pressure gradient forces in ocean surface waters. Surface ocean flow is controlled by a combination of atmospheric circulation and continental configuration. Ocean gyres are major features in North and South Atlantic, North and South Pacific, and Indian Oceans. (A) Atmospheric circulation cells generate trade winds and westerlies; cells operate across the whole ocean but are shown only on the right-hand side of this diagram for clarity. (B) Surface winds (dark arrows, 1), driven by atmospheric circulation, drag the ocean surface by friction (with 3% efficiency) and cause ocean currents (white arrows, 2); Coriolis deflection operating via Ekman transport drives the currents to the right of the wind in the Northern Hemisphere, used here to illustrate the process; ocean current flow is constrained by location of continents and flow is deflected (3); Coriolis deflection forces water to flow to the right of currents towards the centre of gyre (4); deflection continues to the right, to push water towards the centre, building a mound in the gyre centre; mound growth is counterbalanced by gravitational flow away from centre (5), which is driven to the right by Coriolis deflection; the downward flow is the Pressure Gradient Force (PGF); this explanation is incomplete, and the next stage of the process is shown in (C). (C) Here the processes shown in (B) are continued but the Southern Hemisphere is used to illustrate them: flow lines into and out of the gyre are the same, but flow in the opposite direction; the system shown in (B) develops dynamic stability in (C), so that ocean gyre currents actually flow in a set of concentric lines with pressure in the gyre decreasing away from the mound centre. Concentric flow is called geostrophic flow; the winds above the gyre are omitted for clarity. (D) Schematic cross-section of ocean along a latitude line through the gyre's centre, showing the central gyre water mound shown exaggerated for clarity.

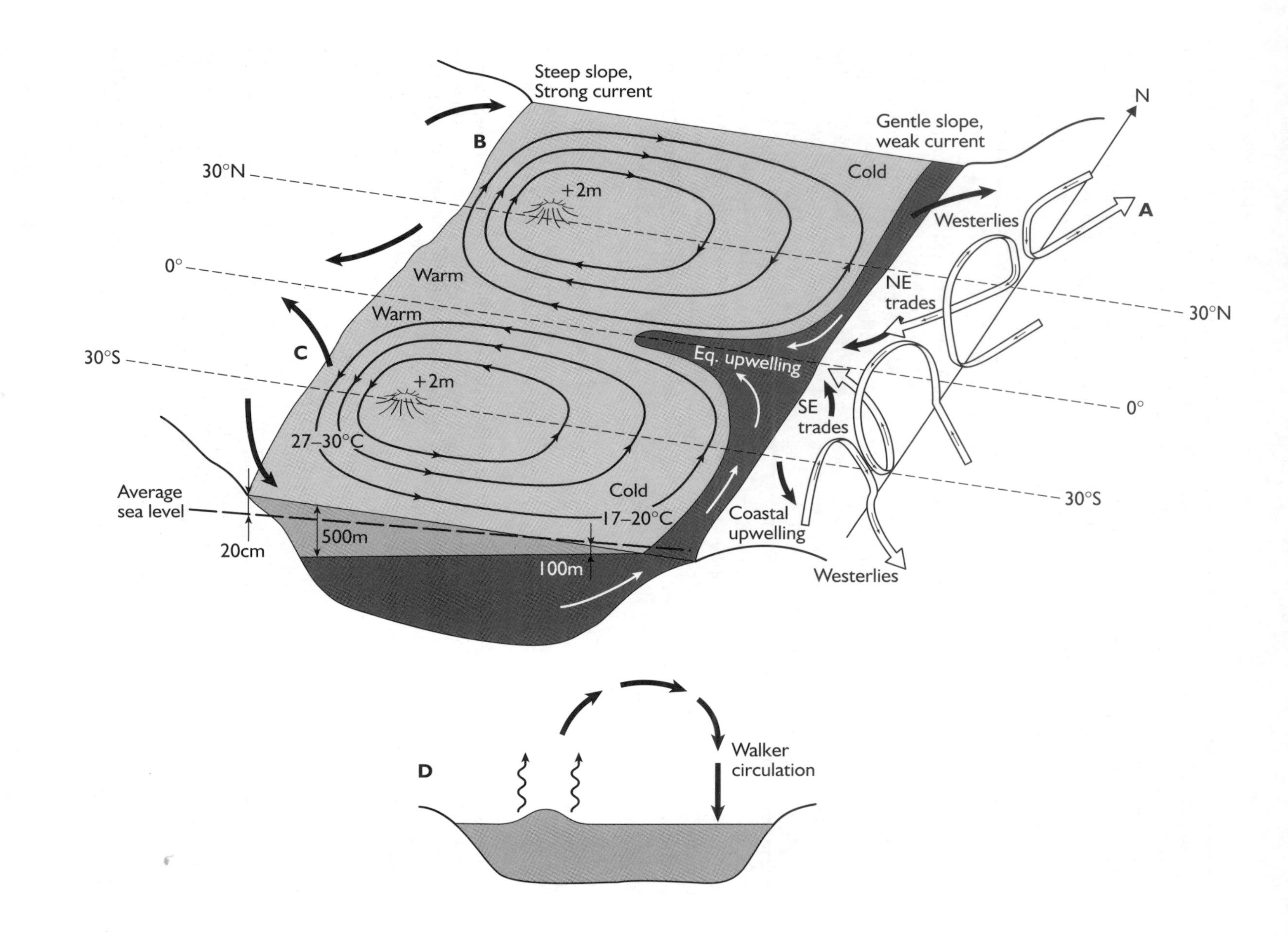

Steep slope,
Strong current
N
Gentle slope,
weak current
B
30°N
Cold
+2m
Westerlies
A
0°
Warm
NE
trades
30°N
Warm
C
30°S
Eq. upwelling
+2m
SE
trades
0°
27–30°C
Cold
Average
sea level
17–20°C
30°S
Coastal
upwelling
500m
20cm
100m
Westerlies
Walker
circulation
D

brief survey of ocean surface water movement. A wave is an undulating disturbance that moves across the sea surface, propagating energy. Waves occur on a variety of scales, and are caused by wind stress, severe storms, earthquakes, and the attraction of the Sun and Moon. The latter are longer period waves, i.e. tides. Waves and tides are most importance in shallow water areas, and influence the development of shallow marine and estuarine sedimentary deposits. Nevertheless, they are arguably relatively minor features in open and deep ocean circulation as a whole, and a detailed examination is not appropriate for the scope of this book. Important shallow water areas of oceanic significance are estuaries, critical in discussions on pollution; estuarine processes are examined in Section 11.5.3, particularly in relation to the estuarine filter concept.

3.3 DEEP OCEAN CIRCULATION

Section 3.2 dealt with the rudiments of surface circulation; in contrast, deep circulation involves most of the ocean volume, and is of greater overall importance. However, knowledge of its mechanisms is incomplete, and the debate continues as to how the system really works; here we examine the processes and problems.

Below the thermocline, ocean circulation is powered by density gradients. Both temperature and salinity exert strong controls over density, although temperature is more important than salinity. Migrating water masses will move to a level where their density is equal to that of the surrounding water, so denser water will sink, and less dense water will rise.

3.3.1 FORMATION OF NEW WATER MASSES AND THEIR PROPERTIES

Changes in temperature and salinity of seawater occur when heat or water are gained or lost at the air–sea interface. Once water sinks and is no longer in contact with the atmosphere, salinity and temperature become remarkably stable, and can be used to identify and trace discrete parcels of water with uniform temperature and salinity relationships, termed water masses (Figure 3.12). Water masses thus acquire their physical characteristics while they are in contact with the atmosphere. In time, however, both salinity and temperature change slightly as mixing occurs with adjacent water bodies, so that any particular water mass eventually loses its identity over a period up to about 1500 years for the maximum known age of deep waters (Crowley and North, 1991).

Temperature measurements on the ocean floor show that ocean basins contain a lot of very cold water (c.2°C). The only places on the Earth's surface where such cold water forms are at the surface near the two poles, and clearly indicate that most of the water in the deep sea originates from shallow water in polar latitudes, where it is chilled by losing heat to cold air masses (Figure 3.13). The two best-known sites are: (1) the Weddell Sea, Antarctica, which produces the Antarctic

Figure 3.9 Schematic model of ocean surface circulation. Figure 3.8 demonstrates the processes of flow in ocean gyres, but the process involves more. (A) Atmospheric circulation cells controlled by heat flow and Coriolis deflection, as in Figure 3.8. (B) Northern Hemisphere geostrophic flow is asymmetric because the Earth's movement piles water up on the west side of the oceans in combination with the NE trade winds. (C) The flow is of opposite direction in the Southern Hemisphere. (D) Schematic cross-section of sea surface showing water mound is actually located nearer the western side (cf. Figure 3.8(D)). In the Pacific Ocean, there is also increased evaporation of water on the western side because that side is warmer; latent heat is transferred to the atmosphere and passes to the low pressure zone on the eastern side, generating an atmospheric flow called Walker circulation (see Figure 3.15); ocean topography shown exaggerated for clarity.

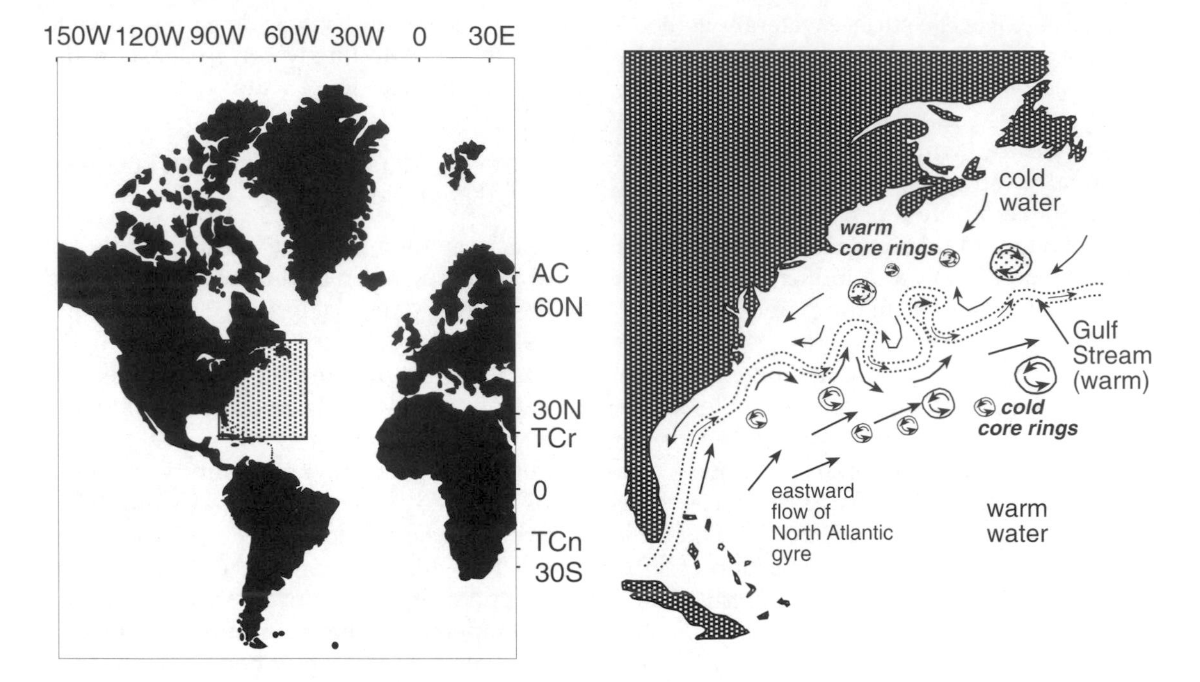

Figure 3.10 Complexity of ocean surface circulation as illustrated by the Gulf Stream of northwest Atlantic Ocean. The flow develops river-like patterns of meanders, with cold and warm water pinched off to form cold-core and warm-core rings, respectively. The system is dynamic and the precise pattern of flow changes with time.

Bottom Water (AABW); and (2) the northern North Atlantic, especially east of Greenland, where North Atlantic Deep Water (NADW) forms. A third site which is quoted more rarely is the Ross Sea of Antarctica, but which has all the attributes of a cold water generation site. Cooling in these sites generates water masses, which are referred to as new water. Of course the water is not new, in the sense that it has been in existence since the Earth formed; but water masses acquire density properties at the present time, which are new.

It is often explained that surface waters flowing into the North Atlantic from equatorial regions (i.e. the Gulf Stream then the North Atlantic Drift – Figure 3.3) cool around the latitude of Iceland, at about 60°N, release heat to the atmosphere, and sink to the deep ocean by virtue of being denser. Another view is that water rises from intermediate depths (of *c*.800 m) to the surface before it reaches polar latitudes, and then cools from 10°C to *c*.2°C (Broecker and Denton, 1990b). The image of NADW formation south of Iceland, however, seems to be an oversimplification. Figure 3.3 illustrates the passage of the warm North Atlantic Drift current into the Norwegian and Greenland Seas, where deep water also forms (Henrich, 1998). It is interesting that the North Atlantic Drift Current manages to pass northwards through the polar front at 60°N without being deflected by the polar easterlies (Figures 3.3, 3.4), discussed later.

It is heat released by NADW formation that keeps western Europe warmer than it should be, not the (incorrect) assertion that the heat is carried in the Gulf Stream, which does not reach Britain (Broecker and Denton, 1990b). The amount of heat released in the North Atlantic is enormous, equivalent to 30 000

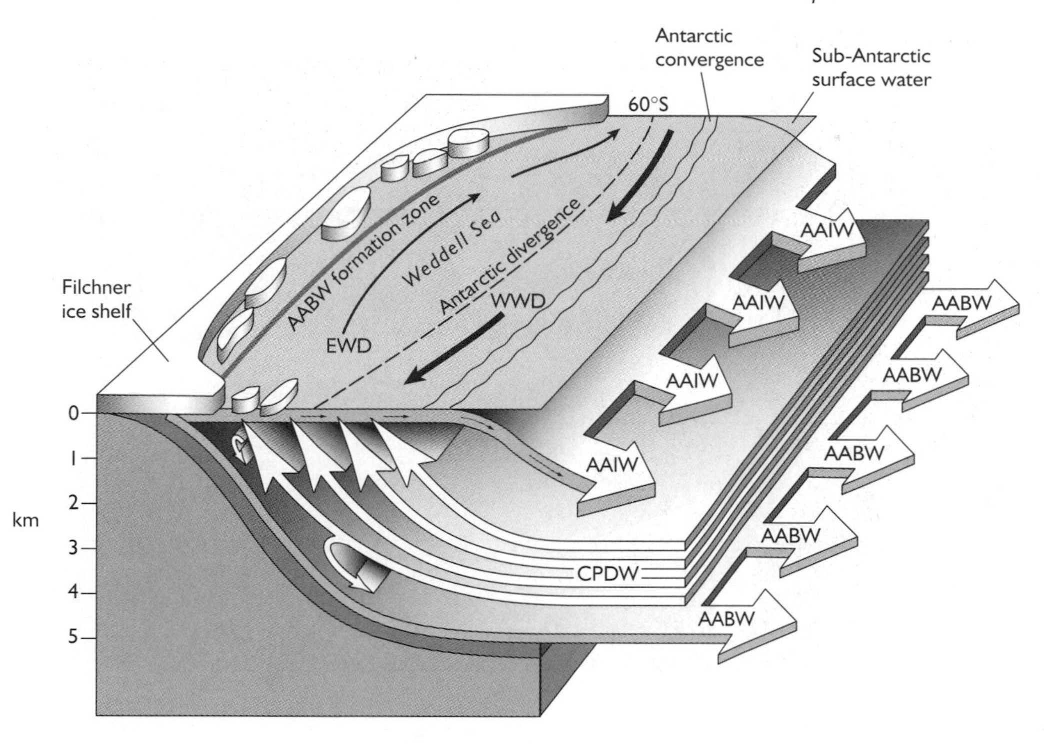

Figure 3.11 The region of the Southern Ocean around the Weddell sea (cf. Figures 1.3. and 3.3(A)) is an important site of deep-ocean water flow. Cold water adjacent to the Filchner Ice Shelf sinks down the continental margin into deep water to form Antarctic Bottom Water (AABW); further north, cold surface water meeting the warmer subantarctic surface waters of the South Atlantic gyre (at the Antarctic Convergence) sinks to form Antarctic Intermediate Water (AAIW). These two sinking water masses are replaced by upwelling of Circumpolar Deep Water (CPDW), which brings nutrients to the surface to stimulate the high productivity of these cold waters. Ocean bottom waters are oxygenated, and so little organic storage occurs in sediments in this region (discussed in Chapter 9). Surface water flows in opposing directions either side of 60°S, because of the polar high pressure zone (see Figure 3.4(C)), and forms the West Wind Drift (WWD) and East Wind Drift (EWD) (see Figure 3.3(A)).

times the power-generating ability of Britain (New, 1998), or *c*.30% of the annual solar input to the region, and involves the vertical sinking of a water mass equivalent to 20x the total volume of the world's rivers (Broecker and Denton, 1990b).

The increase in density by cooling aids sinking of the water mass, because it is denser than the underlying water. An additional feature is the formation of polar sea ice, because when seawater freezes, it excludes *c*.70% of its dissolved salt, so that the water is more salty and hence denser still. More recent work suggests that the cooling of poleward-flowing water in the Atlantic is only part of the control on the formation of new water. Although it is well known that cooling is accompanied by evaporation in the polar latitudes to increase the salt concentration (Pinet, 1992, p. 183), of seemingly critical importance is the location of cooling in semi-enclosed marginal seas (Price, 1992). In fact the densest polar waters are

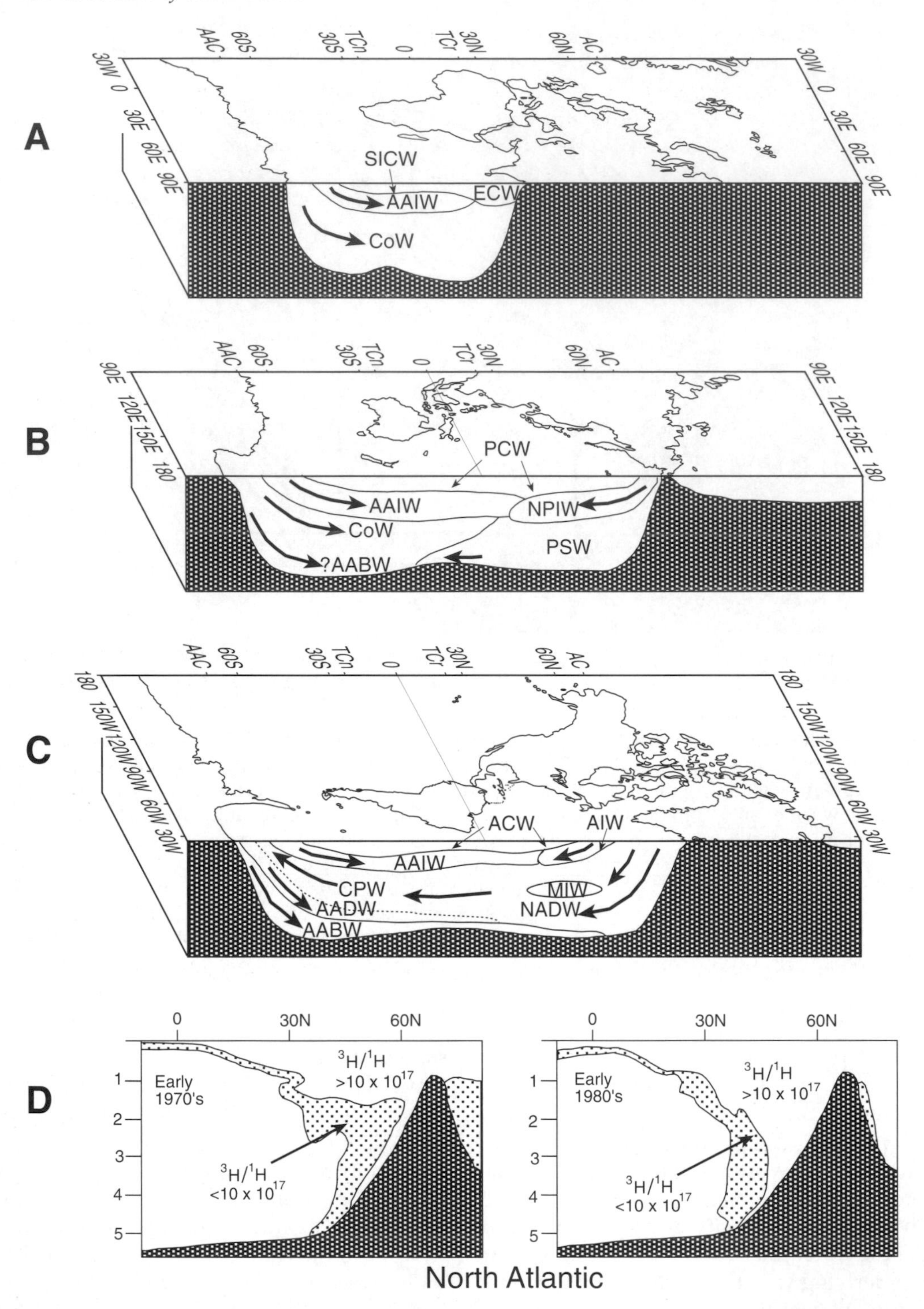
A
SICW
AAIW
ECW
CoW
B
PCW
AAIW
NPIW
CoW
PSW
?AABW
C
ACW
AIW
AAIW
CPW
MIW
AADW
NADW
AABW
D
Early 1970's
$^3H/^1H$ >10 x 10^{17}
$^3H/^1H$ <10 x 10^{17}
Early 1980's
$^3H/^1H$ >10 x 10^{17}
$^3H/^1H$ <10 x 10^{17}
North Atlantic

created in the Norwegian Sea between Iceland and Norway, accompanied by dense water formed in the Denmark Strait (between Iceland and Greenland), and in the Labrador Sea (between Greenland and Canada). All three cold-water generation sites are partly enclosed by land, as are the Antarctic Weddell and Ross Seas wherein lie the Filchner and Ross Ice Shelves respectively. The large size of the Weddell Sea contributes to its importance as a source of new cold water. Cold waters do not form in the Bering Strait between Russia and Alaska, despite there being a surface flow through the gap. The North Pacific region is not cold enough to form deep water, even though it is at a similar latitude as the North Atlantic. The dominant northward flow from the Pacific to the Arctic Ocean has been a feature of that region for *c.*4 million years, due to ocean circulation reorganisation following formation of the land bridge between North and South America (Marincovich and Gladenkov, 1999); this flow probably prevents cold deep water formation in the North Pacific.

Apart from the generation of cold deep waters in the Atlantic Ocean, an observation of some importance is that the Atlantic Ocean has a greater salinity than the Pacific (Bearman, 1989b, p. 35), because of the greater cooling and evaporation of Atlantic polar waters. The salinity gradient may aid the flow of deep waters to the Pacific, and may be a key driving force in the global conveyor, described next.

3.3.2 THERMOHALINE CIRCULATION

Now we have seen how new water forms, what happens to it? Salinity and temperature differences of various water masses amongst the oceans is thought responsible for causing a large-scale flow pattern called thermohaline circulation, which appears to be a worldwide system, also called the global ocean conveyor (Broecker and Denton, 1990b). A popular simple diagram is reproduced in many oceanographic publications (Figure 3.13), but the progress of flow is clearly complex.

In the Atlantic Ocean, deep and bottom waters are dominated by AABW and NADW, and the NADW is the principal driving current. Estimates of production rates suggest that the AABW forms at 38×10^6 m^3/s and NADW at 10×10^6 m^3/s (Murray, 1992). It is curious, therefore, that NADW should be the more important, considering it is produced at only about one quarter of the rate of AABW. This may be related to the constraint provided by sea bed topography for the AABW (because it flows along the South Atlantic floor), but as yet not enough is known about the controls on the currents. The two water masses sink until they each reach a level where their density matches the surrounding water. Continued sinking from above pushes the water horizontally along this depth so that they slide under mid-depth (intermediate) and surface (central) waters, with AABW being overlain by NADW. The effect of sea bed topography is important

Figure 3.12 Deep-water mass structure in the oceans; temperature and salinity characteristics are maintained for considerable lengths of time, to ensure that each water mass is an identifiable feature. (A)–(C) show the deep-water masses of the Indian, Pacific and Atlantic Oceans. (D) shows evidence for sinking cold water in the north Atlantic (NADW), as shown by differences in distribution of the isotope of hydrogen called tritium, released by anthropogenic nuclear sources, between the early 1970s and 1980s (depth in km; adapted from Segar, 1998).
AABW = Antarctic Bottom Water; AADW = Antarctic Deep Water; AAIW =Antarctic Intermediate Water; ACW = Atlantic Central Water; AIW = Arctic Intermediate Water; CoW = Common Water; CPW = Circumpolar Water; ECW = Equatorial Central Water; MIW = Mediterranean Intermediate Water; NADW = North Atlantic Deep Water; NPIW = North Pacific Intermediate Water; PCW = Pacific Central Water; PSW = Pacific Subarctic Water; SICW = South Indian Central Water.

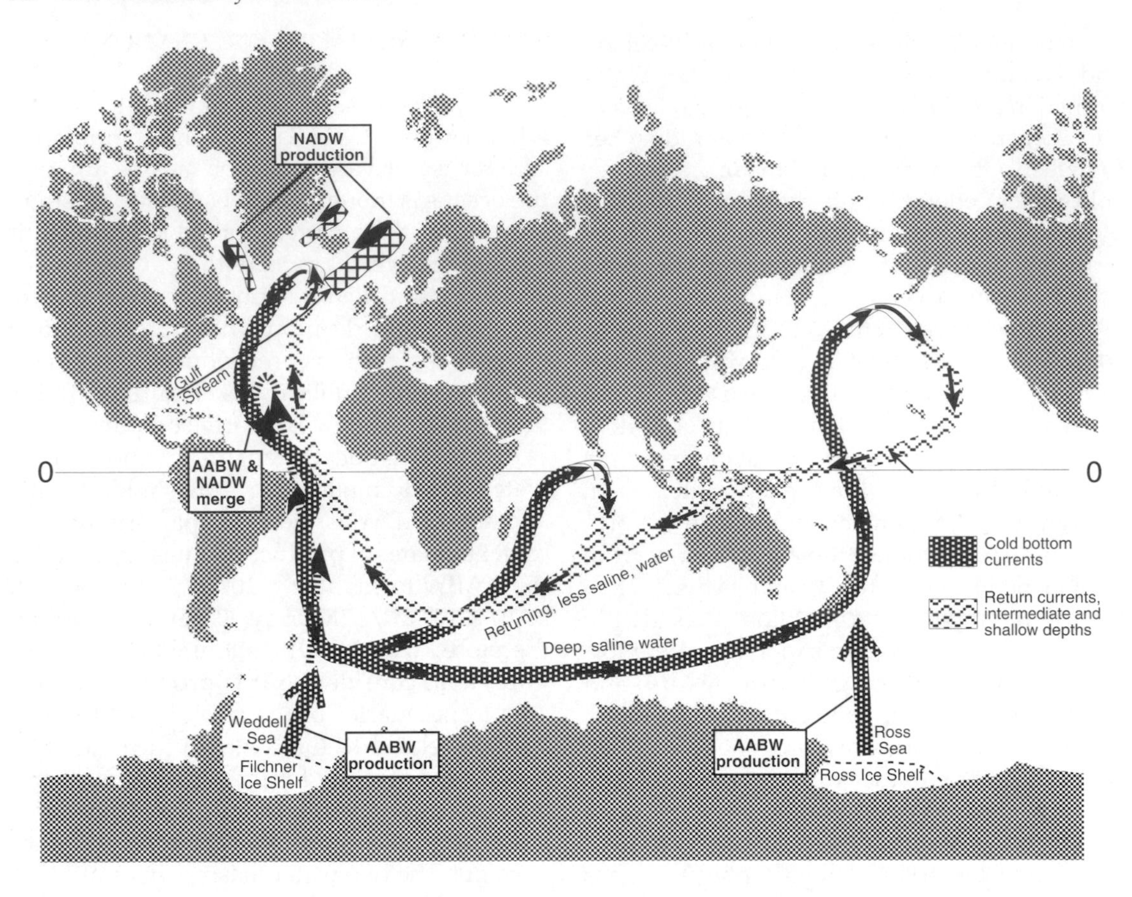

Figure 3.13 Global deep ocean circulation is often simplified as a loop-like conveyor. Evidence of North Atlantic Deep Water (NADW) formation in three sites shown in the North Atlantic region is more complex. Southern Ocean deep water formation of Antarctic Bottom Water (AABW) is well known from the Weddell Sea, but also occurs in the Ross Sea. (See text for discussion.) The exact mechanism of the global conveyor system is not understood, so care is needed in its application to ancient oceans! Adapted from Broecker and Denton (1990b).

to the path of the deeper AABW, because it flows north along the western South Atlantic, being prevented from passing to the east by the presence of the mid-Atlantic Ridge, except in small amounts permitted by narrow valleys formed by transform faults on the ocean floor. AABW passes into the western North Atlantic through the Vema Channel (in the Rio Grande Rise, a large obstacle in the Atlantic floor), and does eventually gain access to the East Atlantic via the Romanche Fracture Zone, the largest of the transform faults, located at the tightest part of the bend in the mid-Atlantic Ridge (Figure 3.13). This zone has probably been active as a bottom current conduit connecting the East and West Atlantic since the Late Cretaceous.

Although the AABW and NADW move by advective processes, they gradually mix by turbulence, until AABW reaches about 40°N, when it ceases to be an effective current (Murray, 1992). Southward-flowing NADW mixes with AABW beneath it, and also incorporates both salty warm Mediterranean Intermediate Water (MIW) exiting from the

Gibraltar Straits (Figure 3.14), and less dense Antarctic Intermediate Water (AAIW), which itself flows along with and above AABW. These mixed waters passing south through the South Atlantic are then flushed across the Southern Indian Ocean and on into the Pacific Ocean by the Antarctic Circumpolar Current, forming a deep water mass known as Common Water (CoW), which travels along the Tonga Trench and the Samoan Passage into the North Pacific. AABW from the Ross Ice shelf adds to the mass of cold water in the Pacific. The central Pacific location of the Hawaiian Islands serves to divide the CoW flow but it recombines north of the island chain, and passes into the North Pacific where the currents rise to mid-depths and flow southwards. Why they rise here is unclear, but having travelled a long distance from the South Atlantic, it is possible that the water no longer maintains its original integrity and loses some of its high density character. Channels between the Indonesian Islands provide the access for the water back into the Atlantic via the Indian Ocean. CoW also enters the Indian Ocean directly from the Weddell Sea source area, and provides a subsidiary loop, rising south of the tip of India, flowing south to join the stream returning to the Atlantic.

The Antarctic Circumpolar Current forms near the Antarctic Convergence (Section 3.2.9: this generates the West and East Antarctic Drift surface currents because of its 60° latitude, see Figure 3.4). Circumpolar flow is presumably related more to the lack of continental obstruction at those latitudes for the fluid ocean, and less related to the atmospheric circulation patterns which drive surface currents. Both AABW and AAIW sink because of their high densities, and are replaced at the surface by upwelling of Circumpolar Deep Water, which draws nutrients to the surface. This unusual setting for upwelling provides the circum-Antarctic seas with a rich food supply, making waters surrounding Antarctica a highly productive region (Figure 3.11); high productivity here is thought to lead to significant removal of CO_2 from the region's atmosphere (Di Tullio *et al.*, 2000).

Earlier work on ^{14}C ages of deep waters (Stuiver *et al.*, 1983) indicates that the AABW spreads to Pacific, Indian and Atlantic Oceans from the Weddell Sea, and gets older as it moves away from the Weddell Sea area. The oldest ages in each branch of the water masses are: Pacific 510 a, Atlantic 275 a, Indian 250 a; so, water currently in the deep Pacific began its journey from Antarctica about the same time as Columbus made his epic voyages to the Americas.

The global conveyor transports heat and dissolved/suspended matter around the global water mass, and the supply of oxygen to the seafloor. Furthermore, it removes dissolved CO_2 from the atmosphere, thereby providing a potential moderating effect on climatic fluctuations. If such circulation were to stop or reduce in intensity, parts of the deep sea would become oxygen-starved, removal of CO_2 would cease, and heat transport around the Earth would be significantly affected. There is a growing body of evidence that the conveyor system is sensitive to change, and is potentially disrupted by both glacial episodes and periods when the climate was much warmer than at present; the latter is recognised from geological evidence as being the state of the world's oceans for much of geological time – modern cool climates are unusual.

3.3.3 THE GLOBAL CONVEYOR IN PERSPECTIVE

It is clear that although much is known about deep ocean flow, a lot remains to be learned, and despite the simple models, no single universally agreed flow system is established. The complexity lies in the fact that there are several water masses in motion throughout the ocean system, all subject to change and difficult to study because of their inaccessibility. The following two paragraphs indicate the sort of problems that exist with under-

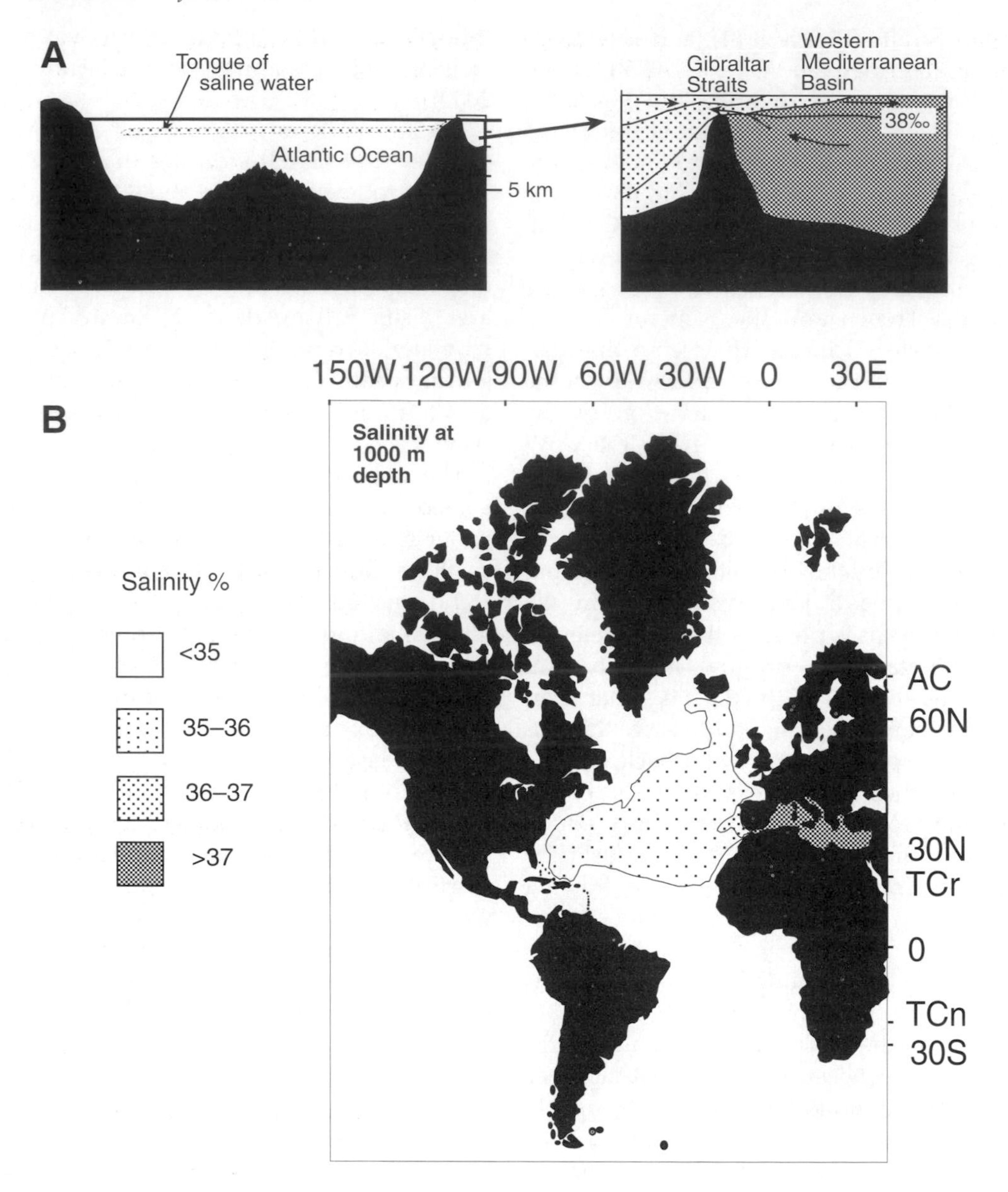

Figure 3.14 Features of the Mediterranean Intermediate Water in the Atlantic Ocean. (A) Cross-section to show the outflow of dense saline warm water from the Western Mediterranean Basin through the Gibraltar Straits into the Central Atlantic, counterbalanced by flow of normal marine water into the Mediterranean Sea. In this system, evaporation from the warm enclosed Mediterranean Basin drives this system, by increasing water density within the basin. (B) Map of distribution of water with salinity of 35–36‰ at 1000 m water depth in the Atlantic Ocean, compared to the salinity immediately west of the Mediterranean (36–37‰) and in the Mediterranean Basin (> 37‰). Most of the water in the 35–36‰ range in the Atlantic at that depth can be attributed to Mediterranean Intermediate Water (MIW) (see also Figure 3.12(C)). Dense salty MIW ia a modern analogue for the proposed formation of past warm saline deep and bottom waters (WSDW, WSBW) in low latitiudes; thought to help drive ocean circulation during greenhouse phases of Earth's past climate, this is halothermal circulation (see Section 9.3).

standing the global conveyor, which converge on the difficulty of explaining how the water masses form a complete global circulation system, which, logically, they must do.

The simple ocean conveyor diagram (Figure 3.13) indicates that water flowing into the North Atlantic on the return loop from the Pacific is cooled and resinks as new NADW when it reaches Greenland once more, suggesting that it must have been warmed up in order to be cooled again! If that logic means the water must reach the surface by the time it gets to the North Atlantic, then there is a problem of determining exactly the path of this water. If it is at the surface in the South Atlantic, then presumably it gets caught up in the South Atlantic Gyre, and passes to the North Atlantic in the Caribbean region, and then becomes the Gulf Stream. One idea views return flow as part of surface circulation, but because this water must be caught up in ocean gyres, its path to the North Atlantic must be very convoluted indeed!! This is also at odds with the observation that the water rises from intermediate depths in the North Atlantic (Broecker and Denton, 1990b). Therefore perhaps the water returns at both shallow and intermediate depth, i.e. above and below the thermocline. However, if water passing into the South Atlantic on the return flow of the conveyor is travelling at intermediate depths all the way, beneath the thermocline and therefore decoupled from surface gyres, then there is another problem of its identification. Published cross-section schematic diagrams of deep flow in the Atlantic do not illustrate the return flow of the conveyor, and there seems to be no space in such diagrams for the water to flow (Figure 3.12).

There is a further curiosity of the ocean circulation system in the north Atlantic. Figure 3.4 shows the North Atlantic Drift passing northwards between Iceland and Europe, seemingly against the flow patterns of the polar atmospheric cells, which would be expected to drive water *south* at those latitudes. Thus it seems that while the Gulf Stream behaves as it should and leads to a southward-flowing current on the eastern side of the Atlantic (Canary Current), the North Atlantic Drift breaks the pattern. Poleward surface flow is at least partly counterbalanced by southerly cold water flow in deeper waters from the Arctic and Nordic regions, and is a key element of NADW formation (Bianchi and McCave, 1999). The seafloor is interrupted by (i) the prominent Denmark Strait Sill (Dickson *et al.*, 1999) between Greenland and Iceland, and (ii) the Iceland–Scotland Sill; Arctic water spilling over these ridges, and through narrow passages cutting them, plays a part in NADW formation. Recent work on Arctic Ocean processes reveal that North Atlantic surface waters are carried as far north as the Barents Sea, and interactions with Arctic waters is part of the recently discovered North Atlantic Oscillation of climate, demonstrating the great complexity in these northern regions (Dickson, 1999).

3.3.4 CIRCULATION IN THE MEDITERRANEAN, AND BASIN OUTFLOW

In addition to the sinking of cold, dense water masses described above, density-driven vertical movement of warm, salty surface waters also occurs to intermediate depths around the Mediterranean region. Deep circulation in the Mediterranean is controlled by surface processes. Over the Mediterranean basin as a whole, evaporation is greater than precipitation and river runoff, causing high salinity (up to 39%). Intense evaporation in the east of the basin lowers the sea surface, creating a sea surface slope and PGF, which causes a surface inflow of Atlantic water via the Straits of Gibraltar. The hot, more saline water in the east of the basin sinks, and flows out of the basin over the shallow sill at Gibraltar as a deeper flow. Residence time of water in the basin is around 80–100 years. Sinking water removes nutrients from the surface, causing a generally low biological productivity, giving the Mediterranean its intensely blue colour (although fertile areas occur locally, near to

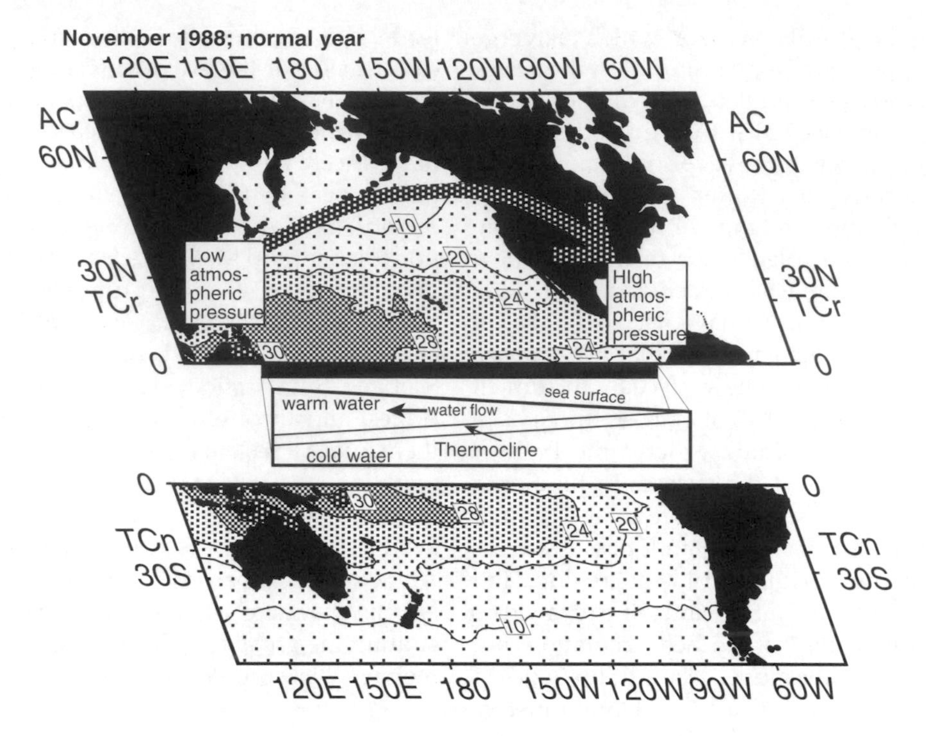
November 1988; normal year
120E 150E 180 150W 120W 90W 60W
AC
60N
30N
TCr
0
Low atmos-pheric pressure
High atmos-pheric pressure
10
20
24
28
30
sea surface
warm water
water flow
cold water
Thermocline
TCn
30S

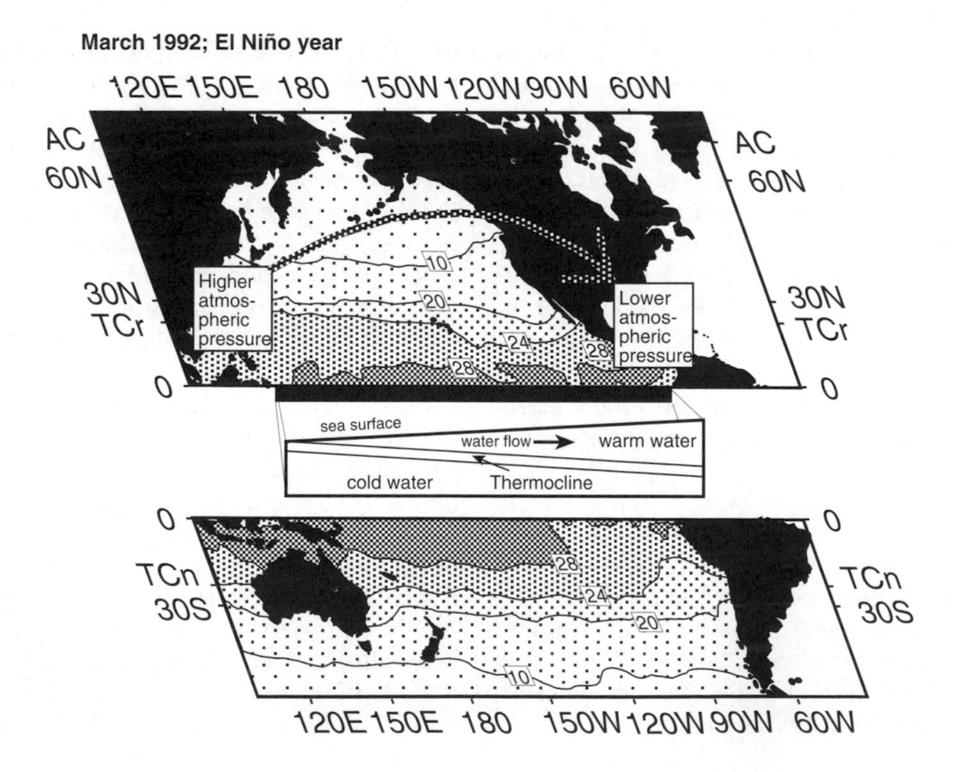
March 1992; El Niño year
120E 150E 180 150W 120W 90W 60W
AC
60N
30N
TCr
0
Higher atmos-pheric pressure
Lower atmos-pheric pressure
10
20
24
28
sea surface
water flow
warm water
cold water
Thermocline
TCn
30S

river discharges or following storms which stir up the water column).

The warm, saline deep water which flows out of the Mediterranean sinks to *c.*1000 m depth and then spreads out horizontally across the North Atlantic, forming a distinctive water mass of temperature 13°C and salinity 38.4–38.5‰, which can be detected 1000s of km from its origin (Figure 3.14). The Mediterranean basin is thus an important source of North Atlantic intermediate water. This warm, saline water is not dense enough to displace cold deep waters in the present ocean and so forms a mid-depth water mass, but in the geological past formation of warm, saline deep water masses in equatorial areas may have been an important control on deep ocean circulation.

3.4 EL NIÑO, THE ULTIMATE INTERACTIVE SYSTEM?

Described *ad nauseam* in contemporary texts, El Niño is important because it embodies all that is not clear about global ocean circulation. As a result of the surface water circulation patterns we examined in the first part of this chapter, water in the Pacific basin is usually piled up on the western side because of the rotation of the Earth and the northeast trade winds (Figure 3.15). In normal years, the South Pacific gyre causes water to flow northwards along the eastern margin (South American side), inducing Ekman transport to the left (west) as it should do, and consequently stimulates upwelling of cold nutrient-rich water from the deep Pacific, and a thriving Peruvian fish industry from the upwelled nutrients (Figure 3.7). Furthermore, because there is a piling of surface water on the west side, there is a thicker layer of warmer water in that region, so that the thermocline is deeper on the west, and slopes upwards to the east.

Each year around Christmas time (El Niño, the child), the pattern changes. The southeast trade winds weaken, and normality breaks down; warm water flows eastwards down the water gradient across the Pacific to interrupt the usual pattern off the west coast of South America. The thermocline flattens out as the warm ocean sloshes eastwards and there is a severe decrease in the upwelling off South America, depriving surface water of nutrients and interrupting the fishing industry; a time for repairing nets and painting boats.

In certain years the change is more profound, leading to a set of events now commonly referred to as an El Niño *event*. In an El Niño (event) year, the whole process is intensified, resulting in a catastrophic decrease in the upwelling off South America and decimation of the Peruvian fishing economy.

It gets worse. In normal times, the southeast trades flowing across Eastern Pacific waters draw cool, dry air from the subtropical high pressure zone. Remember that in the Southern Hemisphere, the descending subtropical air is cool, dry and is deflected to the west by the Coriolis effect (Figure 3.4); this air is not warm enough to rise to produce rainclouds until it gets across to the Western Pacific region. Because the southeast trades weaken during El Niño, the rainclouds extend eastwards across the Pacific, causing considerable increase in rainfall in the central Pacific, and also in central South America. Shifting of the rain focus to central Pacific waters can leave Western Pacific

Figure 3.15 The Pacific Ocean surface circulation in normal and El Niño years. The upper diagram shows the temperature distribution and the piling of warm water along the western Pacific; the thermocline slopes up to the east, so that upwelling cold water on the eastern margin is able to reach the surface, stimulating ocean productivity in that area. The patterned arrow in the atmosphere is the compensating flow of warm moist air, with the whole system being called Walker Circulation. The lower diagram shows an El Niño year, when Walker Circulation diminishes, warm water flows east along the equatorial regions, suppressing eastern margin upwelling, and promoting dryness in the western Pacific. Temperature fields in the ocean are in °C. Adapted from Bryant, E. 1997. Climate Process and Change. Cambridge University Press, Cambridge, UK.

areas short of rainfall, giving droughts in Australia and Indonesia, and forest fires. Finally, the fluctuations of normal and El Niño times led to the term El Niño-Southern Oscillation (ENSO); how much of this can be attributed to oceanic processes is only guesswork, but the ENSO system forms a complex ocean–atmosphere–land interaction. The oceanic part seems to involve mostly surface currents (= Central Waters), but also the deep ocean (upwelling). The big question, of course, is why the southeast trade winds weaken to initiate the El Niño event. Although the heat transport system of the atmosphere, plus Coriolis deflection, creates these trade winds, the controls are clearly more complex, and are suspected to include recently recognised large-amplitude waves in the ocean body (Kelvin waves). How much of the disturbance is related to heat transference between the atmosphere and ocean awaits further research, but more refined models indicate that, if future greenhouse warming predictions come true, then El Niño conditions will become more frequent and powerful (Timmerman *et al.*, 1999). Because water has *c.*30× the heat capacity of air, it may not come as a surprise if it is eventually established that El Niño has an oceanic origin!

3.5 CONCLUSION

Once again we have seen the importance of the physical properties of water, especially its heat capacity relative to land, and the maintenance of discrete water packages because of their density characteristics. The global flow of ocean water controls the heat budget of the planet's surface. The applications of the principles discussed here are widespread, modelling both the modern and ancient ocean flow patterns, and we shall see how they are used in Part B.

3.6 SUMMARY

1. Ocean movement may be divided into two discrete systems, surface and deep, their separation being effected by the thermocline, which acts as a barrier to vertical migration of water.
2. The thermocline limits the depth of surface ocean circulation to *c.*200 m. Surface ocean flow provides *c.*20% of short-term heat distribution, the rest being provided by atmospheric flow.
3. Surface currents are dominated by gyres generated by a complex of controls governed by the Coriolis deflection effect.
4. Coriolis deflection is an illusionary force caused by the fact that the Earth's speed of rotation differs with latitude, while a freely moving object (in this context, an air or water mass) travelling north or south does not have this constraint, and will either be faster or slower than the Earth it passes over, thus appearing to an Earth-bound viewer to be deflected.
5. Coriolis deflection is transmitted to the ocean, to generate Ekman transport and pressure gradient effects, key processes in water flow and upwelling; upwelling has implications for marine ecosystems.
6. Deep ocean circulation is governed by density differences of discrete water masses which have distinct temperature and salinity features, both of which control density.
7. Deep ocean water involves the bulk of the ocean's volume and so controls the long-term heat distribution of the planet via the global thermohaline circulation.
8. Both surface and deep ocean circulations are ultimately constrained by continental positions, and, over geological time, evolve with continental drift.
9. The complexities of ocean, atmosphere and land interactions are epitomised by the ENSO, but the degree to which the oceans are involved awaits a better database.

CRUST–OCEAN INTERACTIONS 4

Having examined seawater structure and motion in Chapters 2 and 3, it is now appropriate to take account of the relationships between the oceans and the Earth's geology. Here the aim is to examine the key features of crust–ocean interchange, which is largely related to the role played by plate tectonics in oceanography.

4.1 INTRODUCTION: THE NATURE OF CRUST–OCEAN INTERACTIONS

Present-day ocean waters' composition and structure is inherited from a long geological history of interaction with the crust. This history began when the oceans first condensed from volcanic degassing in early Precambrian times, and the oceans have been recycled through the crust many times over. Ocean water evolution over geological time is discussed in Part B; here we examine the current state of crust–ocean interactions, which further demonstrates the integrated nature of the ocean and solid Earth. There are three principal areas of interaction:

1. The shape of ocean basins is inherited from millions of years of continental drift, leading to the present continental configuration, of critical importance to ocean current flow.
2. Ocean floor topography has an effect on deep ocean circulation, because of the physical obstacles to bottom-water flow, as illustrated in Chapter 3. Ocean floor mountains, such as those formed at spreading ridges, are the highest peaks on Earth.
3. Ocean water cycles through the crust, interchanging dissolved and particulate matter between water and rock, and having powerful effects on the composition of the water.

The result of interchange is sometimes called the crustal-ocean factory (Libes, 1992), and involves that often-quoted, mind-bending concept of biogeochemical cycles; mind-bending because, although the concept of cycles seems simple enough (e.g. the carbon cycle or phosphorus cycle), cycles are actually complex and involve a combination of:

- organic control at various stages (*bio-*);
- geological control by a variety of rock properties over a long time (*geo-*); and
- inorganic *chemical* change.

The idea of biogeochemical cycles is therefore truly holistic, completely integrated with oceanography, and consequently enormous in its complexity (Butcher *et al.*, 1992). We begin by examining the nature of the crust.

4.2 OCEANOGRAPHIC RELEVANCE OF THE EARTH'S STRUCTURE AND BEHAVIOUR

4.2.1 BASIC FEATURES

Continents exist because of differentiation of the Earth's mineral composition, the lighter components having risen to the surface, and denser parts concentrated deeper down. Continents began as small features (cratons), which grew by accretion of additional light rock as the Earth differentiated over geological time. Problems of understanding the nature of early continents are enormous, because they have been involved in several collisions, modifying them, and because they are covered by later sedimentary deposits. The modern continents are concentrated mostly in the Northern Hemisphere, but through much of geological time evidence indicates that the Southern Hemisphere held most of the land. Continental movements have modified the patterns of ocean circulation through time. The Earth's structure is divided into three depth-related portions: crust, mantle and core (Figure 4.1). It is the crust and upper mantle that mainly concern us.

The Earth's surface elevation shows a bi-modal distribution (higher and lower zones), enabling two forms of crust to be recognised:

1. continental crust, typically granitic material comprising rocks made principally of O, Si and Al, with a density 2.7–2.8 g/cm^3;
2. ocean crust, typically basaltic material, comprising rocks made principally of Si, O and Mg, with a density of 2.9 g/cm^3.

Both sit on the upper mantle, rock made of peridotite (Si, O, Mg, Fe), density around 3.3 g/cm^3, and the interface between crust and upper mantle is termed the Mohorovicic disconformity (Moho for short). The Moho lies at a greater depth beneath continents than beneath ocean basins (Figure 4.1). Continental and oceanic crust act simply as two rock masses of different densities sitting on an even denser mantle, with continental crust sitting at a higher level than thinner, more dense ocean crust. Note that the mantle is *solid*, but has the ability to deform in a ductile fashion, important for the process of isostatic adjustment, discussed below.

Ocean crust is almost everywhere covered by ocean water, and is only accessible by ocean floor drilling, except for Iceland, and small slices of old ocean crust emplaced on top of continental crust by collisional forces. The weight of seawater on ocean crust adds to the topographic depression of ocean basins.

4.2.1.1 Isostasy

The analogy of crustal masses as floating objects is supported by the reaction of continental crust when loaded with ice during glaciation. That the land was physically pressed down is revealed by raised shorelines following deglaciation, in Northern Europe, for example. Considerable uplift occurs; e.g. parts of the Baltic Sea area have risen 250 m due to postglacial crustal rising following the last glacial maximum. This illustrates the principle of isostasy (Greek *iso* = equal; *stasis* = position) which in this context means that, under equilibrium conditions, any portion of the Earth's crust is subjected to equal pressures from opposing directions, so that there is no movement (Figure 4.2). It is related to the application of Archimedes' principle that an immobile object at rest, lying on a solid surface for example, is experiencing an equal force opposite to gravity; that must be true, otherwise it would move vertically.

To demonstrate how isostasy works: step into a boat (a rigid object, representing the Earth's lithosphere) floating on water (representing the asthenosphere), and the boat will be depressed a certain distance, with the water displacing sideways under it. When the boat stops moving downwards, isostasy is achieved; get off the boat, and it will rise upwards, undergoing isostatic rebound, in a similar fashion to the Baltic area after the ice

Figure 4.1 Upper diagram: cross-section of the Earth showing the thicknesses of the crust, mantle and core. Note that the lithosphere comprises solid rocks of the crust and upper mantle, and that the asthenosphere consitutes the remainder of the upper mantle. Lower diagram: a three-dimensional schematic representation of the upper Earth showing constructive, destructive and conservative plate margins, and their various features. Note that the rigid crust and upper mantle rocks (lithosphere) ride on the more deformable asthenosphere (Greek *asthenos* =lacking strength); but both lithosphere and asthenosphere (and the lower mantle) are solid.

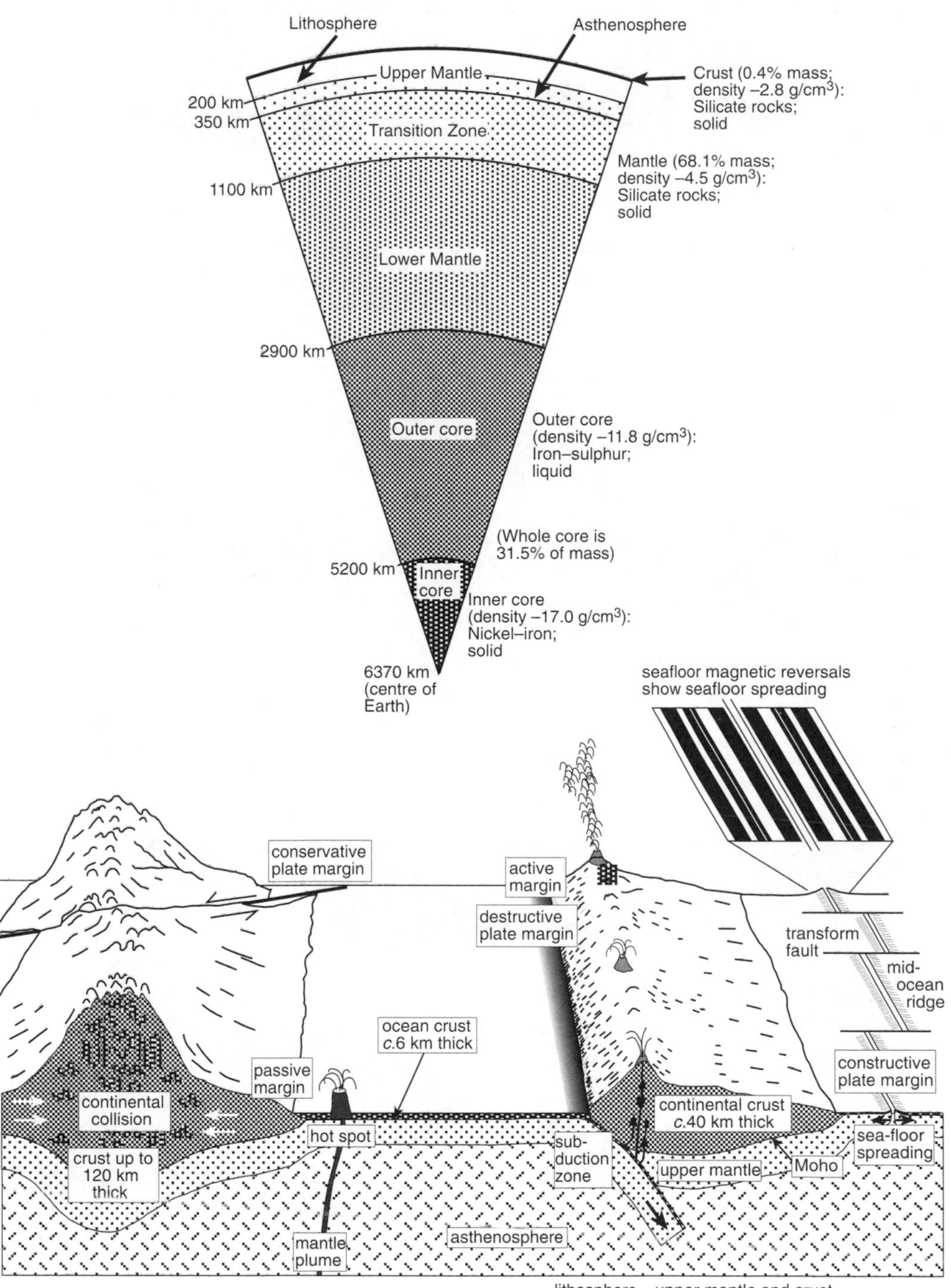
Lithosphere
Asthenosphere
Upper Mantle
Crust (0.4% mass; density –2.8 g/cm³): Silicate rocks; solid
200 km
350 km
Transition Zone
Mantle (68.1% mass; density –4.5 g/cm³): Silicate rocks; solid
1100 km
Lower Mantle
2900 km
Outer core
Outer core (density –11.8 g/cm³): Iron–sulphur; liquid
(Whole core is 31.5% of mass)
5200 km
Inner core
Inner core (density –17.0 g/cm³): Nickel–iron; solid
6370 km (centre of Earth)
seafloor magnetic reversals show seafloor spreading
conservative plate margin
active margin
destructive plate margin
transform fault
mid-ocean ridge
ocean crust c.6 km thick
passive margin
constructive plate margin
continental collision
continental crust c.40 km thick
hot spot
sub-duction zone
sea-floor spreading
crust up to 120 km thick
upper mantle
Moho
mantle plume
asthenosphere
lithosphere = upper mantle and crust

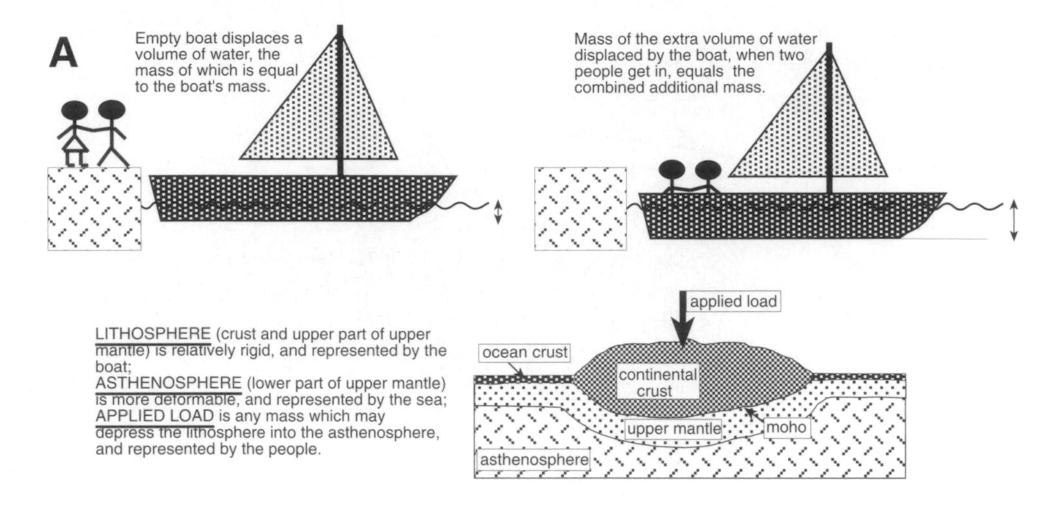

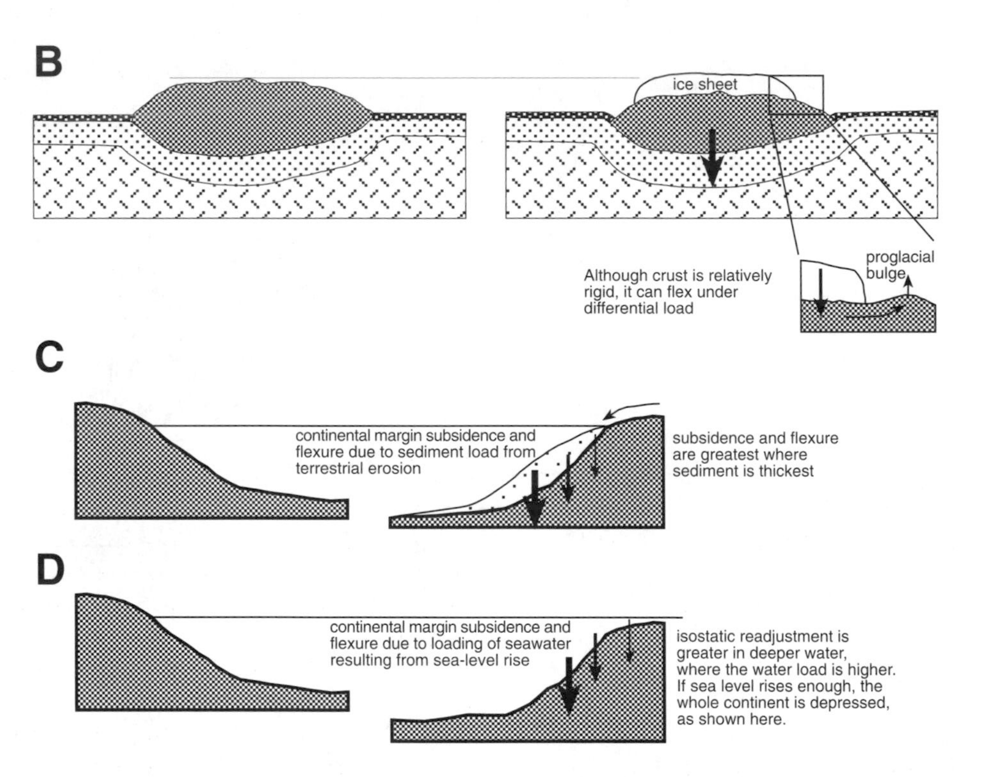

Figure 4.2 Principle and process of isostasy. (A) The mass of the extra volume of water displaced by a boat, when two people get into it, equals the mass of the two people. This relationship is analogous to the lithosphere riding on the asthenosphere. (B)–(D) Not only does the lithosphere sink down isostatically when load is applied to it (either as ice, sediment or water), but it also deforms because of uneven mass distribution, as shown.

age. When the boat stops rising, isostasy is achieved once more. In practice, the Earth's crust is flexible, and itself deforms unevenly; the mantle is very viscous and also moves unevenly, contributing to deformation in the area close to the depressing crust, for example, creating the well-known proglacial forebulge. Northern parts of Britain are currently uplifting because of isostatic rebound following the last glaciation, while southeastern Britain is sinking because it lies in the area of the forebulge, which is gradually (*c*.1–2 mm/a) collapsing back to its preglacial level. Because global sea level is also rising at about 1–2 mm/a, that gives a 2–4 mm/a relative subsidence rate of southeastern Britain.

4.2.1.2 Isostatic rebound in the Baltic area

The Baltic Sea is a relatively small, intracontinental shelf sea in NW Europe and illustrates the relationships between isostasy, plate tectonics and oceanography. The sea is mostly land-enclosed, but mixing with North Sea and Atlantic waters occurs through the narrow Kattegat and Skagerrak channels. The entire Baltic Sea area underwent multiple glaciation during the Quaternary epoch. During the last glacial maximum (*c*.18 ka BP) the area lay within the margins of the Fenno–Scandinavian ice sheet, the weight of which caused depression of the continental crust. Maximum depression occurred around the centre of the ice mass, over the Northern Gulf of Bothnia. Ice retreat started at *c*.12.5 ka BP, and the subsequent development of the sea was largely governed by the interplay between isostatic rebound and the global sea-level rise which resulted from melting land ice (note that melting sea ice will not cause sea-level change). A number of stages in the evolution of the Baltic Sea have been interpreted by researchers, and summarised (Lambeck, 1999), as follows.

1. 12.5–10.3 ka BP, the Baltic Ice Lake developed, during intensified crustal uplift and slow eustatic rise, so at that stage there was no connection with the open ocean, and the Baltic 'Sea' contained a freshwater lake.
2. 10.3–9.5 ka BP, further retreat of the ice mass gave a new outlet across lowlands of central Sweden, resulting in a sudden drop in lake level and a new connection with the open ocean (strictly speaking, the North Sea), turning the ice lake into the Yoldia Sea.
3. 9.5–8.5/8.3 ka BP, the sea level rose, but crustal uplift was still greater than the global sea-level rise. Connection with the open sea was severed by the uplift of the Swedish landmass, leading to the Ancylus Lake stage. At this stage the continental ice sheet had retreated beyond (north of) the Baltic Sea basin. Uplift of land was greater in the north (centre of former ice sheet) than in the south, so that the lake gradually transgressed southwards, and a new contact was established with the rising ocean through the Danish sounds. The transition to brackish water at *c*.8.5 ka (Mastagloia Lake) was the beginning of the final stage.
4. 8.5/8.3–7/6.5 ka BP, brackish and marine faunas of the succeeding Litorina Sea, after which levels fell to the present sea level.

From 3 ka BP to the present day, salinity has decreased due to limited mixing with North Sea waters, and river input exceeding evaporation. Crustal uplift continues in the Baltic Sea area, at up to 9 mm/a in the northern Gulf of Bothnia. The sequence of isostatic readjustment and sea-level change is also recorded in Britain (Figure 4.3).

Isostasy is important in oceanography because it controls not only the amount of space the oceans have available, but also the shape of that space. The accumulation of sediment (and water during a sea-level rise) in an ocean basin over a long period will depress the basin floor, affecting its shape and volume.

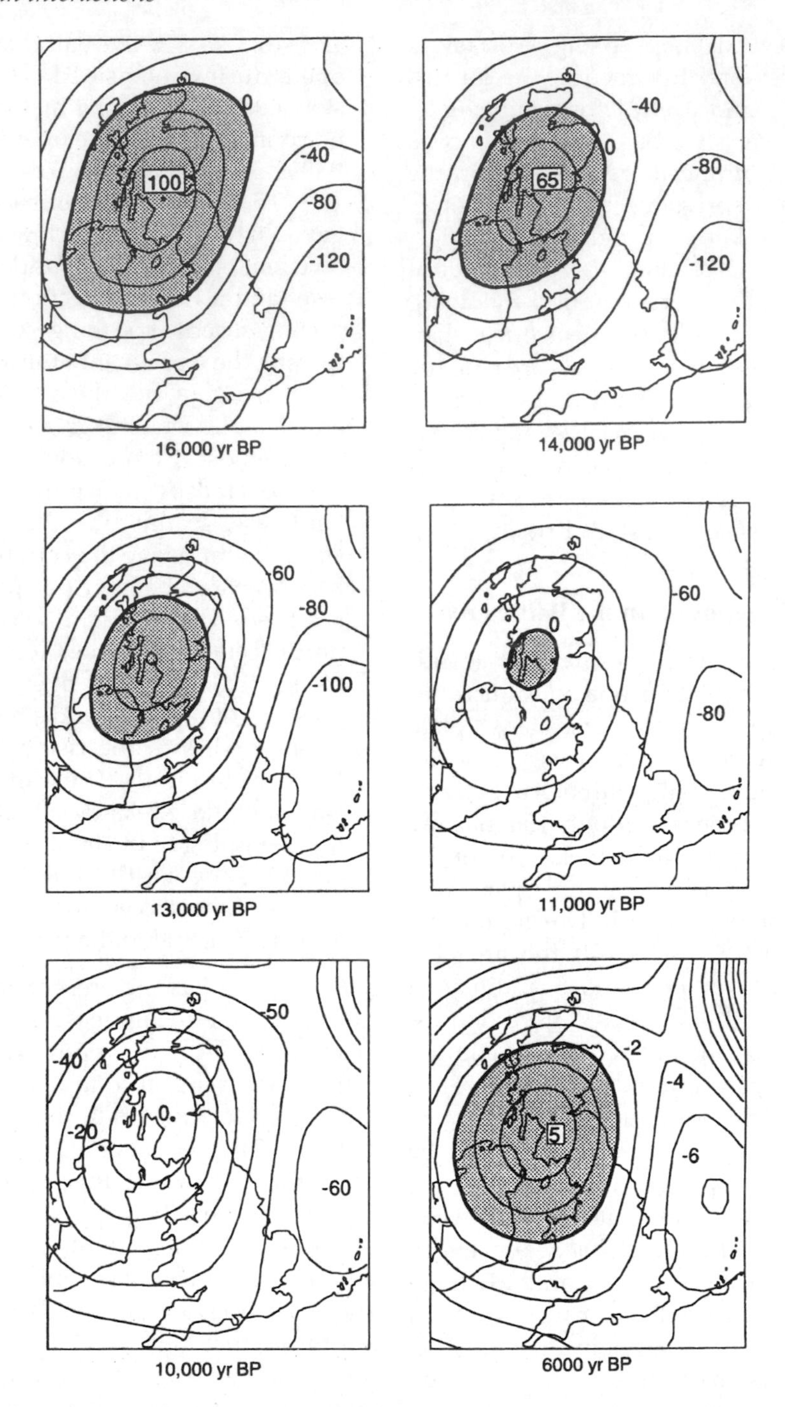

Figure 4.3 Sequence of diagrams of the evolution of Britain during deglaciation, illustrating the effects of isostasy. Reproduced from Terra Nova, **3**, Lambeck, K., *Glacial rebound and sea-level change in the British Isles*, pp. 379–389 (1991), with permission from Blackwell Science.

4.2.2 INTERACTIONS OF CONTINENTAL AND OCEANIC CRUST

Movement of continents was recognised early in the 20th century by Alfred Wegener, who created great controversy with the idea of continental drift that we are now so familiar with. The later model of plate tectonics involves interactions within and between ocean crust and continental crust, and is the major control on ocean basins. Activity occurs at plate margins, of which there are two types: active and passive (Figures 1.2 and 4.1). On passive margins, the oceanic crust simply abuts against the continental edge (such as on the east and west margins of the Atlantic Ocean plate). On active margins, there is either construction, destruction or conservation of crustal matter; the construction and destruction largely happening to ocean crust.

Ocean crust is dense, and originates as magma from the rising portions of convection cells in the mantle. On reaching the ocean floor, hot peridotite (Fe- and Mg-rich silicate, with olivine as its principal mineral) is added along the constructive margins of ocean floor. This seafloor spreading was proposed by Hess in 1962, and proved by Vine and Matthews in 1963, from the symmetrical mirror-image pattern in the variation of magnetic orientation of ocean crust parallel to the ridges. The Earth's magnetic field switches orientation every 0.6–1 million years, and magma contains elements susceptible to magnetism. When solidified, ocean floor basalt contains accessory magnetic minerals, e.g. magnetite (Fe_3O_4) and ilmenite ($FeTiO_3$), which become aligned with the current magnetic field orientation (thermoremanent magnetisation) (Figure 4.1). The continents, being less dense, are dragged along with the moving ocean floor. The age of basalt crust increases as it moves away from ridge axis at a rate of 1–10 cm/a.

The ocean floor is crossed by a pattern of mid-ocean ridges (Figure 1.5), with active global motion of the crust diverging from the ridges. The motion of ridges can go round corners by a set of short offset faults (transform faults) accommodating the irregular movements of the plates. Not all ocean spreading ridges are on the ocean floor; Iceland has received so much mantle material that it reached the surface of the Atlantic, and the mid-ocean ridge is observed as a fracture zone across the island. Iceland is expected to continue growing.

Ocean trenches identify sites of subduction zones (destructive margins) in the ocean crust, either along continental or neighbouring oceanic plate margins; cold dense ocean crust passes down, to be remelted and recycled. Melting is assisted by some water which is carried down the subduction zone and reduces the temperature at which the crust melts. Deep earthquake activity originates within an inclined zone dipping towards an island arc or mountain belt (called the Wadati–Benioff zone), indicative of a dipping slab of ocean crust. Folding and faulting demonstrate powerful compression in these regions; volcanic arcs and mountain belts associated with deep sea trenches contain a range of igneous and volcanic products.

Ocean crust is destroyed at subduction zones by sliding under the adjacent plate, at rates of 15–45 cm/a. The process appears to be faster than the production rate at MOR, but there are more spreading sites than subduction zones, leading to a steady state of production and destruction. This steady state must exist because the Earth is apparently not expanding or contracting. Along conservative margins, such as the San Andreas Fault, no material is added or removed. The plates move episodically, release accumulated stress, and cause earthquakes and differential ground movement.

Subduction zones identify sites of plate collision between ocean crust and either continental or other ocean crust. In each case, a chain of volcanoes (usually curved and therefore called a volcanic arc) results above the subduction zone, as some melted material from the descending plate reaches the surface.

Continental volcanic arcs build mountain ranges (e.g. the Cascades of western North America, the Andes of western South America), while oceanic arcs form ocean volcanoes (e.g. the Aeolian islands north of Sicily, the Japanese islands). One result of continued creation and destruction of ocean crust is that the oldest ocean floor is only of Jurassic age (*c.*200 Ma). In contrast, continental crust is so light that it is not subducted, so the oldest continental rocks are around 3900 Ma.

Destructive margins are complicated places. The portion of continental crust beyond the volcanic arc is stressed by the presence of the descending ocean plate beneath it, and a secondary convective cell is suspected to form in many cases. This is because there may be crustal extension and results in the opening of a back-arc basin; this not only becomes a site of sediment accumulation, but also the crustal activity may have an effect on regional, and even global, sea level. On continental margins, subsiding fore-arc basins may also exist, due to crustal subsidence on the outer continental margin. These various subsided areas of continental margins become marginal basins, filled with seawater, and accumulate sediment. They can play a role in organic evolution, because they may become isolated, leading to divergence in communities in time.

From the above, and from the discussion of isostasy in Section 4.2.1.1, it is clear that continental plates are deformable. Thus, although the Earth's crust may be regarded as brittle when subjected to sudden high energy events (e.g. earthquakes fracture the ground), periods of extensive stress causes plates to flex. Flexure is one of the factors influencing regional and global sea level.

Finally, certain places on Earth have ocean crust moving over sites underlain by hotter mantle (mantle plumes), leading to hot spots, the site of long-term volcanism. The best known resulted in the Hawaiian island chain in an intraplate setting; island volcanoes created above the hot spot are carried away from the area as the underlying plate migrates, which subsides as it passes away from the hot spot because of cooling (a feature now called thermal relaxation, relating to subsidence of all cooling crustal masses). Eventually the island submerges and becomes a seamount. Submerged volcanoes may be called guyots, but that term strictly is applied to flat-topped submarine mountains (Figure 4.4). In the tropical Pacific Ocean, the island volcanoes are ringed by reef growth. If the submergence rate is slow enough, the reefs keep their growth at the surface, and when the volcano submerges, the reef remains as an atoll; this process was recognised by Charles Darwin. Mid-ocean islands also play a large part in biotic migrations, not only of land creatures, but also marine forms, as island hoppers. Ancient mid-ocean volcanic islands are recognised in fossil assemblages, and are believed to have assisted in the evolutionary migrations of biotas.

4.2.3 THE OCEANOGRAPHIC SIGNIFICANCE OF PLATE TECTONIC ACTIVITY

This falls into two broad areas.

1. *Contintental configuration and change controls ocean basins*. There is plenty of evidence that the creation and destruction of ocean basins has happened through much of geological time, based on a range of information from continental positions, fossils and rock associations. Rifting of continents is taking place today in the Red Sea and East African Rift Valley. The Red Sea is already flooded, and in time it is expected that East Africa will divide. The reasons why rifting is occurring in those sites is probably due to mantle plumes, splitting the crust apart where they abut against the base of the crust. Recent work suggests that mantle plumes are the main control which causes continents to follow particular collision courses with others. The fact that these events happen is what has shaped modern oceanography, and has a huge impact on climate and evolution. The occurrence of greenhouse and

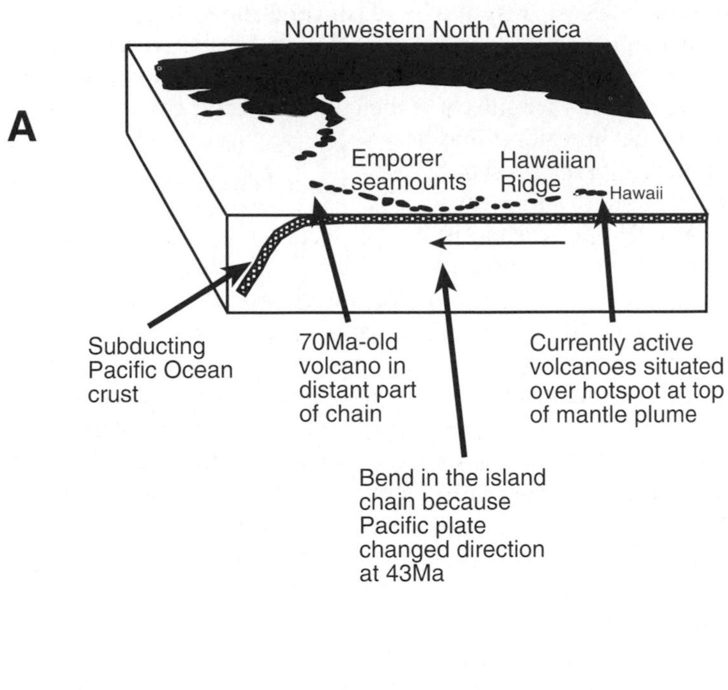

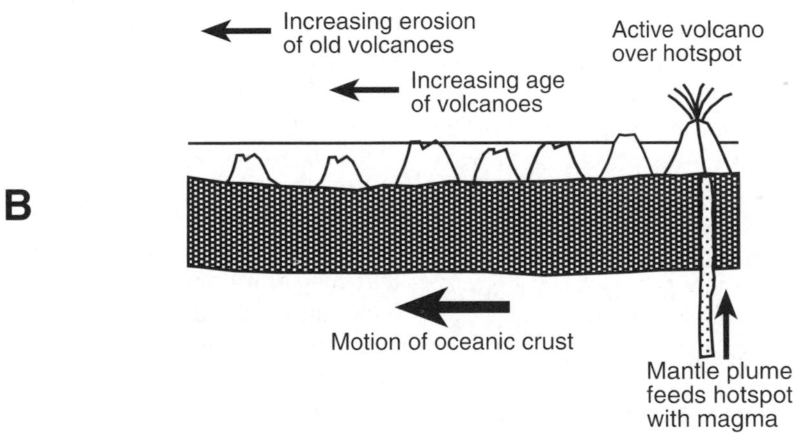

Figure 4.4 Features of volcanic island chains formed as a result of a tectonic plate over a hotspot. In this well-known example, the Pacific Plate moved over a stationary hotspot, where a mantle plume feeds submarine volcanoes. The current location of the hotspot feeds a developing volcano just southeast of Hawaii. The age of successive volcanoes in the chain increases away from the hotspot; the oldest volcano is 70 Ma, indicating the hotspot has persisted at least that long. A bend in the chain at 43 Ma shows the plate changed direction then.

icehouse worlds at particular times in Earth history, described in Part B, is largely attributed to the consequences of plate motion. The action of plate tectonics seems not to have been constant, and the degree of activity affects global sea level. Active plates cause the ocean floor to swell and displace water onto the continents, as in much of the

Palaeozoic and Mesozoic Eras; inactive plates lead to subsidence of ocean floor and large sea-level fall, such as in parts of the late Palaeozoic Era. Ocean basins are sites of storage of carbon, both as limestone and organic carbon, without which the presence of free oxygen in the atmosphere and oceans could not be maintained. 75% of carbon in the Earth's crust is preserved as carbonate (Holmén, 1992, p. 248).

2. *Seawater interacts with crust*. The input of large amounts of rock material to the Earth's surface settings, the drawdown of crust at subduction zones, and the pattern of fractures across thousands of miles of seafloor, all make the crust, and possibly the upper mantle, somewhat 'leaky' and accessible to ocean water. As noted elsewhere in this book, the cycling of ocean water through the crust and back again is so ubiquitous that over any 10 Ma period, a volume of water equivalent to the sum of all ocean water has passed through the crust. Not only has this cycling resulted in the present-day composition of seawater, but it also has great economic implications. The extensive Precambrian Banded Iron Formation (Chapter 8), derived from input of iron from the ocean floor, supplying most iron for industry, is probably the most important example, while the formation of ferromanganese nodules on the present ocean floor is a potential source of metals.

On the large scale of geological time, ocean basins can be seen to have opened and closed in many places. This has led to an idealised sequence of ocean basin evolution called the Wilson Cycle. In reality, ocean basins do not follow a predictable pattern of opening and closing again; instead there is a long-term trend of continents separating and then colliding with other continents. A major pivotal point in Earth history was in the late Palaeozoic Era, when all continents became fused to a single landmass (Pangaea), a detailed discussion of the implications of which is given in Chapter 9.

From the foregoing, we can easily see how the relationship between crust and ocean is a fully integrated factor in oceanography, giving the concept of the crustal-ocean factory.

4.3 THE GLOBAL CRUST–OCEAN SYSTEM

4.3.1 THE OCEAN STEADY STATE

Marine chemistry involves the study of mechanisms and rates by which materials are exchanged between land, atmosphere, ocean water and ocean sediments, leading to the present concentrations of dissolved material in the ocean (Chapter 2). Materials exchange between these components through the geological, physical, chemical and biological processes which comprise biogeochemical cycles (Butcher *et al.*, 1992). The sum of biogeochemical cycles is the global crustal-ocean factory, which involves all manner of interchanges between oceans and crust, and also incorporates the atmosphere, summarised in Figure 1.2. There is a continual input of elements to the oceans, mainly from continental weathering and erosion of silicate rock material carried by rivers, with small amounts input as wind-blown dust, but also from hydrothermal activity from beneath the ocean floor. Geological evidence, however, suggests that the salinity and composition of seawater have not changed significantly for at least the last 1500 Ma (Murray, 1992), which indicates that over geological timescales, rates of removal and input have been equal. This constancy of composition (also called the steady-state ocean) is a powerful factor which governs the modelling of ancient oceans, because the properties of seawater control how it circulates (note the density characteristics of water of different salinities and temperatures discussed in Chapters 2 and 3). Nevertheless, geological evidence indicates that, while the idea of a steady-state ocean is a good approximation, short-lived change in the chemical composition of the oceans appears to have occurred at certain times in the past

500 Ma. For example, oceanic crises in the past have included important excursions from the steady state affecting whole regions (e.g. the Messinian salinity crisis in the Mediterranean, *c.*6 Ma). Major changes in ocean composition occurred in earlier Earth history, and there is good evidence for minor cyclic change throughout Phanerozoic time, as detailed in Part B.

4.3.2 DYNAMICS OF SEAWATER–CRUST INTERCHANGE

Input of material to oceans cannot be summarised easily, because seawater is a complex solution containing at least some of every naturally occurring element. However, a useful idea of the key process of terrestrial weathering and erosion, and their products which are carried by rivers to the sea, may be gained by examining major components of the system. The results of the following three weathering reactions of common minerals cover important components of seawater. For each, (s) = solid, (g) = gas, (aq) = water or material dissolved in water.

1. Albite ($NaAlSi_3O_8$), one of the feldspars, is an alumino-silicate common in many surface rocks; it weathers to produce material common in ocean water and sediment:

$$NaAlSi_3O_8(s) + 2CO_2(g) + 3H_2O(aq) \rightarrow Al_2Si_2O_5(OH)_4(clay, s) + 2Na^+(aq) + 2HCO_3^-(aq) + 4SiO_2(s).$$

2. Anorthite ($CaAl_2Si_2O_8$), a calcium feldspar is an important source of Ca^{2+}:

$$CaAl_2Si_2O_8(s) + 2H_2CO_3(aq) + H_2O(aq) \rightarrow Ca^{2+}(aq) + 2HCO_3^-(aq) + Al_2Si_2O_5(OH)_4(clay, s).$$

3. Magnesium olivine (Mg_2SiO_4), a non-alumino-silicate component of basic (= alkaline) igneous rocks, produces key dissolved components:

$$Mg_2SiO_4(s) + 4CO_2(g) + 4H_2O(aq) \rightarrow 2Mg^{2+}(aq) + 4HCO_3^-(aq) + H_4SiO_4(aq).$$

4. Calcium carbonate ($CaCO_3$) produces two of the most significant dissolved materials in the oceans:

$$CaCO_3(s) + CO_2(aq) + H_2O(aq) \rightarrow Ca^{2+}(aq) + 2HCO_3^-(aq).$$

The chemical formulae seem forbidding for those not familiar with chemical symbols, but they are included here in their complete form, so that you can see the full results of weathering of crustal materials. There are several important features to note in the above reactions:

- 1 produces silica (SiO_2), a component of sand, and 1 and 2 yield clays; both are important sedimentary deposits in oceans (see Sections 5.2–5.5); 1 also produces dissolved sodium (Na^+), a major source of the oceans' vast content of that element, a component of salt (NaCl).
- 1–4 produce HCO_3^- which is passed to the oceans and becomes *the* critical control on the oceans' pH system in surface waters (see Section 5.6).
- 2 and 4 produce calcium (Ca^{2+}) and 3 produces magnesium (Mg^{2+}), important in the formation of limestone (see Section 5.4).

Other major components, such as Cl, P and N, arrived in the oceans mainly from volcanic degassing. Some material is carried by wind, either as dust particles or in aerosols, and by rainwater directly into the ocean (see also Figure 1.2). From these relationships it is easy to see how the oceans became rich in the common materials they contain.

Because there is a steady state of concentration of components, they must be removed from the oceans as well. Elements are removed from seawater by the following processes:

1. trapping of water in sediment pore spaces, with subsequent growth of cements on the sediment grains, either as marine or burial (= later, maybe millions of years) cements;
2. evaporite formation in hot arid shallow marine settings, drawing salt from the

oceans (not to be confused with evaporites in continental interiors, which are not ocean-related);
3. reaction of water with basalt lavas on the ocean floor, chiefly in ocean spreading ridges;
4. reaction of water with clays (or with inorganic and organic coatings on clays) which rain down through the water column;
5. formation of authigenic minerals on the seafloor (e.g. ferromanganese nodules) (authigenic minerals are those which form *in situ*, and are therefore not transported into the area);
6. organic processes (uptake by organisms or reaction with sinking organic material), principally resulting in removal of CO_2, but also may remove critical nutrient elements, especially P, N, Si and Fe.

These processes remove elements to ocean sediments, and sediments then follow alternative pathways, over long timescales:

1. Sediments subducted along with the ocean crust at ocean trenches pass to the mantle to be recycled, presumably to spreading ridges after millions of years, but the CO_2 they contain may well be released back to the ocean by volcanic eruptions at the surface.
2. In continent–ocean collisions, volcanic mountain regions (e.g. Andes) are eroded, with sediment being returned to the ocean trench or, in the case of rivers flowing east from the Andes, to the opposite side of the continent via the Amazon River, to be deposited on the passive Atlantic margin on the huge Amazon fan.
3. Some sediments are scraped off the descending oceanic plate and become part of the continental crust (called underplating, this process leads to the development of a wedge-shaped pile of sediments and volcanics, called an accretionary prism, at subduction zones).
4. Small parts of oceanic crust involved in powerful collisions may be thrust on top of the continental margin, in a process called obduction. The resulting rock associations (consisting of ocean crust, volcanic rocks and oceanic sediments) are called ophiolites, and are critically important in the identification of the sites of ancient subduction. Note that in rocks older than Jurassic age, ocean floor has all been destroyed, except for ophiolite fragments stranded on continental margins.
5. In continent–continent collisions, subduction does not occur because the rock masses are buoyant, although one block may partly override the other. The resulting mountains (e.g. Himalayas, European Alps) are rapidly eroded, the sediments accumulating either internally in the mountain chain, or externally, on coastal and interior plains. Coastal plain sediment reaches the sea by river transport. Sediment eroded from orogenic zones is often called molasse, and, incidentally, contributes to plate flexure by loading the continental plate, with consequent isostatic readjustment.

Therefore the fate of ocean sediments and volcanics, and consequently components drawn from the water into them, is variable. The average length of time that an element remains in solution in ocean waters before being removed to the sediment is defined by its residence time, expressed as A/R (where A is the mass of each element in the oceans, and R is the rate of removal or input, in the steady-state ocean model).

Residence times range from a few hundred years or less to tens of millions of years, and depend on an element's reactivity in seawater. Examples of residence times are shown in Figure 4.5. Highly reactive elements are rapidly scavenged (removed) to the sediments by (a) inorganic processes, by interactions with sinking particles or the sediment itself; and (b) organic uptake by organisms or sinking organic material. Unreactive elements tend to stay in solution and so have comparatively long residence times. The major ions in

Figure 4.5 Residence times of dissolved elements in oceans

Element	*% of crust (by mass)*	*% of ocean (by mass)*	*Mass in ocean* $\times 10^{18}$ *kg*	*Mass entering ocean per* 10^8 *years* $\times 10^{18}$ *kg*	*Number of times cycled through oceans per* 10^8 *years*
Na^+	2.4	1.077	14.4	20.7	1.4
K^+	2.1	0.038	0.5	7.4	15
Mg^{2+}	2.3	0.129	1.9	13.3	7
Ca^{2+}	4.1	0.041	0.6	48.8	81
HCO_3^-	0.02(total C)	0.002	0.19	190.2	1000
SO_4^{2-}	0.026(total S)	0.09	3.7	36.7	10
Cl^-	0.013	1.95	26.1	25.4	1
SiO_2	27.7(total Si)	0.0002	0.008	42.6	5300
H_2O	?	96.67	1370	3333 000	2400

Compiled from Segar (1998, p. 116), Bearman (1989b, see Figure 2.5) and Murray (1992).

seawater have residence times that are much longer than the mixing rate of water in the oceans, and so these substances are distributed uniformly around the oceans, exemplifying the principle of constant proportions, mentioned in Section 2.4.2. The next sections examine the processes that influence the composition of seawater in a little more detail.

4.3.3 VERTICAL AND HORIZONTAL DISTRIBUTION OF DISSOLVED OCEAN CONSTITUENTS, AND REMOVAL MECHANISMS

A classification of seawater components was given in Chapter 2. Elements in seawater show two main types of distribution: conservative-type, with stable levels over time, and non-conservative-type, such as nutrients, which vary from place to place and time to time (Pinet, 1992). In addition, as discussed in Section 2.4, there are major and minor element concentrations, with minor defined as less than 1 ppm.

Major dissolved ions are conservative, because although they are involved in chemical and biological processes, these generally have little effect on the elements' overall concentration. The most conservative are Na, K, Cl, S, Br, B and F. Ca is an exception because of its link with the CO_2 equilibrium reactions (see Chapter 5), and is slightly enriched in deep sea, along with HCO_3^- (Murray, 1992). Minor constituents are non-conservative, because their concentrations vary significantly from area to area. However, because they are in low concentration, there is little effect on overall salinity. Involvement in chemical and biological reactions significantly affects the vertical distribution of minor constituents. Examples of minor constituent activity are Cd and Pb. Cd shows similar behaviour to nutrient elements, being depleted in surface waters by phytoplankton uptake and regenerated at depth by decomposition of sinking organic material. Pb shows a profile that is enriched in surface layers due to atmospheric input. The dissolved Pb is rapidly scavenged by particles causing a decline in concentration with depth.

For both major and minor elements, the detailed chemistry strays beyond the scope of this book, but it is worth noting that, although many elements exist as being simple dissolved ions, much dissolved matter is held in complex associations. The major ions of Na, Ca, K, Mg and Cl are mostly present as simple ions, but are partly involved in complexes. Other major constituents (SO_4^{2-}, CO_3^{2-} and HCO_3^-) are mostly complexed, as are most minor constituents (Murray, 1992). As noted below in Section 4.3.4, the principal nutrient elements P, N, Si and Fe are at low concentration in shallow water due to biogenic removal, and are therefore non-conservative.

Of the Earth's gases, 95% are held in the atmosphere, reflecting the relatively low solubility in water. N_2 is the most abundant dissolved gas (reflecting its atmospheric levels), with O_2 and CO_2 and others in much lower amounts. Note that the solubility of CO_2 in particular is affected by temperature (more soluble in colder water, with deep ocean cold waters acting as a significant CO_2 store).

4.3.4 EXCEPTIONS TO THE RULE OF CONSTANT PROPORTIONS

Globally, the conservative major ions in seawater are present at constant ratios to each other, and to total salinity. Important exceptions to this rule do occur, particularly in the following areas:

1. In marginal seas and estuaries, large river inputs can considerably affect the major ion composition of seawater.
2. Anoxic basins contain water that is largely devoid of oxygen. Organic matter breakdown proceeds under these conditons using the oxygen contained in the sulphate ion, converting sulphate (soluble) to sulphide (insoluble), causing depletion of dissolved SO_4^{2-}, and leads to the formation of pyrite (iron sulphide). Reactions involving sulphides and sulphates account for most of the chemistry of sulphur (one of the most important elements in the sea).
3. As sea ice forms in polar regions, only small amounts of salt are incorporated into the ice, leaving the remaining seawater enriched in salt. This is part of the control process on formation of deep cold water in the thermohaline circulation (Chapter 3).
4. Hydrothermal vents are zones where acid waters, enriched in Mg and S, mix with alkaline seawater.
5. Evaporite formation plays a major role in geochemical cycles of the major ions in the regions where evaporites form. Precipitates (limestone, halite, anhydrite and possibly dolomite) are deposited in stages, causing the various major ions in seawater to be removed at different times and rates. Consequently, the composition of the remaining solution (or brine) differs considerably from that of average seawater.
6. Sediment pore (= interstitial) waters are usually involved in early diagenetic reactions considerably affecting the ion content of interstitial waters.

In these six cases, the sites are peripheral to the main mass of ocean water, and involve partial or complete isolation of the location from the oceans. This leads to a principle that has been used by palaeoceanographers; areas of ocean that become isolated (usually basins stranded by sea-level change or subject to regional tectonics) undergo unique seawater evolution following isolation, because of the lack of interchange with the open ocean. Examples of this have been proposed for Mediterranean basins associated with the Messinian salinity crisis, a problem which is much more complex than the traditionally accepted story of a simple desiccating region once the Atlantic gateway at Gibraltar had been closed (see Section 10.9).

4.3.5 SEAWATER OXYGENATION; THE IMPORTANCE OF THE BLACK SEA MODEL

Oxygen (and the processes that control its concentration in seawater) is critically related to photosynthesis in the photic zone, the principal source of oxygen in the sea. Plankton are the dominant source of marine organic matter, via photosynthesis in surface layers of the ocean, and remove dissolved CO_2, N and P, and produce O_2, in a reaction summarised as:

$$106CO_2 + 122H_2O + 16HNO_3 + H_3PO_4 \rightarrow (CH_2O)_{106}(NH_3)_{16}H_3PO_4 + 138O_2.$$

$(CH_2O)_{106}(NH_3)_{16}H_3PO_4$ is not a real compound, but is an average composition for plankton, including both phyto- and zooplankton (Murray, 1992), and applies to the open ocean. Although this is a rather forbidding formula, note that for the $138O_2$ molecules produced, $106CO_2$ molecules are used, a ratio of 1.3:1. A more simple version of the photosynthesis equation is commonly expressed as:

$$6CO_2 + 6H_2O \rightarrow C_6H_{12}O_6 + 6O_2$$

($C_6H_{12}O_6$ is glucose, the simplest carbohydrate) which suggests a O_2/CO_2 ratio of 1:1, indicating that each molecule of free oxygen is matched by an atom of carbon buried in the Earth's crust. Both reactions are approximations of complex global systems, and show that, although the amounts of free O_2 and buried C are approximately equivalent, balancing the amounts of each in the various reservoirs is imprecise. Some O_2 is used up in crustal weathering; for example, the oxidation of iron.

However, in the oceans, rates of transport of constituents between reservoirs is a critical feature, exemplified by the Black Sea setting. Thus, if the rate of O_2 removal from seawater (by bacterial decay of organic matter) exceeds the rate of supply from thermohaline circulation (in which cold surface waters sink and take oxygen to seafloor), oxygen deficiency builds up in deeper waters and may lead to anoxic conditions. This has a large influence on seawater chemical composition, particularly with respect to biologically cycled elements, and the best-known site is the Black Sea, the largest anoxic basin in the world.

A stratified, silled basin, the Black Sea is almost landlocked; it is connected to the Mediterranean via the Straits of Bosporus (only 30 m deep in places), so exchange with Mediterranean water is limited (Figure 4.6). It is located in mid-latitudes, and its inputs from river discharge and precipitation exceed outputs through evaporation, resulting in low salinity, 18–22%, compared to 35% for the open ocean. Summer heating generates a sharp density gradient at *c.*100 m depth, so that warm fresher water overlies cooler, more saline water. The result is little mixing of water between layers. Furthermore, the presence of shallow sills gives a long residence time for deep waters (up to *c.*3000 years), during which there is no turnover, so that lower layers become stagnant, because any oxygen is used up in the oxidation of organic matter.

Surface layers are O_2-rich, with large populations of phytoplankton due to nutrient inputs from rivers (Figure 4.6). As plants die, organic matter sinks into the lower layers and is decomposed (oxidised) by bacteria. Bacteria use dissolved oxygen in lower layers which is not replenished because there is no mixing, so the deeper water is permanently anoxic. In those depths, chemosynthetic bacteria use oxygen from SO_4^{2-} to decompose organic matter and release much highly toxic H_2S. Occasionally (every century or so) a severe winter storm may mix some of this into the upper oxygen-rich layer, causing massive fish kills.

In the anoxic zone, sediments are laminated, because the absence of bottom fauna means there is no burrowing to destroy the layers (a feature used to interpret seafloor anoxia in ancient oceans). Black, organic-rich sulphide mud is deposited, consisting of fine siliciclastic turbidites interbedded with organic-rich, microlaminated limestones, with some similarities to ancient black shales (Figure 4.6). Between 6600 and 3200 a BP, a finely laminated organic-rich shale (sapropel) with silt laminae was deposited (3–20% total organic carbon). We have little idea of water column conditions during its formation, but because the sediment is finely laminated with no *in situ* benthic fossils, it demonstrates a persistent anoxia. Most organic-rich sediments are found in topographic lows on the basin floor, because low-density organic particles and muds are focused in areas where currents are slowest. Sluggish Black Sea deep currents circulate

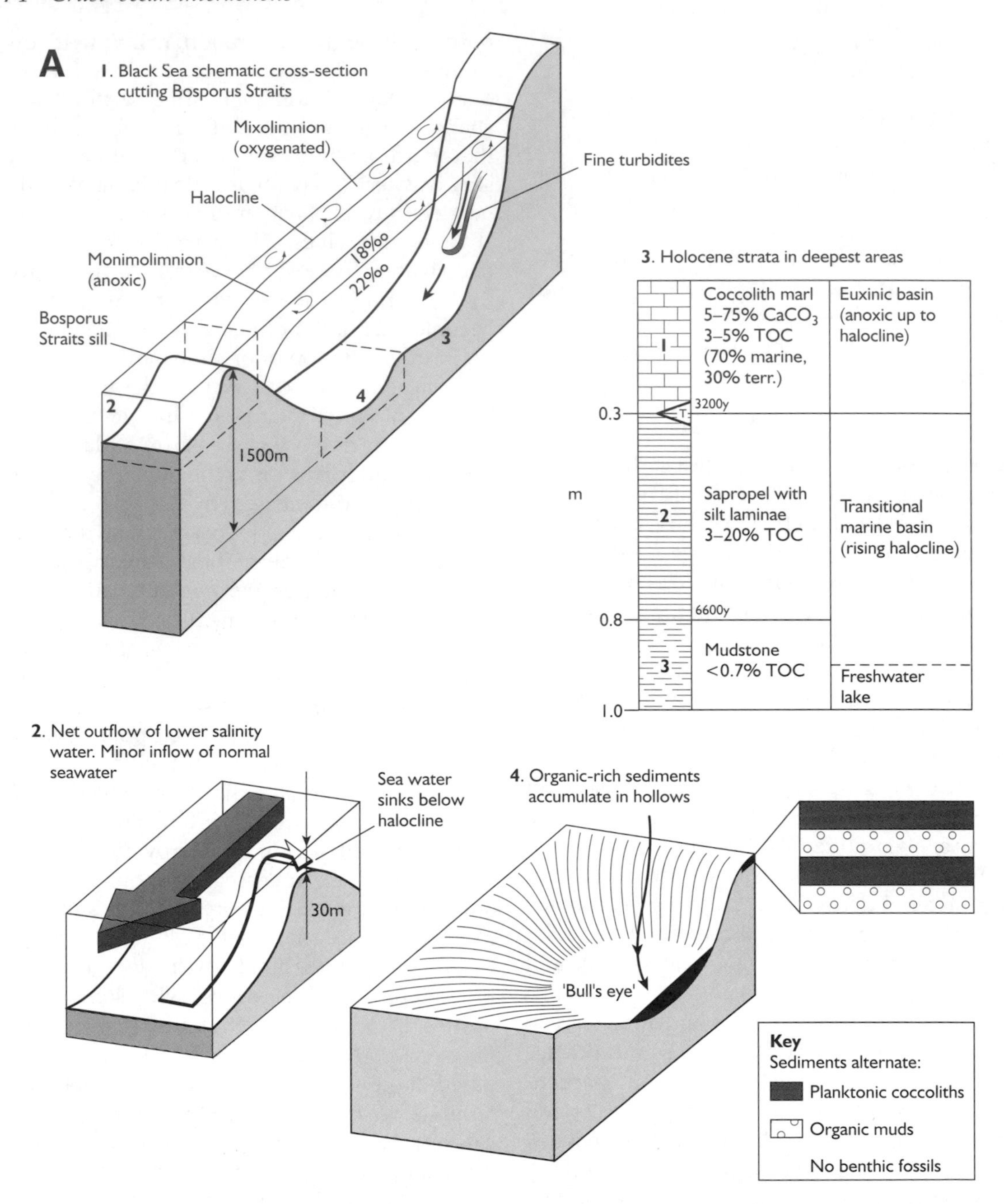

Figure 4.6 Processes in a restricted basin, using the model of the Black Sea. (A) shows the features of the Black Sea. In the schematic cross section (1), the Bosporus Straits sill restricts circulation, with a net outflow of low salinity water (2), and only the upper 200 m of Black Sea water is actively circulated. Turbidity flows deposit fine sediment into the basin from its margins. Holocene strata (3, adapted from Wignall, 1994) demonstrate the transition from a fresh water lake to an euxinic basin, and the deepest hollows of the sea floor (4) accumulate the greatest concentration of organic matter. A3 adapted from Wignall, P. (1994) *Black Shales*, Oxford University Press, Oxford.

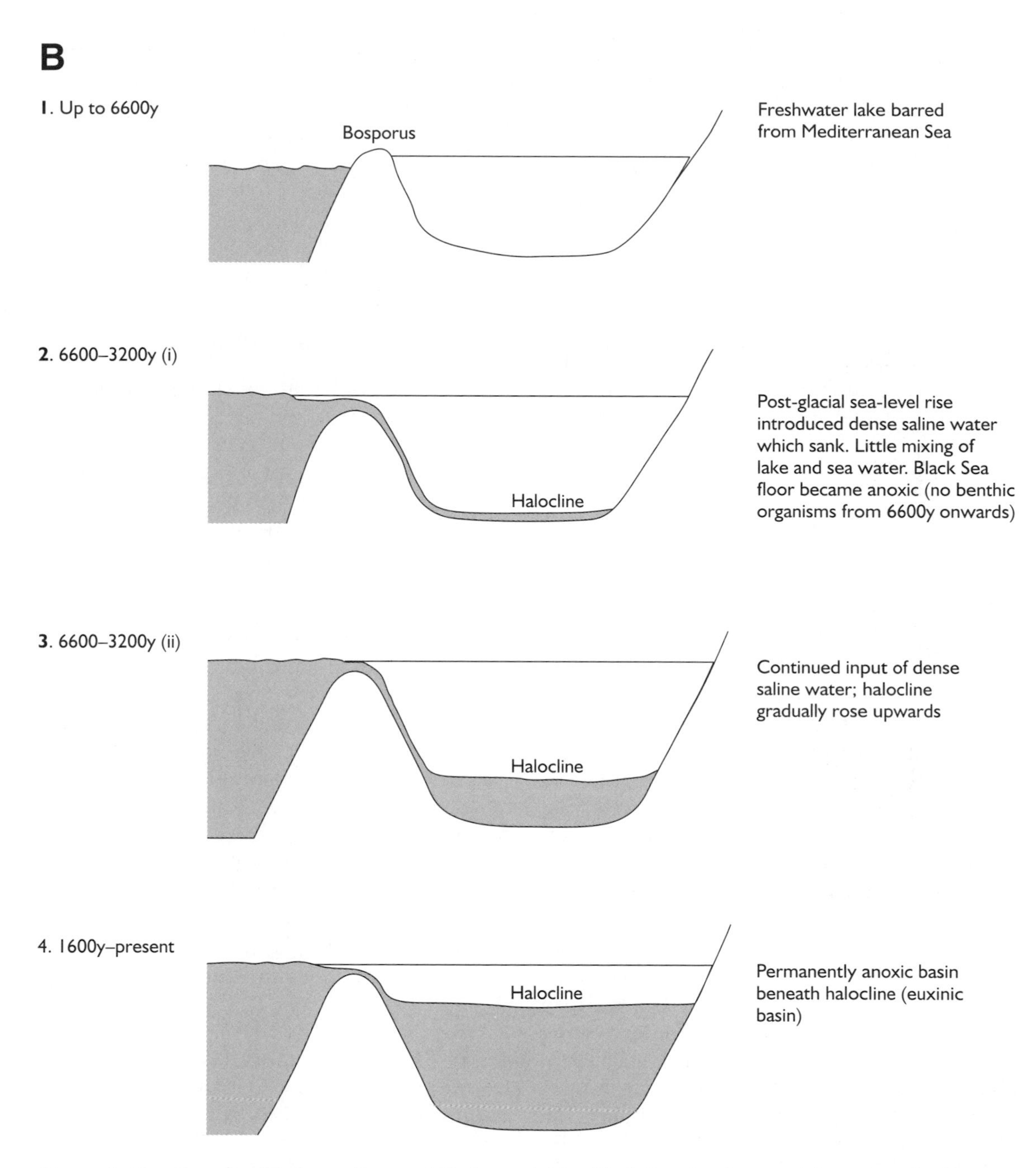

Figure 4.6 (*continued*) (B) shows the most likely model for Black Sea development, based on the Deueser model, whereby gradual replacement of most of the original freshwater lake water by denser saline water also led to the progressive anoxia of water below the halocline.

around hollows, sometimes referred to as the centripetal effect, or bull's eye. Centripetal effects have been suggested as explanations for variation in seafloor oxygenation in the British Jurassic Oxford Clay.

The conditions of the Black Sea provide an analogue for some ancient black shales, although not fully representative, because it is deeper and more brackish than ancient black shale environments.

4.3.6 THE SEAWATER CO_2 SYSTEM

An important component of seawater–crustal interaction, CO_2 is stored on the ocean floor as sedimentary material, largely limestone; therefore, the bulk of evidence for CO_2 features in modern and ancient oceans is in sediments and rocks, and is discussed in Chapter 5 rather than here.

4.3.7 DIAGENESIS

4.3.7.1 Principles

Once elements have been carried to sediments by the processes described in Section 4.3.2, they become subject to diagenetic changes. Diagenesis is the term used to describe post-depositional changes in sediments, brought about by physical, chemical and biological reactions, and includes both cementation of the sediment into rock (lithification) and recrystallisation of existing cements (sometimes called neomorphism); the results are the formation of altered or new mineral phases. The most important result of diagenesis is that the concentrations of all substances precipitated (sometimes also called scavenged) into ocean floor materials can change with time, creating the biggest problems for reconstructing ancient seafloor and ocean conditions, especially in geochemical studies. Chemical constituents can change mobility according to both pH and Eh (measure of oxidation state) of the fluids passing through sediments and rocks. Worse still is that change can take place at any time after the sediment has formed, either when it is still in an unconsolidated state, or millions of years later when it has been lithified. Thus older marine sedimentary rocks are likely to show the most change, but even recent material can be altered to such an extent that reliable reconstruction of seafloor and seawater are severely compromised. Burrowing organisms also mix or ingest sedimentary material, and influence the post-depositional changes.

Early diagenetic reactions occur during the burial of sediments to depths of a few hundred metres.

4.3.7.2 Early diagenesis (Figure 4.7)

Organic material that rains down onto the seafloor is buried, and oxidised by bacteria present in the sediment. The most efficient way for bacteria to break down organic material is to use oxygen itself (aerobic respiration). In pelagic deep sea sediments (mostly clay and $CaCO_3$), where the supply of organic material and the rate of burial are low, diagenesis does not generally continue beyond this stage until the sediment is buried deeper. Where the rate of organic carbon supply or burial is higher, the oxygen concentration in the sediment interstitial water can fall to such a low level that diagenesis must proceed using secondary oxidants (anaerobic respiration). One of the most important reactions in the shallow burial setting is sulphate (SO_4^{2-}) reduction. SO_4^{2-} is very abundant in seawater, and therefore passes through the upper layers of sediment. Anaerobic SO_4^{2}-reducing bacteria *Desulfurovibrio* and *Desulfotomaculum* use SO_4^{2-} as an energy source in the following reaction:

$$2CH_2O + HSO_4^- \rightarrow H_2S + 2HCO_3^- + H^+$$

where CH_2O represents carbohydrate derived from the rain of organic matter collecting in sediment from the death of phytoplankton, and HSO_4^- is a derivative of SO_4^{2-} formed in the water by reaction of SO_4^{2-} (which is an abundant component of seawater) and hydrogen ions (H^+). The role of microorganisms can-

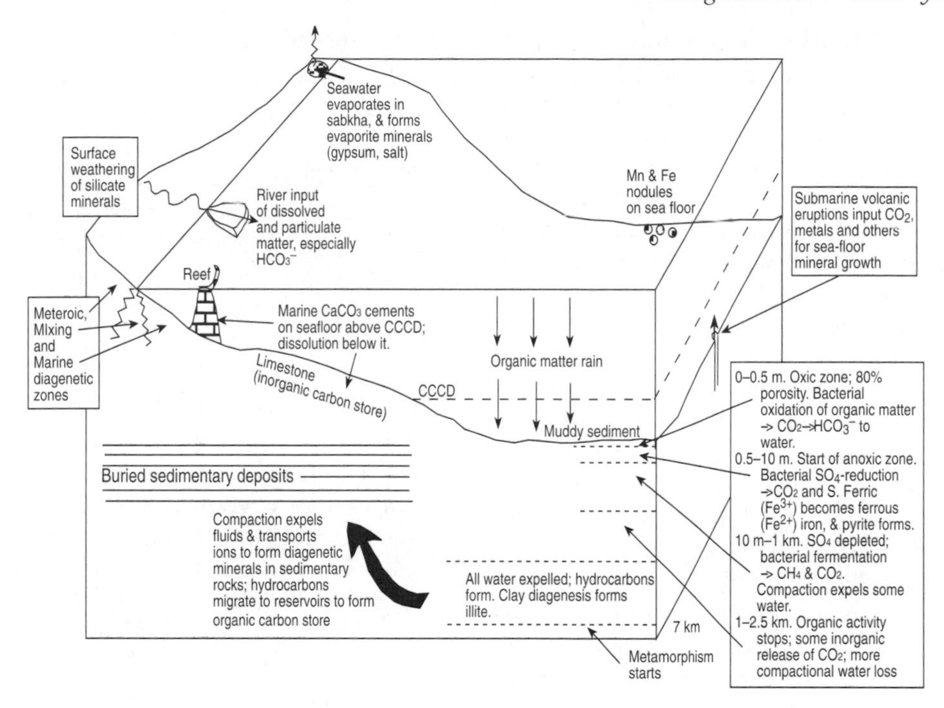

Figure 4.7 Diagenesis of ocean floor materials, emphasising the main processes. Note that the key components are clays and limestones; diagenesis occurs in sediments on the sea floor, in fresh water environments (particularly affecting limestones) and in various stages of burial of sediment. The sum total of these changes leads to important modification in the appearance of marine sedimentary rocks, and such change frequently needs to be interpreted in order to understand ancient ocean processes.

not be overstressed in shallow burial environments; because they are mobile and very abundant, with estimates of 10 million individuals per cm^3 of pore water (Hallberg, 1992), they result in considerable mixing of sediment pore waters and facilitate reactions in the sediment. Thus, during early diagenesis, elements are mobilised into solution and are able to migrate through the interstitial waters. These may be reprecipitated and incorporated into authigenic materials, such as glauconite, various carbonates, and iron and manganese oxides in oxidising conditions, while pyrite (FeS_2) is the common product in reducing conditions. Alternatively, the elements may escape from the sediment entirely and be released into the overlying water column. As the sediment is buried more deeply the sedimentary material begins to dewater under the effects of compaction and primary cementation of sediment may begin, leading to lithification. Calcareous sediments (composed of $CaCO_3$) can become partly cemented to solid rock directly on the seafloor, and form hardened patches of seabed called hardgrounds.

Much oceanographic information is drawn from the early cements in sediments, especially limestones, because these cements were precipitated while the pore waters were in contact with the ocean water. Of particular importance, stable isotopes of carbon and oxygen (^{12}C, ^{13}C, ^{16}O and ^{18}O) are used in palaeoceanography.

4.3.7.3 Burial diagenesis

During deep burial, dewatering and compaction continues, and pore waters are modified further by reactions with clay minerals, dissolution of unstable material and precipitation of authigenic minerals. Because clay is compacted, many soluble ions are expelled and pass through the sediment to form minerals in less compressible sands or calcareous material. The sediment is therefore cemented by material derived from pore waters and grain dissolution. Note that the pH and Eh conditions of deeply buried sediments are not easily predicted. Methods using a variety of petrographic characters (using rock sections so thin that they are transparent under the microscope) and geochemical analysis can be used to reconstruct diagenetic changes; these methods are beyond the scope of this book, but are available in a range of publications (e.g. Leeder, 1982; Tucker and Wright, 1990; Wignall, 1994). Deep burial processes operate over tens of millions of years and affect sediments to depths of around 10 km. At greater depths reactions occur at elevated temperature and pressure and are termed metamorphic, and the boundary between the later stages of diagenesis and the early stages of metamorphism is gradational.

4.4 CONCLUSION

Interchange between crust and ocean occurs in several forms, and has developed over geological time. The action of plate tectonics in distributing continents, and in generating a 'leaky' ocean floor which allows recycling of elements through the 'crust–ocean system', demonstrates the close integration of geological and hydrological processes needed to understand the oceans. Such information is of critical importance both to modern oceanography, and to reconstruction of ancient oceans. Here we have introduced the general principles, but the study of the interchange of material is a huge one, and is continued in appropriate places in Part B of this book. More details of the ideas outlined in this chapter are available in a number of publications (e.g. Bearman, 1989b; Butcher *et al.*, 1992; Chester, 1990; Leeder, 1982; Pinet, 1992; Wells, 1997).

4.5 SUMMARY

1. The crust and oceans interact because plate tectonics controls ocean basin form, and the interconnection between water and the plate mechanisms leads to interchange of material. Therefore the crust–ocean interaction is heavily interdependent.
2. Plate tectonics is probably the most important feature which controlled and modified ocean basin shapes, topography and connections with other oceans.
3. Isostasy has a critical effect in that any form of loading on the continents and ocean crust has the effect of depressing the crust, and modifying the volume available for oceans. Isostasy may be affected by ice, sediment or water load on the continental crust.
4. Plates are brittle under short-term forces, and therefore fracture in earthquakes. Longer-term pressures cause flexing, with consequent changes in ocean volume and therefore sea level.
5. The global cycling of elements through the ocean system can be summarised by the 'crust–ocean system' (which also includes the atmosphere!), whereby elements are input into the ocean via continental erosion and seawater interactions with new ocean crust. These elements then undergo reactions in seawater and are ultimately removed to the sediments. The average length of time that an element remains in solution in ocean waters before being removed to the sediment varies considerably, depending on the element's reactivity and involvement in biogeochemical cycles.
6. Diagenesis is a set of early processes affecting both the solid and dissolved constituents of ocean-floor sediments. It has great significance for palaeoceanography because reconstructions of past ocean and climate conditions rely on knowledge of chemical reactions in the ocean floor.

MARINE SEDIMENTARY DEPOSITS 5

5.1 INTRODUCTION

Sedimentary deposits are those formed at the Earth's surface by the action of water, wind and ice, and encompass the full range of environments, from land to ocean. Oceans and seas have large-scale accumulation of sediment in all marine settings, from shallow shelf to abyssal plain, as unconsolidated particles that cover the bedrock of the underlying crust. However, it seems that plate tectonic action has been necessary to maintain both a source for erosion of sediment (by mountain building), and accommodation space for sediment carried to the sea (by removal of ocean crust with its sediment veneer in subduction zones). Otherwise, global estimates indicate that the oceans would fill up with sediment in 100 Ma and the mountains would erode flat in 50 Ma (Holland, 1978). The average marine deposit is 0.5 km thick, being thinnest at the youngest ocean crust of mid-ocean ridges, and thickest at continental margins, due to both a higher sedimentation rate at the margins (nearer the sources of supply from land), and the greater age of underlying ocean crust due to seafloor spreading (more time for sediment to accumulate). Marine sediments are important reservoirs in the crust–ocean system, as discussed in the previous chapter.

Observations of sedimentation processes today are used to reconstruct past environments and events. Ancient sediments are the dataset for interpretation of past oceans, because the oceans themselves no longer exist, and the oldest ocean crust is Jurassic age, limiting the maximum age of sediments laid down on the ocean crust. Sediments deposited on continental crust, however, date back to the oldest rocks (Chapter 8), plus some information from ocean floor material emplaced on continents by obduction, as ophiolites. A great deal of information from ocean floor sediments, derived from drilling the seabed, provides data for models of ocean and climate change since the Jurassic. However, it is important to note that modern sea levels are much lower than they have been for most of the geological record, so continents are more exposed, and there is plenty of ancient marine sediment available for study on land. This chapter provides a general survey of the nature and controlling processes of marine sedimentary deposits, important for the study of sedimentary rocks in Earth history, and planning for modern ocean resource management.

5.2 TYPES OF SEDIMENTARY DEPOSITS

Marine sedimentary deposits are composed of two main fractions: inorganic and organic, which are intermixed to varying degrees. Sediments are unconsolidated particles, while sedimentary rocks are made of cemented particles. However, some sedimentary deposits are composed of material precipitated from water into the site of deposition, and are essentially crystalline. However, the classification of sedimentary deposits is based (like so many other classifications in the earth sciences) on the processes of formation, with four broad categories as follows (Figure 5.1), in which both organic and inorganic components may be present. The organic component of sedimentary deposits is present either as calcareous or siliceous shell material (called allochems), or as organic tissue and products (which, in marine

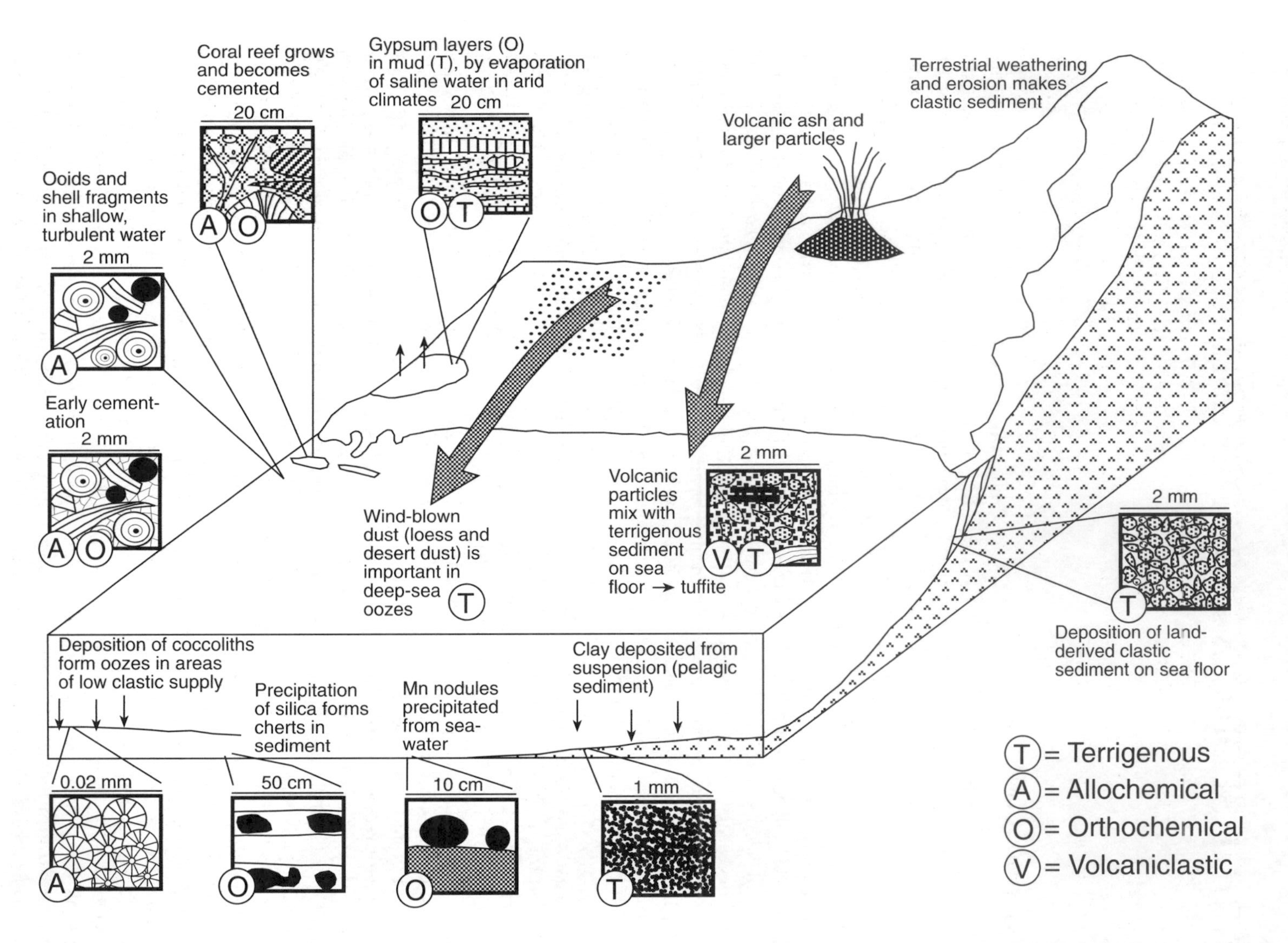

Figure 5.1 Diagram of the four types of sedimentary material – terrigenous, allochemical, orthochemical and volcaniclastic. The setting of formation of each is shown, with typical examples. Note the various means by which sediments arrive on the seafloor. The processes of diagenesis are indicated by Mn nodules, cherts and calcite cements.

sediments, may form hydrocarbon deposits under the appropriate burial conditions). Carbon is present in particulate organic matter on the modern seafloor as 20% $CaCO_3$ and 80% organic carbon, but over geological timescales, $CaCO_3$ stores most crustal carbon. The following are the commonly used groups.

1. *Terrigenous* (Greek for earth-formed), also called clastic (Greek *klastos* = to break), deposits are those derived from weathering and erosion of igneous, metamorphic and even older sedimentary material from land, and carried to the sea, principally by rivers. They consist mostly of quartz and clay minerals, because these are the most stable forms of crustal material on the Earth's surface. Weathering converts common rock-forming minerals (e.g. pyroxenes, amphiboles, feldspars, see Chapter 4 for examples of chemical reactions) into clays, the main component of mud and shale, and releases soluble ions. On the other hand, quartz (common in granites, for example) is chemically virtually inert, and passes through weathering and erosion cycles to form the principal component of sand and silt, and their lithified versions, sandstone and siltstone. Quartz is so durable that grains are known to have been recycled several times over millions of years of erosion and deposition.
2. *Orthochemical* (Greek *orthos* = straight (as in direct), *chyma* = fluid) deposits are crystallised *in situ* from seawater; the commonest are evaporite minerals (e.g. halite, gypsum) formed by evaporation in arid climates, but also included are metallic minerals around volcanic vents, manganese and iron nodules, and glauconite on the ocean floor. Orthochemical deposits are also called authigenic. The lithification of unconsolidated sediment either on the seafloor, or buried beneath it, is achieved by the growth of cement crystals of clays, silica, iron oxide or calcite, which are also orthochemical, during early diagenesis.
3. *Allochemical* (Greek *allos* = other) deposits form by crystallisation from seawater, followed by local movement of the material, so that although the material was formed *in situ*, it has been transported short distances. This concept applies almost entirely to limestones, which are composed largely of organically secreted $CaCO_3$, as shells, which then become locally moved detritus on the seafloor.
4. *Volcaniclastic* deposits are principally igneous material ejected from volcanoes, including volcanic bombs and ash. These are deposited as sedimentary particles, either on land or in water, where the particles may mix with pre-existing sedimentary material.

Another concept applied to ocean sediments relates to transport distance and time, and borrows terminology from biology. *Pelagic* (Greek *pelagos* = the sea) sediments are those deposited in deep water by the gradual settling of particles suspended in water, and therefore similar in meaning to the pelagic mode of life of fish and plankton, which live in the water, not on the substrate. Important pelagic sediments are clay minerals, particles of which are too light to settle in any quantity in turbulent shallow waters; but pelagic also applies to biogenic particles of carbonate and silica from planktonic microorganisms floating in the upper ocean, which on death settle to cover large areas of the modern seafloor. Inorganic pelagic particles are largely derived from river input to oceans, but wind-blown dust and cosmogenic particles are also involved. *Hemipelagic* refers to fine-grained sedimentary deposits which are re-mobilised, either by downslope movement by turbidity currents, or by bottom currents, and therefore redeposited. Clearly, deep-water sediments can contain mixtures of such deposits, and useful discussions of these sediments and their processes of formation are published (Kemp, 1996; Scholle *et al.*, 1983). Modern deeper water sediments are dominated by carbonates, in contrast to muds in the case of many ancient examples. This is partly because of the higher percentage of pelagic carbonate-secreting organisms in recent geological times, but there is also the effect of deeper water on the stability of $CaCO_3$, which dissolves in the deeper water due to its higher pressure (remember that in past geological times the sea level has been generally higher than today).

5.3 DISTRIBUTION OF SEDIMENTARY DEPOSITS IN OCEANS

Figure 5.2 shows the patterns of distribution of marine sediments throughout ocean systems. Note the following four points.

1. Shelf sediments are dominated by coarser clastics than deeper sediments because water energy decreases with depth. Therefore this leads to deposition of sands and muds as clastic shelves in areas of low-angle slopes. On higher-angle submarine slopes, bulk emplacement of terrigenous sediments (derived from slumping and turbidity currents, and in polar latitudes, from ice-rafting) carries coarser sediment into deep water. Where remobilised sediments are preserved on continental margins in ancient rocks, they assist in the identification of ancient fold belts and continental margins, with excellent examples in the UK Cambrian Welsh Basin and the north Cornwall area of the Carboniferous Variscan belt. Furthermore, during times of high sea level,

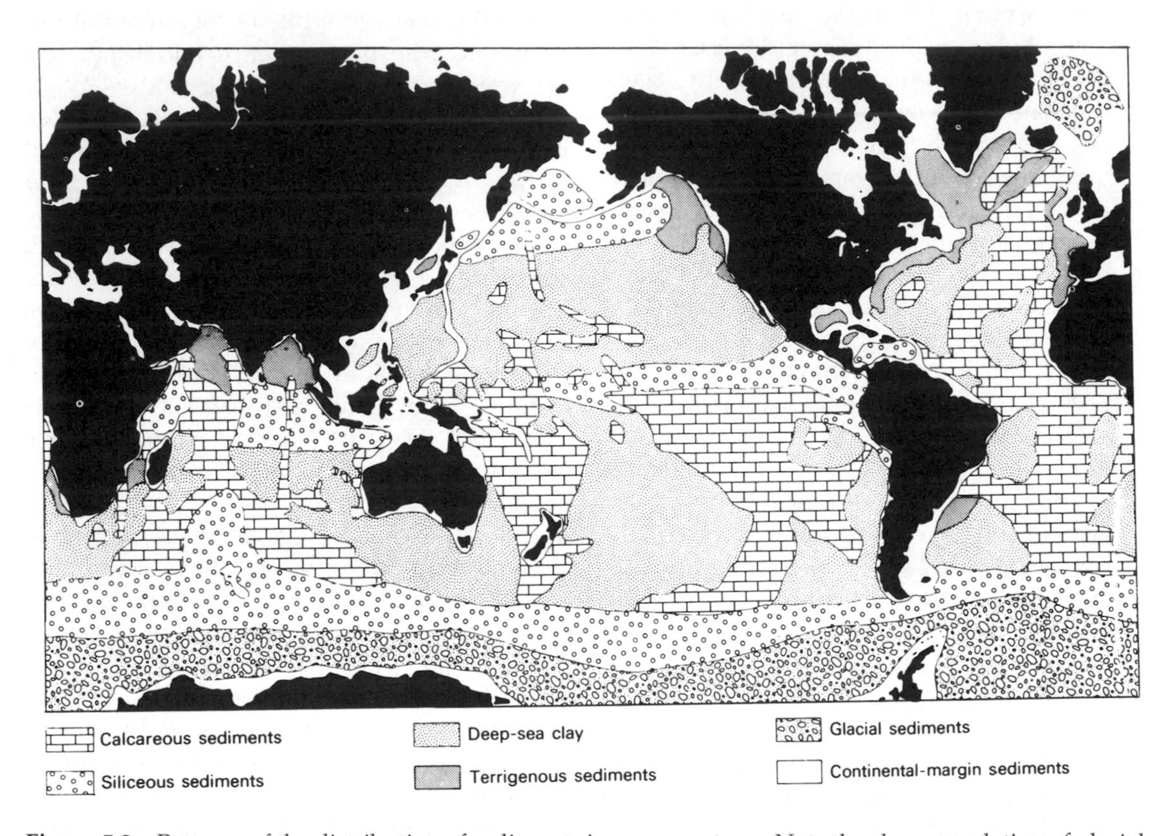

Figure 5.2 Patterns of the distribution of sediments in ocean systems. Note the close correlation of glacial sediments with the North and South polar regions, and of terrigenous sediments with neighbouring mountainous land areas. Deep ocean sediments are dominated by clays and carbonate oozes. Reproduced with permission from Tucker and Wright, *Carbonate Sedimentology* (1990), Blackwell Science, Oxford.

most clastic material is deposited on the inner continental shelf, while at low sea-level stands, rivers extend their channels seawards and material is deposited on the outer shelf and upper continental slope.

2. Clay minerals are spatially controlled by climate, and therefore their occurrence in ancient settings assists reconstructions of past climates. For example, kaolinite is formed in low-latitude humid climates, especially near land areas with abundant granite, because the kaolinite is derived from the weathering of feldspars; it is deposited in ocean settings adjacent to those areas (Figure 5.3). Similarly, illite is more related to temperate and semi-arid settings, and little is found near the equator. Smectites (including the common montmorillonite) are Mg-rich, and so relate to areas of weathering of volcanic rocks, both on land and in the oceans, and are associated with mid-ocean ridges; this is because seawater weathers fresh basalt, which mid-ocean ridges are largely composed of, and draws Mg out of the water into the forming clays. Finally, chlorite contains much ferrous iron (i.e. the reduced form of iron, which is soluble, and so is easily weathered and oxidised); therefore chlorite is found in cold (largely high latitude) areas where chemical weathering is less (Leeder, 1982, p. 14).
3. Carbonate sediments are composed mainly of organically secreted shell material, and therefore develop best in places where the temperatures are higher, but also depend on the rate of accumulation of clastic sediment. If clastic input is high, carbonates are suppressed, so their occurrence near rapidly eroding continental areas such as major orogenic belts is limited (Figure 5.2).
4. Orthochemical deposits form in low-sedimentation settings, both on ocean margins (evaporites) and deep ocean (manganese nodules), because, like carbonates, their formation is diluted by the presence of clastic sediments. Evaporites, like clay minerals, are therefore indicators of climate in past sequences of sediments. Orthochemical materials also include diagenetic cements, and other curios like the siliceous material called chert. Figure 5.4 shows the modern distribution of economically important orthchemical deposits.

The rates of sediment deposition vary with proximity to continents and ocean ridge systems, shown in Figure 5.5.

5.4 CONTINENTAL MARGIN MARINE SEDIMENTARY DEPOSITS

These deposits constitute the marine shelf environments, largely made of carbonates and siliciclastic material (sand and clay). Also, *c.*90% of deposited marine organic matter accumulates on continental shelves, of importance for the petroleum industry (overall, sedimentary rocks contain an average of 0.5% organic matter). Shelf settings are therefore the most important petroleum resources, and it is important to note the major role played by organic matter in sediments, with microbial activity in the surface sediment being a key feature. It has been noted by numerous authors that the presence of free O_2 in the atmosphere is due to burial of carbon in the crust, where it cannot be oxidised, but actually there is less O_2 than should be present if it was all equivalent to stored carbon; in fact, weathering draws much O_2 down to the crust, oxidising surface materials (not just carbon), and aids the maintenance of the steady state of the atmospheric O_2 level (Hallberg, 1992).

5.4.1 SHALLOW WATER MARINE CLASTICS

Estuaries, deltas and clastic shelves collect much sediment released by rivers, and represent the coarser fractions, laid down on ocean margins. A major effect is the isostatic response of continental shelves to the presence of an accumulating pile of clastic material, and this relates to the accommodation space available for the collection of material. Of great

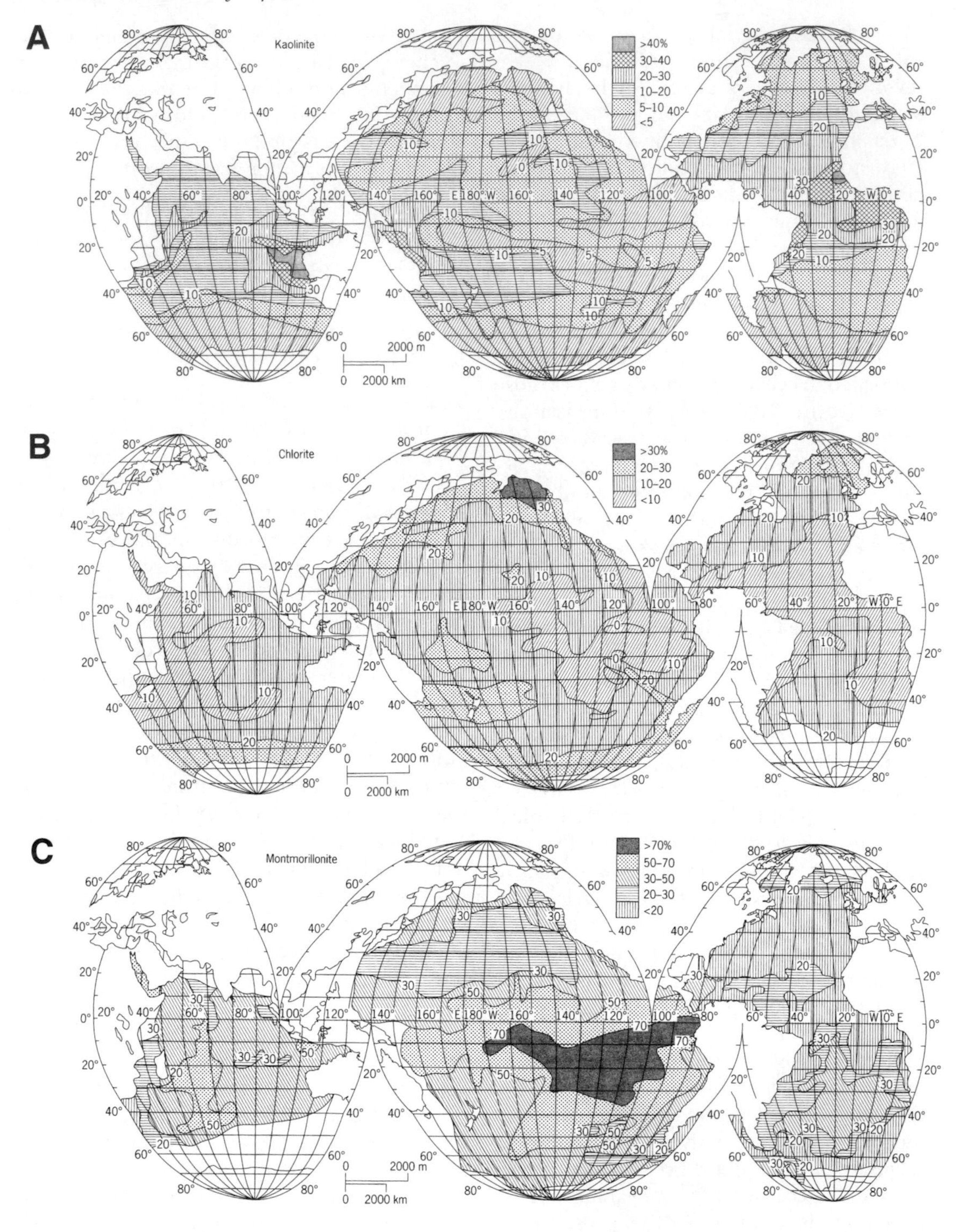

Figure 5.3 Global clay distribution in world oceans. A kaolinite; B chlorite; C montmorillonite; D illite. Reprinted from Griffin *et al.*, *Deep Sea Research*, Copyright © (1968), with permission from Elsevier Science.

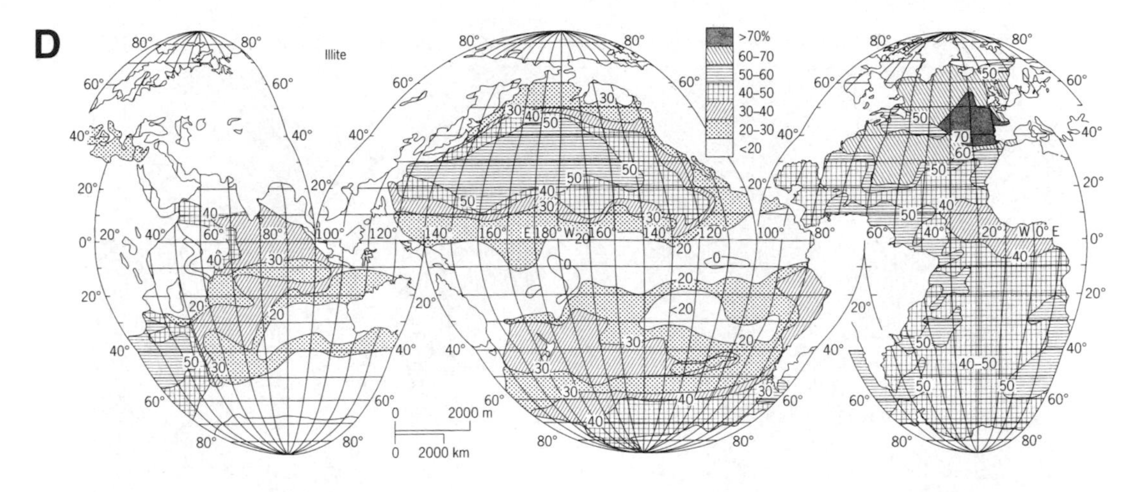

Figure 5.3 (*continued*)

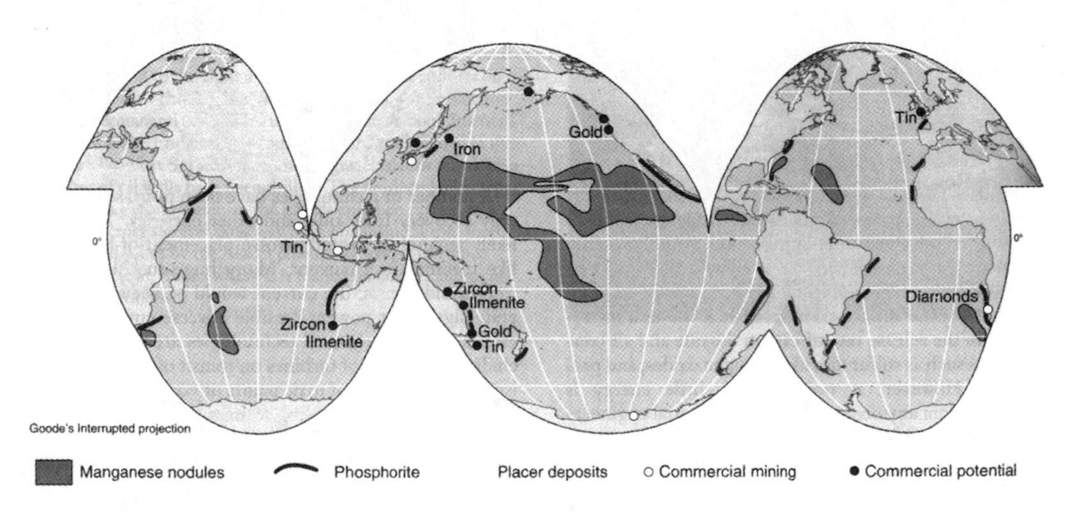

Figure 5.4 The modern distribution of economically important orthochemical deposits is shown. Reproduced with permission from Segar, *Introduction to Ocean Sciences* (with Infotrac), 1st edn (1998), Brooks/Cole Publishing, Thomson Learning, CA, USA.

importance for the geological record is the course of sea-level change. Sea-level rise loads the continental margin with water, causing isostatic depression of the continental crust. This increases the accommodation space, so that more sediment can collect on the shelves, causing further isostatic depression. Epeiric seas under rising sea level also collect sediment, of great value both to hydrocarbon source rocks, but also to associated reservoirs. The North Sea grabens were generated as failed rifts as the North Atlantic Ocean opened, but remained as continental crust when the line of rifting shifted west of Britain. The grabens became collection sites, not only for deep-water clays in which oil

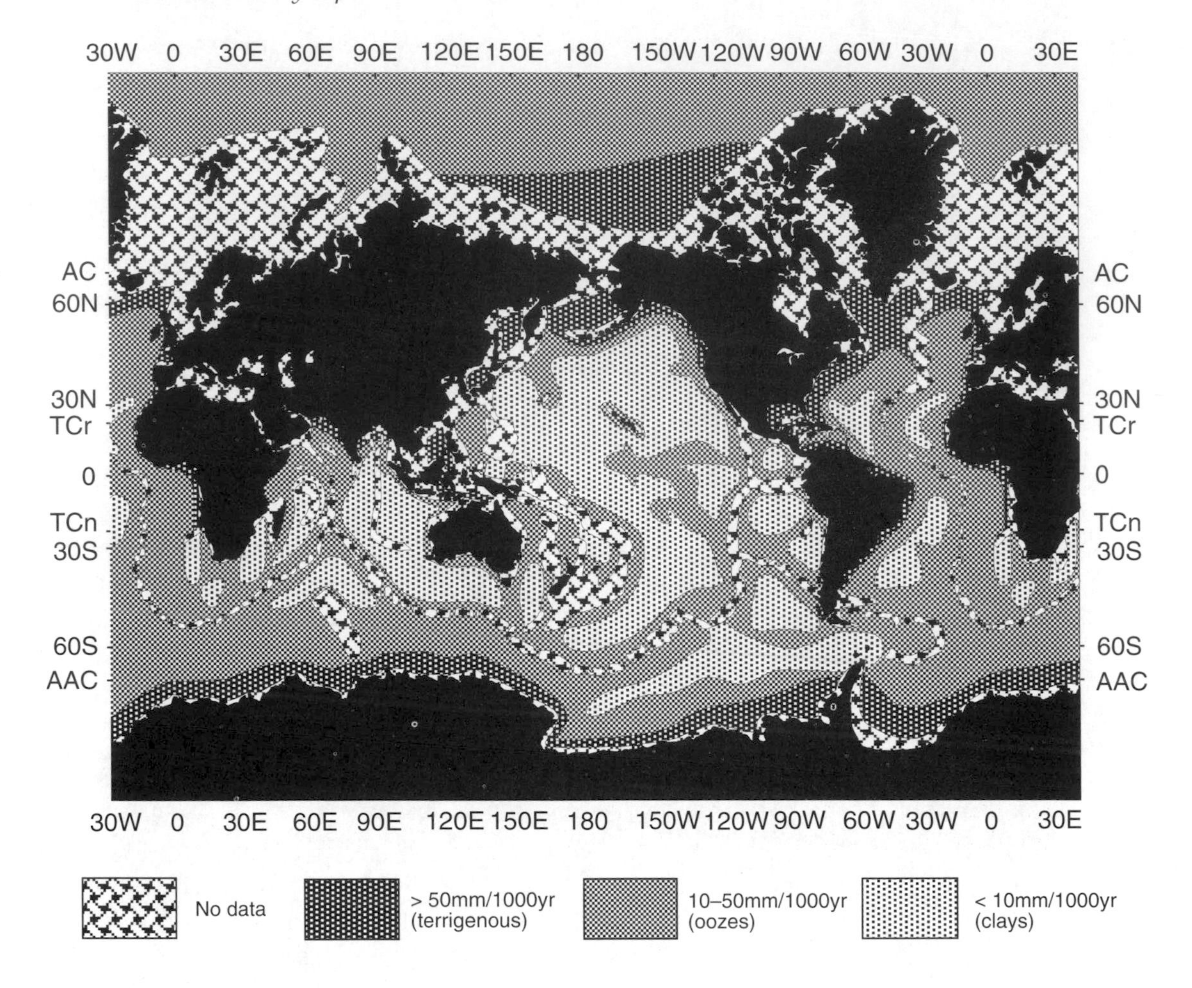

Figure 5.5 Global sedimentation rates. Sedimentation is greatest near continents, where terrigenous sediments collect. Lowest rates are in regions of clay deposition; the terrigenous part of this is wind-derived dust from desert sand and loess, more abundant in the northern hemisphere because there is more land there. There are no data for a) sedimentation rates for continental shelves, because they are so variable; b) mid-ocean ridges, where there is little or no accumulation on these youngest of oceanic seafloor surfaces. Adapted from P. Pinet, *Oceanography: An Introduction to the Planet Oceanus* (1992), Jones and Bartlett Publishers, Sudbury, MA, USA.

and gas could mature, but also of deltaic and marginal marine sediments nearby for petroleum reservoirs (Glennie, 1990a).

5.4.2 SLOPES

Local sediment build-up on continental margins leads to instability and slope failure (which may be triggered by earthquakes), mobilising a fluidised mixture of sediment and water which is redeposited downslope, and some sediment reaches the foot of continental slopes, to form a deep sea fan. Turbidite deposits commonly produced by such flows are characterised by graded bedding formed by the gradual decline in speed of flow with depth, which therefore deposits progressively finer fractions of its load. Particles deposited at the foot of continental slopes may be resuspended and redistributed by contour currents, which flow along bathymetric contours of ocean basins, and when deposited, form contourites.

Ice rafting occurs when continental material is eroded and entrained in glaciers. Icebergs calve into the open ocean and glacial debris sinks to the seafloor as they melt (mostly confined to higher latitudes). Poorly sorted, heterogenous debris may accumulate along with fine-grained sediments on the ocean floor, and critically will contain dropstones (large rocks deposited in fine sediments). Although dropstones are useful evidence of glaciations, they may also be formed by trapped rocks falling from floating trees, so cannot be relied upon after the Carboniferous Period, when large enough plants evolved. However, the end of the recent glacial phase is now well-known to have generated large amounts of ice-rafted debris released from 'armadas' of icebergs flowing from the northern ice mass (called Heinrich layers), and this has resulted in complex theories of climate control in the late-glacial phase, discussed in Chapter 10.

5.4.3 CARBONATE PLATFORMS

Developed in areas lacking clastic supply, these are built by organic secretion of $CaCO_3$ (Figure 5.2). Platforms are traditionally divided into two types: shelves, which are approximately level, and ramps, which usually slope down towards the basin. Platform construction is greater in shallow water where productivity is higher, generating a ramp structure, thinning into deeper water. Evolution of a platform theoretically may then proceed to fill the accommodation space (assuming a constant sea level), and develop a reef rim in its shallow portions. Reef progradation (in a seaward direction) is then possible, as accumulation of carbonate progresses. Reefs have a profound effect on local circulation patterns, and control the development of a range of fore-reef, reef and back-reef environments.

In reality, although such a pattern may explain the historical development of some ancient reefs, platform geometry is much more complex, because it also depends on antecedant topography, and tectonic stability (Wright and Burchette, 1998). The modern Caribbean reefs, for example, are built on the eroded Pleistocene relics, which were killed when the sea level fell during the Quaternary glaciation, and different sites on the Huon Peninsular show different reef development depending on the angle of slope.

5.5 OCEAN BASIN SEDIMENTARY DEPOSITS: CLAYS AND OOZES

These two categories are the major deposits; authigenic materials are discussed in Section 5.7.

5.5.1 PELAGIC CLAYS

These reach the seafloor by sinking through the water column, and comprise largely inorganic sediments from the continents (lithogenous). Some extraterrestrial material (cosmogenous) is also present, including iron spherules (rich in iron oxides), stony spherules (mainly olivine, magnetite and glass) and microtectites (small bodies of green-black glass). Lithogenous deposits are mostly clay-sized particles of weathered continental or oceanic crust transported to the ocean basins by rivers, glaciers and wind action. As noted above, they are composed (in varying proportions) of quartz, illite, kaolinite, chlorite and montmorillonite (smectite clays), and the distribution of the clays is climate-related (Figure 5.3). Not much soft tissue organic matter collects in modern ocean basins because of ecological recycling within the water column, but the skeletons of planktonic organisms dominate ocean floor sediments today (Figure 5.2).

5.5.2 BIOGENIC CALCAREOUS OOZES

The seawater CO_2 system has a relatively simple chemistry, but has a powerful influence on the passage of CO_2 between the atmosphere, ocean and sedimentary deposits of limestone.

Biogenic calcium carbonate is precipitated by surface-dwelling plankton in the form of tiny shells, which today are very abundant in sediments. These include: coccolithophorids (phytoplanktonic algae) and foraminifera (protozoans) with calcite shells; and pteropods (gastropods), with aragonite shells (Chapter 6).

Distribution of calcareous sediments is a function of seafloor topography, rather than primary production, because the stability of $CaCO_3$ shells in seawater depends on depth. Calcite and aragonite will dissolve where seawater is undersaturated with respect to carbonate. The solubility of $CaCO_3$ increases with decreasing temperature and increasing pressure. Consequently, a $CaCO_3$ particle, falling through the water column, experiences increasing hydrostatic pressure from the water above it, and, at a particular level, will show a sharp increase in its tendency to dissolve in the water. This level is called the lysocline, between *c*.3–4 km. Beneath it is the calcium carbonate compensation depth (CCCD, usually just called CCD), below which all $CaCO_3$ is dissolved (Tucker and Wright, 1990, p. 229) (Figure 5.6). Above the CCD, CO_2 dissolved in water can pass through a series of reactions depending on the supply of CO_2, resulting in the deposition of limestones (see Section 5.6).

5.5.3 BIOGENIC SILICEOUS OOZES

Silica is input to the ocean via rivers and hydrothermal emissions, and mostly removed

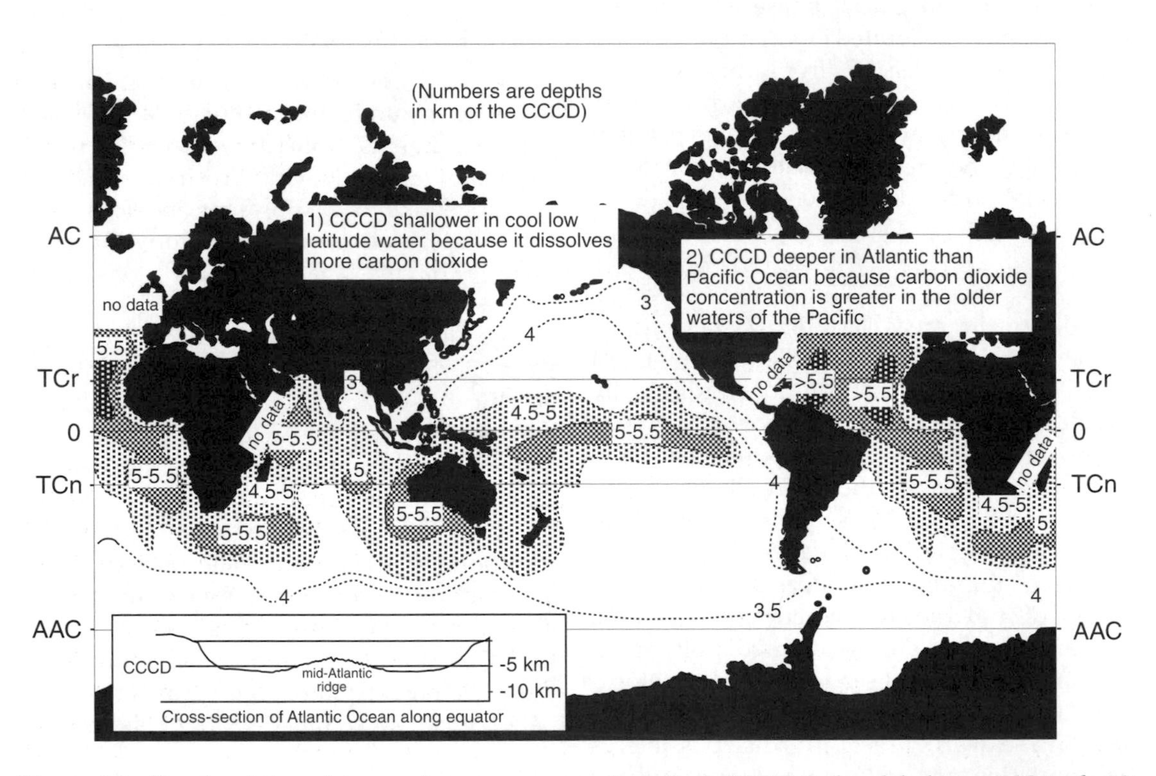

Figure 5.6 Depths of the calcium carbonate compensation depth (CCD) in the global ocean. Note that it is deeper in the Atlantic than in the Pacific, because Pacific deep waters are more mature and are more saturated with carbonate. Adapted from P. Pinet, *Oceanography: An Introduction to the Planet Oceanus* (1992), Jones and Bartlett Publishers, Sudbury, MA, USA.

by the deposition of siliceous tests (skeletons) formed primarily by surface-dwelling plankton, resulting in biogenic silica. The amorphous solid produced is termed opaline silica or opal. In some places this biogenic silica constitutes more than one third of the sediment mass, forming deposits called siliceous oozes.

Biogenic silica is produced by diatoms and silico-flagellates (phytoplankton) and by radiolaria (protozoans) (Chapter 6). A small fraction is produced by sponges, some groups of which secrete silica spicules (small needles). Silica shell productivity is limited by the availability of dissolved silicon. Diatoms are so effective at removing silicon that surface concentrations are virtually zero except in upwelling areas. Silicon is thus a limiting nutrient, which controls diatom growth in most areas. Where dissolved silicon is abundant (i.e. upwelling areas) diatoms are the dominant phytoplankton because they are the most effective pelagic algae at assimilating nutrients and grow rapidly as a result. Globally, diatoms (freshwater and marine) make up *c.* 25% of the mass of all plant matter (Leng and Greenwood, 1999). On the death of the organism, siliceous tests sink. Some reach the seafloor, while others are remineralised in deeper water. Upwelling brings this recycled Si to the surface to fuel further diatom growth, leading to opal-rich sediments. While alive, and after death, the biogenically secreted opaline silica is covered by a coating of adsorbed organic matter, which isolates the shell from contact with seawater, and slows its rate of dissolution. Dissolution rates for Si are highest in surface waters, and decrease with depth.

Highly siliceous sediments are produced when the rain rate of opal is high, and the accumulation rate of other particles is slow, so that silica becomes the dominant sediment component in places. Its concentration in continental margin settings is less, partly because of the dilution by terrigenous material, but also by carbonate oozes forming above the CCD. Furthermore, diatoms are dominant at low temperatures, and when dissolved Si is available, diatom oozes are found at high latitudes. In contrast, radiolarians are dominant at low latitudes, giving radiolarian oozes.

Long-term burial converts biogenic silica into chert or quartz, via the processes of diagenesis outlined in Chapter 4.

5.6 CALCIUM CARBONATE PRECIPITATION, SEAWATER BUFFERING, AND THE CARBONATE COMPENSATION DEPTH (CCD)

5.6.1 CALCIUM CARBONATE PRECIPITATION AND BUFFERING

Observations show that seawater pH stays within narrow limits: between 7.8–8.3, and usually only between 8.1–8.3. As noted in Section 5.5.2, $CaCO_3$ particles in the deep ocean may dissolve if the pressure and temperature conditions exceed certain levels. It is necessary to address the behaviour of CO_2 in water, in order to appreciate controls on $CaCO_3$ at different depths of the oceans, and also the controls on seawater pH. There are two major pathways for CO_2 in ocean waters, as follows.

Organic reactions:

$$6CO_2 + 6H_2O + \text{UV light energy} \Leftrightarrow C_6H_{12}O_6 + O_2$$

($C_6H_{12}O_6$ represents carbohydrates, here as glucose, and $\Leftrightarrow$ means that the reaction may proceed in each direction). This is the photosynthesis reaction if it moves to the right, and the respiration reaction (used by organisms to mobilise their energy resources) if it moves to the left (although the leftward reaction does not use UV). Organic matter not recycled in the ocean is exported to the seafloor and buried to form hydrocarbons. Sinking particles of organic matter play an important part in transferring atmospheric CO_2 to the deep ocean (Kawahata *et al.*, 1998). In Part B we discuss the role of organic carbon burial on climate and atmosphere, and even the recent record suggests a link between them (Kawahata and Eguchi, 1996).

Inorganic reactions and reactions involving $CaCO_3$ skeletons:
There are several reactions given in a range of texts, but they can be rationalised into two major sequences, both reversible:

1 $CO_2 + H_2O \Leftrightarrow H_2CO_3 \Leftrightarrow HCO_3^- + H^+ \Leftrightarrow CO_3^{2-} + 2H^+$
2 $2HCO_3^- + Ca^{2+} \Leftrightarrow CaCO_3 + H_2O + CO_2$.

These two are interlinked by bicarbonate (HCO_3^-) and control not only the pathway of CO_2 in water but also have an important relationship with the stability of pH of seawater. The much-published reaction **1** shows that as more and more CO_2 is added to water, it dissolves to form carbonic acid (H_2CO_3), which becomes partly dissociated into bicarbonate (HCO_3^-), which subsequently forms carbonate (CO_3^{2-}) ions. Reaction **1** is therefore a reaction chain, and the movement of CO_2 along the chain in each direction is the process which holds seawater pH within narrow limits. Each component of the chain is stable at different pH (Figure 5.7), so note that CO_2 is more abundant in acid waters, HCO_3^- in approximately neutral waters, and CO_3^{2-} in alkaline waters, although all three forms co-occur. This reaction occurs in fresh water too, but most stays as H_2CO_3, making fresh waters more acid. By contrast, in seawater, addition of CO_2 pushes the reaction along the reaction line to the right. Although addition of CO_2 increases the acidity of seawater, the fall in pH is limited by conversion of some CO_2 into HCO_3^-, which is slightly alkaline, and prevents the water becoming acid. Some dissolution of $CaCO_3$ on the seafloor is a common feature of carbonate grains, and helps to offset addition of CO_2 by releasing HCO_3^- into the water by leftward movement of Reaction **2**. Similarly, addition of alkaline material to ocean water pushes Reaction **1** to the left, and the H_2CO_3 formed neutralises the added alkali. Thus seawater is *buffered* against wide pH variations. Seawater is saturated with these inorganic carbon ions, and their high concentration provides this major control on pH. In contrast, fresh water lacks the large concentrations of dissolved material, especially dissolved inorganic carbon; there is no buffering, and so fresh waters are much more sensitive to change, of critical importance in water resource management.

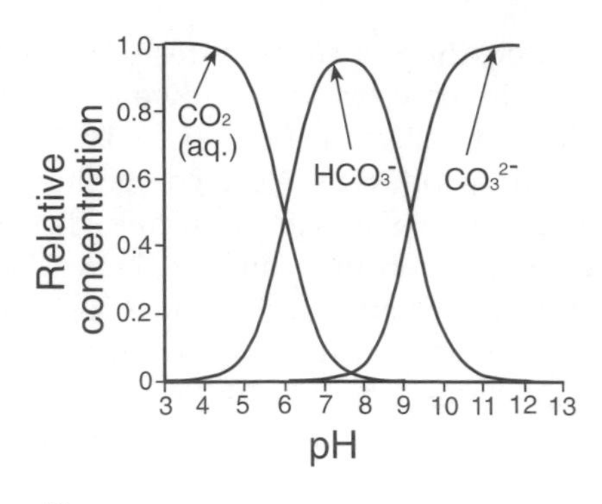

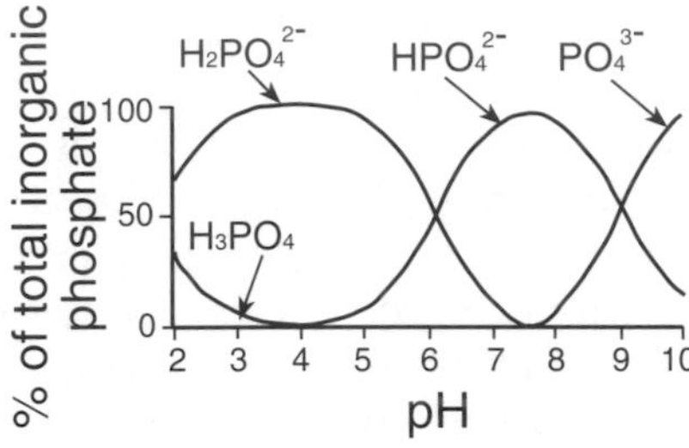

Figure 5.7 Graphs of concentrations of different dissolved components at different pH, of carbonate-related and phosphate-related materials. The implications of these graphs are discussed in the text. Adapted from Butcher *et al.* (1992).

Although you might think that $CaCO_3$ is formed in the sea by the simple combination of Ca^{2+} with the CO_3^{2-} fraction, drawing CO_2 from the atmosphere down the reaction chain to the seabed, that is not consistent with observations and modelling of seawater and the actions of calcifying organisms. There is very little inorganic formation of $CaCO_3$ in the oceans, and the places where it does form (e.g. Great Bahama Bank) show only temporary deposits ($CaCO_3$ crystals floating in the water, called whitings) formed by evaporation of the saturated seawater in shallow tropics; whitings redissolve when it rains, or the temperature falls again. Experimental work with bacteria indicates that $CaCO_3$ precipitates from $Ca^{2+} + CO_3^{2-}$ as a passive process of metabolic activity in the nitrogen and sulphur biogeochemical cycles (Castanier *et al.*, 1999). This is part of the process of microbial mediation of $CaCO_3$ formation, a growing field of study (Riding and Awramik, 2000). The formation of $CaCO_3$ skeletons of corals (and the autotrophic action of some bacteria), however, involves the production of CO_2 (Kawahata *et al.*, 1997); so, rather than removing CO_2, coral reefs generate it! It has been suggested that the burning of fossil fuels to increase CO_2 into the 21st century will cause a reduction in the ability of coral reefs to grow, because the likely enrichment in CO_2 may reduce calcification processes in the sea (Kleypas *et al.*, 1999), presumably because of CO_2's acid nature (seawater pH varies as a result, 7.8–8.3). Consequently, although nearly all $CaCO_3$ accumulation on the seabed is skeletal (i.e. secreted by organisms as shell), at the present and, as far as can be determined, for the majority of geological history, the route of CO_2 from the atmosphere to the seabed and limestone stores cannot be directly from the atmosphere into the ocean, and must be by another mechanism, which is where Reaction **2** becomes important.

In seawater 80% of CO_2 is held as HCO_3^- (Pinet, 1992, p. 150), and it is clear that $CaCO_3$ formation is derived from the HCO_3^- fraction instead of the CO_3^{2-} part. Reaction **2** shows why coral reefs, for example, are producers of CO_2; for every $CaCO_3$ molecule formed into coral skeleton, a molecule of CO_2 is released into the water, from the HCO_3^- fraction. Section 4.3.2 showed how the weathering of terrestrial silicate and carbonate rocks involved the action of acid rain ($H_2O + CO_2$), with HCO_3^- being generated (along with clays and silica, giving mud and sand sediments). H_2CO_3 dissolved in fresh water on the land surface gives it an initial pH of as low as 5.6 (Holmén, 1992). HCO_3^- carried to the sea by rivers therefore is the primary input mechanism for HCO_3^- to give the $CaCO_3$ we now see on the seafloor and in rocks throughout time, using Reaction **2**, and is part of the long-term carbon cycle. That cycle is completed when the CO_2 is returned to the atmosphere after 10s–100s of millions of years by reactions in subduction zones, and subsequent volcanic eruption. The accumulation of HCO_3^- in seawater is also the reason why seawater is buffered, while river water is not.

Reaction **2** also allows us to understand why seawater is alkaline, in contrast to the CO_2-induced acidity of much fresh water. The weathering of terrestrial rocks uses acid rain water to generate HCO_3^-, which is alkaline, and the supply of HCO_3^- in rivers to the sea is the main source of alkaline material for the oceans. This shows why the sea is not acid, because terrestrial weathering converts CO_2 into HCO_3^-. Thus the *alkalinity* of seawater is due principally to HCO_3^-. It is important to realise that alkalinity is not a measure of pH, it is simply the total amount of alkaline matter in the ocean. It can also be viewed as the total number of hydrogen ions (= acid) needed to neutralise the negative charges in alkaline matter. Because about 90% of alkaline material in oceans is due to components of Reaction **1**, seawater alkalinity is often referred to as carbonate alkalinity. A review of the distribution of input of carbon to the oceans (Ludwig *et al.*, 1996) showed that the Atlantic Ocean receives the greatest input, but high concentrations are also transferred to the Pacific and Indian Oceans from southern and eastern Asia.

The alkaline pH of seawater is not maintained by dissolved inorganic carbon alone. Boron (B) is present in abundance in alkaline solutions as hydroxide, as are various forms of phosphate (Figure 5.7), and supplement HCO_3^- in the control of pH (Holmén, 1992, p. 243; Jahnke, 1992, p. 302; Leeder, 1982, p. 18). Because phosphates are important in organic tissues, they are important as a buffer in living tissue. Furthermore it has been suggested that production of ammonia aids the maintenance of alkaline seawater pH (Hallberg, 1992, p. 160), implying that microbial processes may have a hand in the process, because ammonia (as ammonium, NH_4^+) is generated by bacterial mineralisation of organic nitrogen compounds (e.g. amino acids). Furthermore, there is a large amount of dissolved metals in the oceans, especially Na and K, which are highly alkaline metals, and these presumably play a part in maintaining the alkaline pH. Although the modern oceans' pH is controlled by inorganic carbon, there is no guarantee that the earlier Earth's oceans were, or even that they were alkaline at all. We consider in Chapter 8 the possibilities of major ocean chemistry change, with one model suggesting that oceans were even more alkaline than now due to excess Na, which forms a very alkaline solution (caustic soda). This can be demonstrated by adding water to pure sodium metal in the reaction: $2Na + 2H_2O \rightarrow 2Na^+ + 2OH^- + H_2$(gas) (*this is explosive: do not try it at home; do not try it at all !*). The water begins at *c.*pH7 and ends at a more alkaline pH, because the sodium reacts with water to produce NaOH, and donates electrons to the H^+ present in water. Each sodium atom wants to lose an electron to stabilise its structure to form a sodium ion (Na^+); that electron passes to a H^+, making H, and leading to bubbles of hydrogen gas leaving the liquid. Although this is a balanced equation, in that the number of negatively charged OH^- ions balances the number of positively charged Na^+ ions on each side, remember also that water is undergoing a constant dissociation–reassociation at the same time ($H_2O \leftrightarrow H^+ + OH^-$); therefore, in the presence of dissolved Na, at any one time there is an excess of OH^- ions, hence an alkaline pH.

The pathways of dissolved carbon described in this section are shown in Figure 5.8.

5.6.2 CCD

The behaviour of the various forms of dissolved inorganic carbon (called carbon species) in shallower waters maintains the pH as

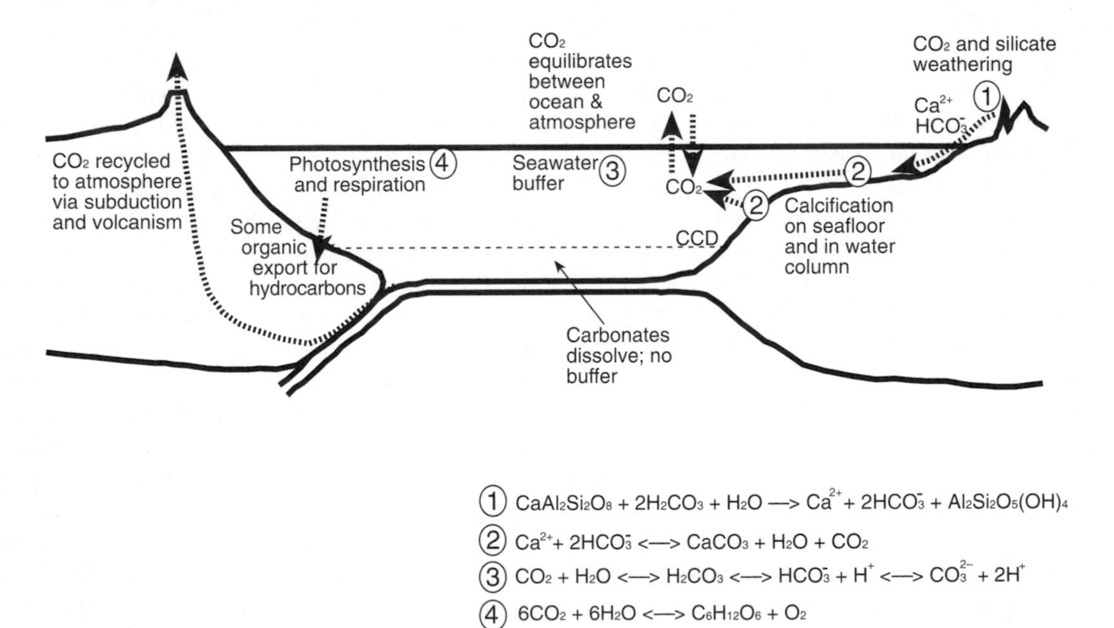

Figure 5.8 A summary of the relationships between different forms of carbon in the oceans, discussed in the text.

slightly alkaline, as described in the previous section. $CaCO_3$ is precipitated on the seafloor as calcareous oozes in areas where the alkaline water is maintained (variable, between 3–4.5 km depth). Beneath a critical depth (the lysocline), waters are colder, and the pressure higher, and under those conditions $CaCO_3$ is more soluble because the water is no longer alkaline, but drifts into the acidic pH; this is because cold water can dissolve CO_2 more effectively than warm water, a feature of monumental importance to the global heat balance. Thus, $CaCO_3$ particles sinking through the lysocline will start to dissolve, and when they reach the CCD, will dissolve completely. Moreover, the CCD is species specific, because shell dissolution rates vary between species. Also, because aragonite is more soluble than calcite, the CALCITE-CD is deeper than the ARAGONITE-CD. The CCD is important because it interrupts the normal reactions of CO_2 in seawater, and the maintenance of alkaline pH breaks down.

So, as pressure exceeds the stability of $CaCO_3$ particles, the water is no longer saturated with respect to $CaCO_3$ and the shells begin to dissolve. This can be recognised by corrosion around the edges of surviving grains, while many are lost completely into solution. Bacterial breakdown of organic matter in the water adds to the dissolved CO_2, and the amount of CO_2 added to a water mass increases as it ages, so that older water masses are more corrosive to calcium carbonate and the rate of dissolution increases. Therefore, because Pacific deep water is the oldest of the ocean waters, the CCD is shallower than in the Atlantic (Figure 5.6), and demonstrates the non-conservative behaviour of CO_2. You might think that this complex relationship between various forms of inorganic carbon shows that the principle of constant proportions does not apply to all parts of the ocean solution, but CO_2 is not included as a major component of seawater, and does not come under the remit of that principle.

5.7 ORTHOCHEMICAL SEDIMENTARY DEPOSITS

Such deposits may be divided into several groups as follows.

5.7.1 EVAPORITES

The evaporation of a given volume of seawater leads to progressive concentration of the dissolved components, and at critical points certain materials are precipitated as inorganic crystals; the commonest are gypsum and salt, because of the high concentration of calcium, sulphate, sodium and chlorine in seawater. Evaporites indicate arid settings, and form in places isolated from the oceans. Examples are flat coastal plains subject to periodic flooding (the Persian Gulf) and marginal basins cut off from the oceans, either by regional tectonism or by global sea-level fall (such as the various parts of the Mediterranean during the Miocene Epoch, see Chapter 10).

5.7.2 METALLIFEROUS SEDIMENTS AND HYDROTHERMAL DEPOSITS

Trace metals introduced to oceans via atmospheric fallout and river runoff may form deposits on the continental shelves, while most of the metals that precipitate in open ocean settings are derived from hydrothermal activity on the ocean floor.

Hydrothermal vents and metalliferous sediments form at mid-ocean ridges. During seafloor spreading, lava is extruded onto the seafloor, and contracts and fractures on cooling. Seawater percolating through fractures down into the basalt is heated by shallow magma, and reaches temperatures of up to 400°C. This dissolves elements from the magma and rises to be ejected at ocean-floor vents, being replaced by cool seawater drawn into the hydrothermal system. This system develops into a hydrothermal circulation cell (Figure 5.9)

Seawater circulates through hydrothermal cells and undergoes chemical exchanges of

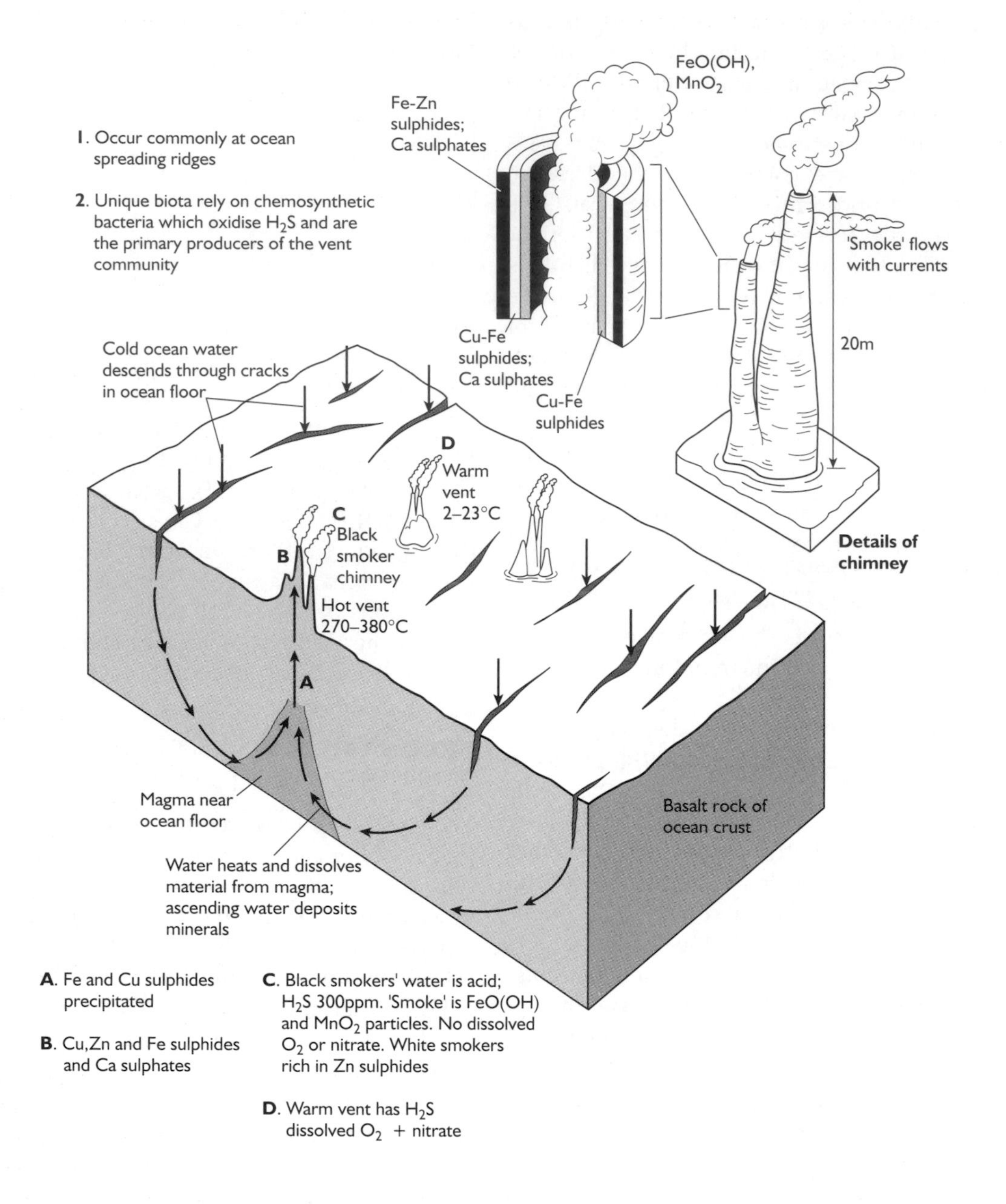

Figure 5.9 Features of hydrothermal vents. Formed largely at ocean spreading ridges, where magma is close to the surface, cold water descending through cracks in the seafloor is heated and rises back to the surface. Superhot water dissolves material from basalt, and precipitates it in the cold conditions of the seafloor; the high hydrostatic pressure prevents hot water from vaporising to steam. Black and white smokers form by mineral particles precipitating in the water as it streams out of the seabed; chimneys of minerals build up on the seafloor. The high levels of hydrogen sulphide drive the ecosystems present around vents, by action of chemosynthetic bacteria oxidising hydrogen sulphide and generating the primary production of the vent ecosystems.

elements with the magma and newly formed basalt rock. Water which is vented out of the rock is hot, acidic, poor in oxygen, but rich in metals and hydrogen sulphide. This water reacts with cold, alkaline, oxygen-rich water and the metals precipitate as sulphides. Because most of the sulphide precipitates are black (i.e. iron sulphide), the water leaving the vent turns into a black smoke (black smokers). Chimneys, several metres high, of precipitated sulphides and anhydrite (calcium sulphate) form around these vents. At lower temperatures, the precipitates are mainly zinc sulphides, which are milky white, called white smokers.

Most of the metal sulphides leaving the hydrothermal vent are deposited close-by, so mid-ocean ridges are blanketed in metalliferous sediment. Very high concentrations of zinc, iron, copper, lead and silver are potential ore deposits if they can be mined in the future. Hydrothermal deposits which have escaped subduction by being uplifted onto land, termed ophiolites, are currently mined for copper, zinc and other metals.

5.7.3 Fe–Mn NODULES AND Fe CRUSTS

Fe–Mn nodules are found in all of the ocean basins, most abundantly in the south central Pacific. Most occur in the size range 1–10 cm, and form around a central nucleus, such as a small rock, bone fragment or shark's tooth. Metals precipitate onto this in concentric bands, accreting at rates between 1 and 200 mm per million years. Most nodules are not buried even though they grow much more slowly than sediment accumulates around them, due to periodic rocking by earthquakes, or possibly by bioturbating organisms. The source of metals for nodule growth may be from precipitation from seawater or from diagenetic remobilisation from sediments. Fe crusts may form in similar situations, but indicate very slow sedimentation rates. Such crusts are recognised in the geological record, and help to identify areas and episodes of reduced sedimentation, which may lead to condensed sequences (i.e. thin sedimentary deposits laid down over long periods).

5.7.4 PHOSPHORITES, BARITE AND GLAUCONITE

These are formed in shallow waters, following total or partial dissolution of other minerals, or input from the crustal-ocean factory or from terrestrial sources by rivers. Phosphorites (phosphate-containing minerals) occur in shallow water upwelling areas; upwelling leads to high biological productivity, and in the shallow waters large amounts of organic P reach the ocean floor following the death of plankton. The fate of this P is critical to ocean productivity, and if the ocean floor circulation is suppressed, the P gets locked up in sediments or stagnant bottom waters. As a result, sites of major upwelling are important productivity localities, and suffer greatly if upwelling is interrupted as in El Niño events (Chapter 3). If the dissolved O_2 content of pore waters is low, buried organic material may be transformed into phosphorite. Barite (barium sulphate) is formed in a similar fashion and barium is another organically mediated element. Glauconite (iron-rich, greenish hydrous silicate) is found as nodules and encrustations, giving rise to 'green muds'.

5.8 VOLCANICLASTIC MARINE DEPOSITS

Airfall deposits, also called tephra (Greek for ash), consist largely of volcanic dust, which settles in oceanic water. Recent eruptions show the distribution of ash from explosive volcanoes reaches stratospheric heights and is globally distributed. Such volcanic deposits can be important mineral sources for diagenesis and productivity. In the geological record, they form plastic clays in deep shelf and ocean floor settings and are variously called tephra horizons, bentonites and tonsteins. Geochemical work shows that particular ash layers may have distinctive trace element suites (related to the composition of the magma they were derived from), and can be used for correlation of sedimentary rocks on a regional basis. In the rock record, it is possible to distinguish water-laid and land-deposited ashes; the latter commonly contain lapilli, small pea-sized balls of ash which nucleated in the rain-clouds developed over volcanoes. If lapilli land in the sea, they dissipate into ash and lapilli are not preserved. Very good examples of lapilli tuff are seen in the Ordovician volcanic suites of the Caledonian orogeny in North Wales and the English Lake District, and reveal the terrestrial nature of many of the deposits, associated with shallow marine sandstones and sand–ash mixtures called tuffites. Such evidence aids the identification of ocean-margin settings of ancient deposits

Pyroclastic flows (high-speed superheated ash, debris and gas from volcanic eruptions) have been recorded entering the sea, and are a minor input of material to ocean waters.

5.9 CONCLUSION

Global distribution of deep sea sediments (including continental shelf and deep ocean) is controlled by the rate of, and proximity to, input from weathering and erosion, biological productivity, and the influence of ocean chemistry on mineral stability. Marginal settings collect much clastic material, but carbonates may be important in areas isolated from clastic supply. The critical importance of marine sediments is that they contain the information used to interpret past oceans, so understanding their nature and controls is a good background for Part B.

5.10 SUMMARY

1. The distribution of marine sediments is closely related to the processes which formed them.
2. Continental margin sediments may be dominated either by clastics in places where the supply is sufficient, or by carbonates where it is not.
3. Continental margin clastics are represented by estuaries, clastic-dominated shelves, deltas and slope deposits. Carbonates are dominated by calcareous shells, and build platforms on the continental shelf of varying morphology and relate to antecedent topography. The standard model of sloping ramps and flat shelves has been shown to be too simple to account for the complexity of these features.
4. Clay mineral suites are sensitive environmental indicators, and give the following general patterns: kaolinite indicates humid tropical weathering; illite, temperate settings; montmorillonite, ocean volcanic sources; chlorite, high latitude environments.
5. The modern seafloor is dominated by calcareous, and to a lesser extent siliceous, oozes, because most of the ocean floor is above the carbonate compensation depth (CCD). Ancient oceans often lack these oozes because the creatures that formed them did not evolve until relatively recent times, and because the higher sea levels of most of Earth history placed more of the ocean floor beneath the CCD so that more $CaCO_3$ was dissolved.
6. Orthochemical deposits of the oceans reflect low sedimentation rates where they occur. Evaporites indicate arid settings, while ocean-floor mineralisation (nodules and crusts) develops from metals input to ocean systems from volcanic vents.
7. Volcanic deposits supply minerals to the oceans, as additional resources for productivity and orthochemical processes; they may also be used to correlate different areas in the geological record.

MARINE ECOSYSTEMS AND OCEANOGRAPHY

6

6.1 INTRODUCTION

Ecology is the study of interactions between organisms, and between organisms and their environments. The Greek root of the term ecology (*oikos* = a house; *logos* = discourse) applies very well to terrestrial ecosystems, because although dispersal and migration are inherent, a large part of their components (e.g. trees, bushes, grass) do not move around (hence *house*). In contrast, oceanic ecosystems are literally fluid; a large portion of the biota is microscopic and suspended (= plankton) in water, and therefore governed by its properties. In all ecosystems, most of the energy follows two lines: *grazing foodchains* (animals eat plants; carnivores count for little) and *detritus foodchains* (small animals, fungi and bacteria eat anything dead, returning nutrients to the environment). On land, movement of matter and energy generally operate locally; but in the sea, the mobile nature of marine ecosystems, and the huge scale of ocean volume, mean that the larger transport distances of ecosystem components are a critical aspect of their operation. Habitats on land are easy to picture, because you can walk amongst and touch them. In the sea, much is microscopic, and everything lies out of sight beneath the waves in a 3D space so large that appreciating the processes is more difficult.

Marine ecology covers the range from coastal processes of land–ocean interaction through to the deepest seas, and therefore occupies a greater volume than terrestrial ecosystems. Global transport of ocean waters, and the widespread nature of biota, means that the study of marine ecosystems can be regarded as biological oceanography. The concept of biogeochemical cycles discussed in Chapter 4 is embedded in biological oceanography.

In recent decades, ecology has become identified with the recognition of environmental degradation and the needs of its management, with a tendency to compartmentalise, and focus on aspects of human impact. While this is of course environmentally and politically necessary, it is easy to overlook ecology as a science of dynamic interchange and flow of energy and matter. So, in this chapter we attempt a balanced view of the components and operation of marine ecosystems, as a background to understanding oceanographic features of geological importance, examined in Part B, and modern ocean problems and management in Part C. This chapter does not provide a detailed description of the types of organisms, which is available in other texts.

Modern marine ecology is governed by the full range of physical and chemical characteristics of ocean water (Figure 6.1), and the geological and geographical parameters of ocean basins; it is the present state of both biological and geological evolution over millions of years, so it inherits past history and is the precursor for future change.

6.2 FUNDAMENTALS OF MARINE ECOSYSTEMS

6.2.1 OCEAN LIFE

Ocean organisms can be either pelagic (consisting of nektonic (swimming), or planktonic

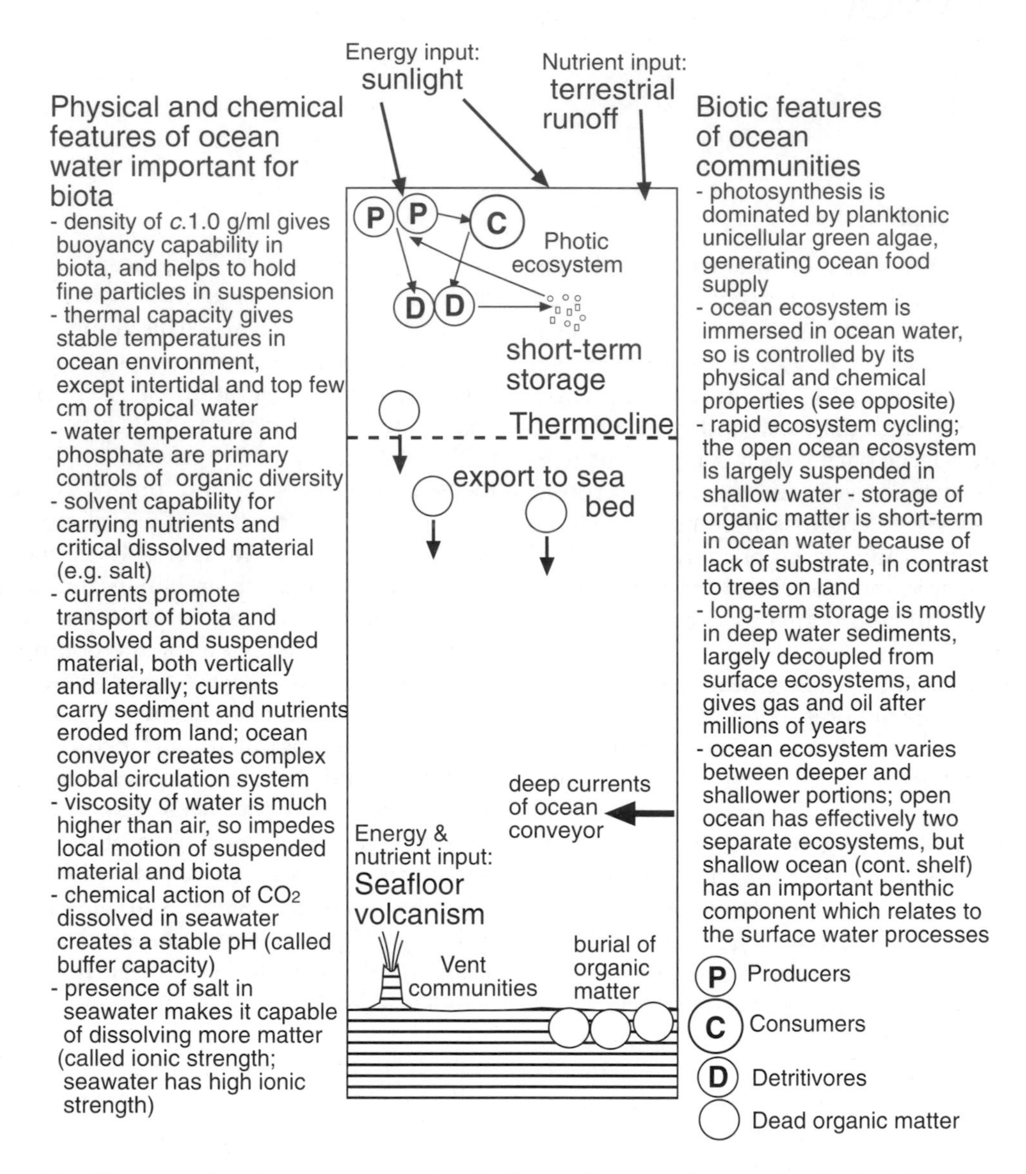

Figure 6.1 Diagram of ecosystem structure, and relevant features of ocean physics and chemistry.

(floating) organisms) or benthic (living either on or in the substrate), and can exist at all depths. Because of the large-scale 3D relationships in the oceans, the ecosystem parameters vary widely (see Pinet 1992, pp. 377–378) and production is highest near the coast, especially where there is significant runoff and/or nutrient upwelling. Furthermore, although the oceans are large enough to have separate ecosystems in different parts, the importance of ocean currents as travel agents, both laterally and vertically, means that the ocean ecosystems are open, and subject to import and export.

Life on Earth can be divided into 5 kingdoms, all of which play a part in ocean ecosystems: **Monera** (usually single cells, bacteria and cyanobacteria), **Protista** (a mixed bag of single-celled organisms, which occupy much of the plankton), **Fungi** (not to be included with plants), **Plantae** (plants, in which most people include algae, e.g. seaweeds, and many plankton species), **Animalia** (animals, including many plankton, all nekton and most benthos). To give you a perspective, a full English breakfast contains three kingdoms (animals, plants, fungi; i.e. bacon/sausages/eggs/black pudding; cereal/tomatoes/baked beans/toast/tea; and mushrooms) and may have monera and protists along for the ride, depending on the cook's attitude to hygiene! All five groups are abundant in the sea, although the key organisms of the oceans are unicellular marine algae, which provide *c*.70% of marine productivity, in the oceanic photic zone. It should be noted that some authorities recognise six kingdoms, by dividing the monera into archaebacteria and eubacteria; such division reflects the great complexity within these relatively simple organisms. For the broader purposes of this text, we use the combined group of monera.

Monera are simpler than the rest, and are prokaryotic; because they do not contain a cell nucleus, the DNA strand floats in the cell fluids. The rest are eukaryotic, having a nucleus containing the DNA replication system (Greek *karyon* = kernel, applicable because the nucleus is a large, dense, separate body in cells). From the geological record, the oldest prokaryotes pre-date the oldest eukaryotes (*pro* = before; *eu* = complete). Within all five groups there is a range of subdivisions, although the diversity amongst monera is lower because they lack sexual reproduction, the key to increased diversity, present in eukaryotes. To put the ecosystems in perspective, the monera dominated oceans throughout geological time; the Earth has been a microbial world for 90% of its history.

Organisms are also divided into two major subdivisions on the basis of their feeding requirements. Autotrophs are organisms which make essential food by either photosynthetic (green plants) or chemosynthetic (many bacteria) processes; heterotrophs are unable to generate food, and therefore must eat other organisms to survive (herbivores, carnivores, detritivores and decomposer bacteria and fungi).

6.2.2 BIOLOGICALLY CRITICAL FEATURES OF WATER

Oceans support abundant life because of the unique properties of water described in Chapter 2. Water therefore has essential features for life, which are:

1. *buoyancy* (allows organisms to live in the water, not just on the substrate);
2. *viscosity* (influences flow capability for transport of organisms, their reproductive structures and the chemicals essential for their lives);
3. *thermal capacity* (temperature changes are slow, good for organic systems because of the time needed to adapt);
4. *solvent action* (critical for transport of dissolved gases and nutrient chemicals);
5. *surface tension* (not important in turbulent waters, but plays a part in still shallow water biology).

These are empirical properties, and have great influence on the operation of the key limiting factors of marine environments, which are: temperature, oxygen, salinity, circulation (ocean currents), nutrient transport (especially upwelling).

6.2.3 OCEAN HABITATS

A simple view of ocean habitats is that they are depth-related (Figure 6.2), and are commonly divided into the following.

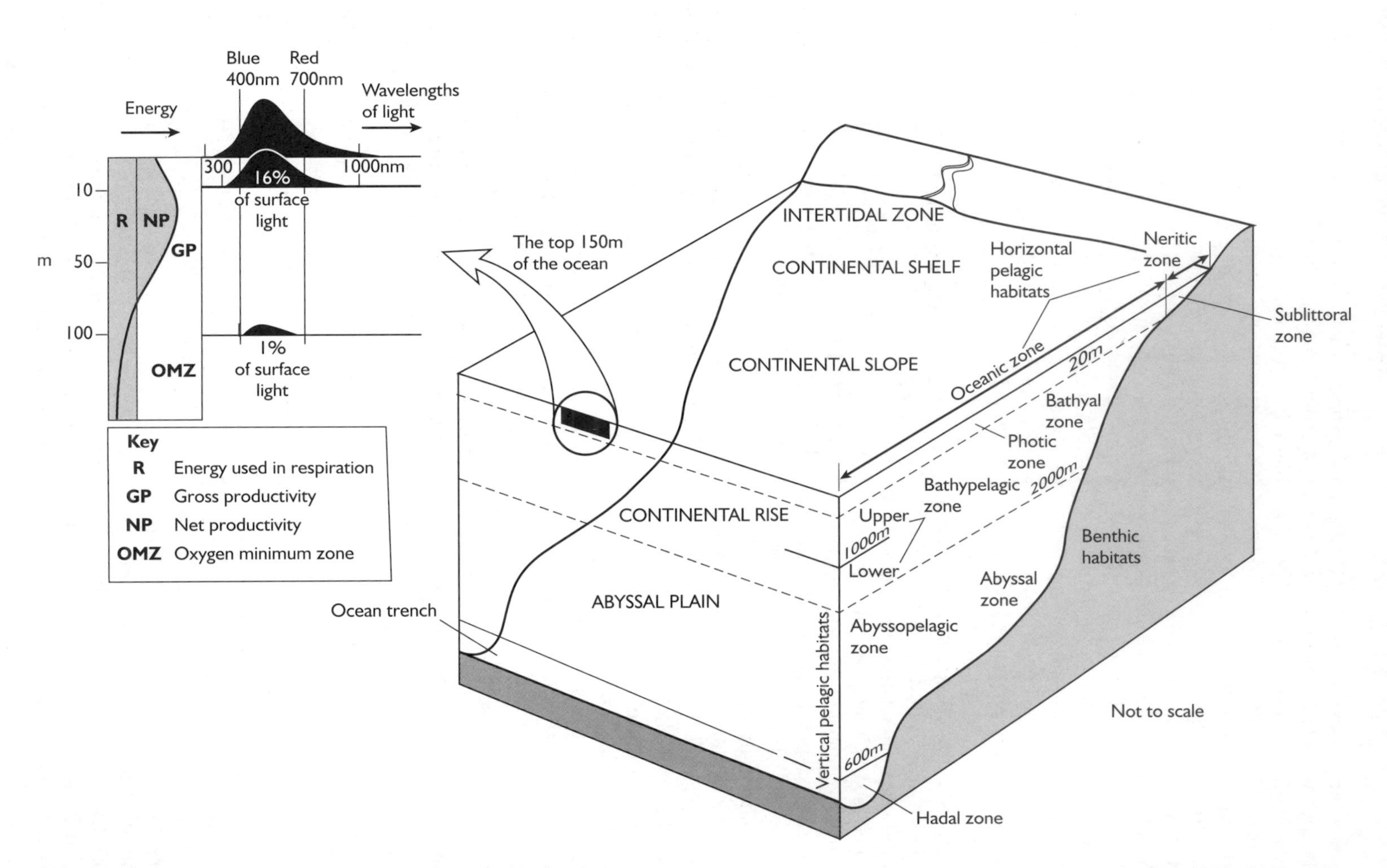

Figure 6.2 Ocean habitat classification. Benthic habitats are named separately from pelagic habitats. Pelagic habitats occupy the full volume of the ocean, and therefore have both vertical and horizontal terminology; benthic habitats in comparison occupy only the upper layer of the sea bed (maximum of a few metres deep). Inset shows that light levels decline rapidly with depth, and correlate to ocean surface productivity and respiration. The OMZ forms beneath the zone of maximum production, where breakdown of organic matter (respiration) exceeds production, with decomposer organisms using dissolved oxygen (see also Figure 10.7).

1. *Intertidal* (littoral) zone is volumetrically insignificant, but important in human perspective: beaches, rocky shores, salt marshes. The key biological feature is salinity, because most organisms cannot survive fluctuating salinity, so diversity of intertidal zones is low. Most organisms on Earth are stenohaline (tolerate only slight salinity change), but intertidal ones are often euryhaline (tolerate wide salinity change), and also have special adaptations to resist twice daily exposure to air and therefore rapidly and widely fluctuating temperatures.
2. *Subtidal* (sublittoral) zone is strictly the rest of the ocean, but usually taken to *c*.200 m depth. Intertidal and subtidal together form the neritic zone, and are the portion of the oceans where most life exists. This is because (a) the top *c*.200 m is lit (or at least partly lit) by the Sun, so that energy input to the heterotrophic components via photosynthesis in autotrophs means that complex food webs are generated; (b) the upper waters are oxygenated (by photosynthesis and by turbulent mixing of atmospheric gases into the sea), so that aerobic respiration takes place, *c*.50× more efficient than the anaerobic respiration occurring in ocean-floor sediments beneath the oxygenated zone. Little mixing with lower waters takes place in many areas because of the stability imposed by the thermocline, and this leads to vertical variations in oxygen levels; and (c) terrestrial runoff and upwelling bring nutrients to the shallower depths, especially near the continental margins.
3. *Bathyal* (200–2000 m), dark, but currently well-oxygenated because of thermohaline circulation, and less populated. Includes continental slopes.
4. *Abyssal* (> 2000 m), includes most of the ocean floor; some texts (e.g. Pinet, 1992) also add *Hadal*, > 6000 m. At bathyl and abyssal depths, hydrostatic pressure is important to organisms. Ocean-ridge vent communities associated with black smokers are the seat of much chemosynthesis, a potential site for the origin of life (see Chapter 8).

These depth relations are quoted in many texts, and are important to the marine ecosystems; however, another aspect of ocean environments is their geometry, which allows us to appreciate the shape of ocean basins and ocean floor, not just the depth relationships, as outlined in Chapter 4. The importance of such 3D features is that they influence water transport and therefore ecological relationships; an extreme example of this is the substantial ecological isolation of Antarctic waters by the Antarctic Circumpolar Current, only possible due to the position of Antarctica.

6.2.4 ECOSYSTEM OPERATIONS

Ecosystems comprise living (biotic) and non-living (abiotic) components, which are thoroughly integrated. Biotic components are the organisms, and abiotic components are the solids, liquids and gases which support and interchange with living creatures. Ecosystem principles are the same everywhere: the double approach of nutrient cycling and energy flow apply equally in the sea and on land. These processes are well documented in a wide range of biological and geographical texts, and only a summary is provided in Figure 6.1 and below.

In all ecosystems autotrophs convert inorganic materials into organic matter. Most autotrophs are photosynthetic, largely plants, and use the chlorophyll molecule as a mechanism for energy transfer; solar energy is captured to split water, combining CO_2 with H_2O to generate carbohydrates, which are the primary foodstore of the rest of the ecosystem. Chemosynthetic autotrophs use chemical energy, largely from organic matter, converting H_2S and CO_2 into small amounts of carbohydrates in dark places where photosynthesis does not operate. Another important chemosynthetic mechanism uses enzymes to convert inorganic atmospheric nitrogen into nitrogenous compounds leading to the formation of proteins and nucleic acids, essential for cell operations and reproductive processes.

The result of these processes is primary production. Some of the carbohydrate and nitrogenous compounds created are used directly by the organisms in operating their systems, and so the reverse process of respiration metabolises a portion of the food store. The remaining material is available for use by other organisms. Thus, ecologists distinguish gross and net primary production, with respiration representing a loss, such that gross = net + respiration, in energy terms, in the same logic as gross salary = savings + expenditure in human economics. The laws of thermodynamics place limits on ecosystems, because matter cannot be created or destroyed (hence matter is cycled) and each stage of the operation of ecosystems involves energy loss (hence energy flows through ecosystems).

Primary production is taken up by other components of the ecosystem, via a series of trophic levels (Greek *trophe* is food). The trophic level concept creates a mental image of an orderly process of passage of production through a chain, in the following sequence: primary production is eaten by herbivores along the grazing food chain, creating a meat store as secondary production, which in turn is passed to carnivores in the next level, and on, to top carnivores. Everything, sadly, dies in its turn, and dead matter is reprocessed via decomposer organisms along the detritus chain, principally fungi and bacteria, and the nutrients are returned to the environment to be taken up by the primary producers once more. At each stage, energy is lost. In reality, although much primary production does indeed follow this orderly path, it is only a model of the general process. For example, humans are omnivores, and knowingly eat meat, vegetables and fungi; less knowingly, we also eat protists and monera, so we eat at different trophic levels and across all kingdoms of life. Thus food chains are really webs, and quantification of ecosystem components and function is beyond the scope of this book.

Life exists in most places in ocean systems, and we must appreciate the geometric parameters of the ecosystems themselves, and their relationships with nutrient supply. Because the ocean ecosystems have a large component of suspended matter, the overall 3D character of ocean habitats interacts with ocean circulation, and is presumably part of the reason why older deep-water masses (the Pacific Ocean) have more dissolved CO_2 and a consequently shallower CCD than younger water (the Atlantic Ocean), discussed in Chapter 5; there has been more time to dissolve more CO_2 in older waters. Most ocean life is in the photic zone, because of the high efficiency of eukaryotic photosynthesis in the plankton (Figure 6.3). The most productive ecosystems are near land and in upwelling regions (Figures 6.4 and 6.6). Mangrove swamps and estuaries are highly productive because of the terrestrial input of nutrients; but these are volumetrically small, so open ocean production is greater, in coastal and ocean upwelling sites. Productivity is largely controlled by nutrient supply especially P, and to a lesser extent N, Fe and Si (Brasier, 1995a,b). Thus coastal upwelling and equatorial upwelling are sites of great productivity, as are the south polar convergence region, and north polar cool zones of the North Atlantic.

6.3 OCEAN COMMUNITIES – PROCESS AND PRODUCTS

6.3.1 OCEAN COMMUNITY DISTRIBUTION AND TIERING

Ocean ecosystems are focused around nutrient supply; thus, as indicated above, productivity is greatest in two main settings: (a) coastal upwelling zones on the eastern edges, and equatorial upwelling areas, of Atlantic and Pacific Oceans, plus Antarctic upwelling associated with cold deep-water formation in the Weddell Sea; and (b) areas influenced by river input, especially surrounding the Northern Hemisphere landmasses, because most land is

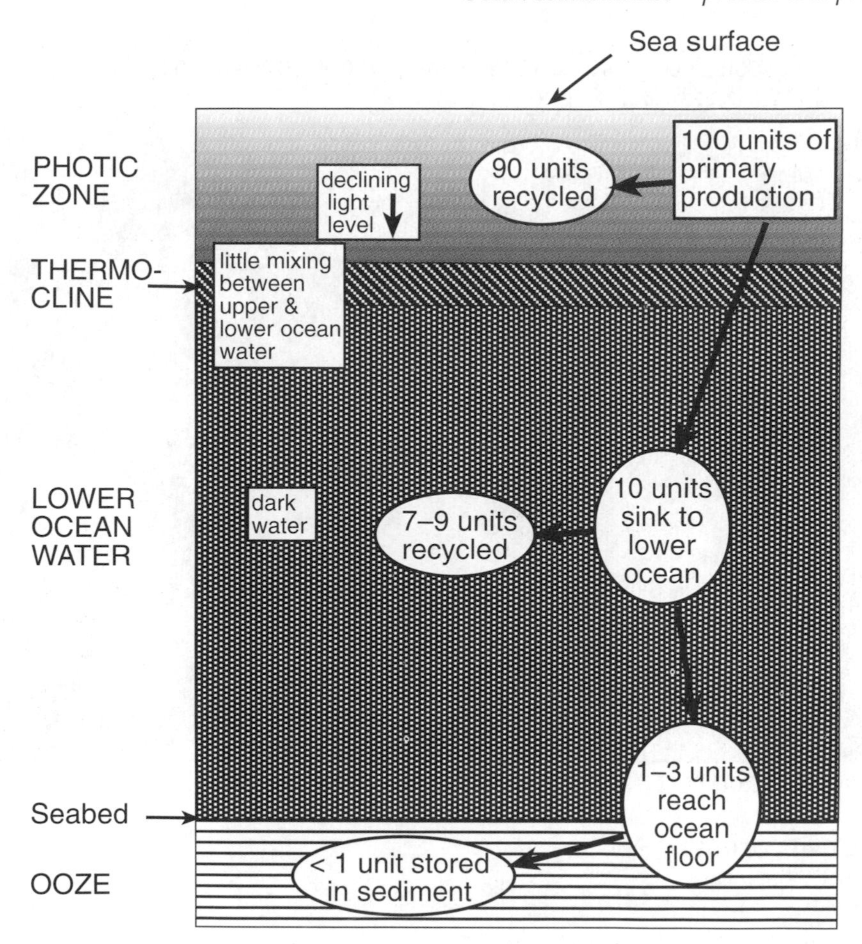

Figure 6.3 Productivity of different depths of ocean water. Note the very small amount of production stored in ocean sediments. While this contrasts terrestrial ecosystems on small time scales, in the long term, the gradual accumulation of organic matter in sediments is a key component in the composition of ocean and atmosphere, especially with regard to oxygen. The removal of CO_2 from the atmosphere by weathering and photosynthesis over geological time has led to progressive burial of carbon in the crust, and allows the atmosphere to maintain a free oxygen content. Adapted from Bearman (1989b) with permission from the Open University.

in the Northern Hemisphere. The open ocean, in contrast, is mostly an ecological desert; despite the transport capacity of ocean currents, nutrients are removed by organisms close to nutrient sources, limiting nutrient transfer throughout the oceans (Figure 6.4). The role of P, in particular, has been shown recently to be very important; low P concentrations can support higher productivity rates than previously recognised, and some bacteria use P to obtain associated nutrients, especially C and N from the oceans (Benitez-Nelson and Buesseler, 1999). Nevertheless, the open ocean is so large that overall production exceeds marginal areas (Figure 6.6).

Ocean biota are tiered both in the water column, and in the benthos. Thus primary production is largely in the upper 100 m of the ocean (Figure 6.3), while decomposer portions are deeper, where they generate an oxygen

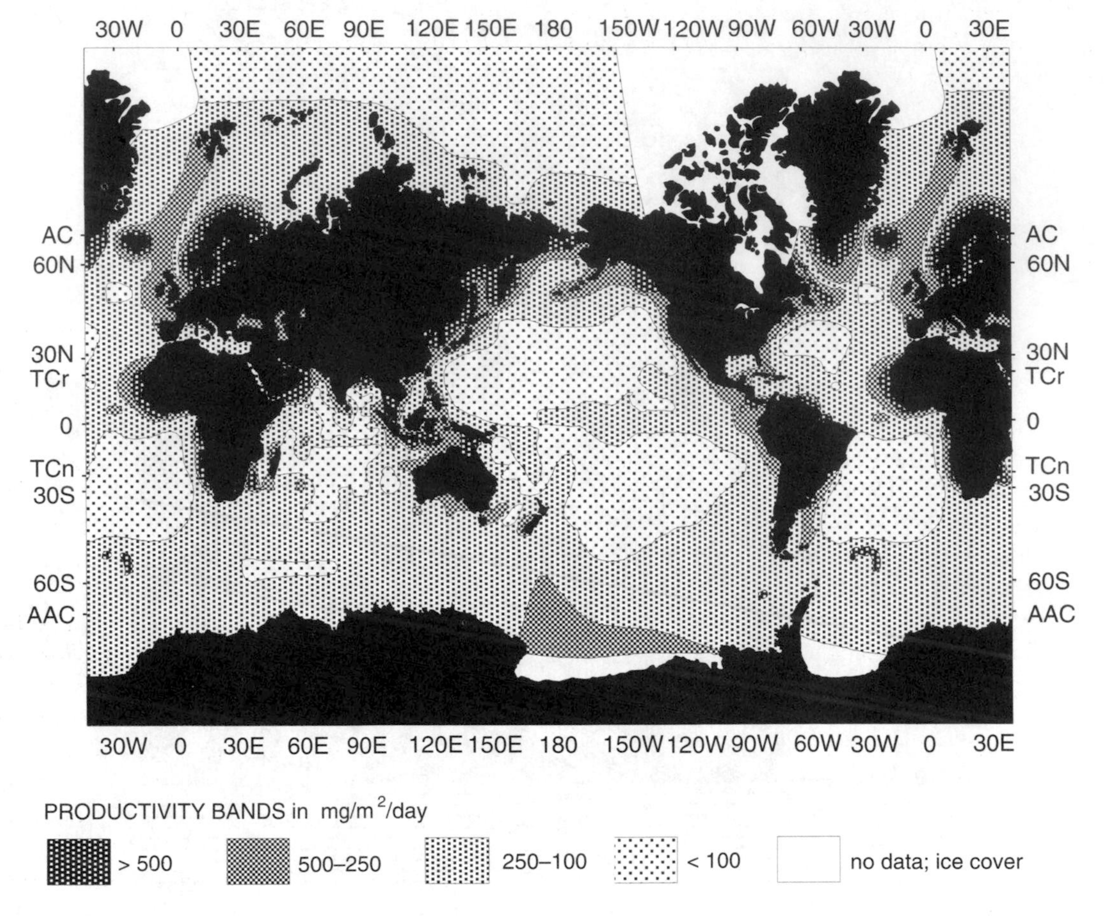

Figure 6.4 World distribution of marine productivity. Productivity is high near regions of nutrient input: continental margins adjacent to mountain areas; continental margins in areas of coastal upwelling; areas of active marine volcanism; and areas of equatorial upwelling. Note that most of the open ocean has very low productivity.

minimum zone (OMZ) through respiration of dead organic matter falling through the water. Therefore the nutrients are mostly recycled within the water column, and only *c*.10% of dead organic matter reaches the seabed. These characteristics are readily shown by the behaviour of limiting nutrients such as N and P, although the relationships are complex in detail. The sinking of organic particles removes substantial amounts of organic matter from the ocean waters, and biomarkers called alkenones and altered chlorophyll compounds are useful to track the accumulation (Herbert *et al.*, 1998; Rosell-Mele and Koc, 1997). Organic particles which survive recycling in the OMZ become a major food source for the seafloor communities, and their input is well known as the seasonal rain of phytodetritus. Such material becomes part of the sedimentary oozes, which contain a mixture of organic carbon plus dead skeletons of skeletonised plankton. Of the sedimentary oozes, calcareous oozes act as important sinks of $CaCO_3$, and siliceous oozes place severe limitations on the availability of silica by locking silica into seafloor sediments. Distribution of high concentration of oozes also fits with areas of upwelling related to Ekman transport and Coriolis forcing of wind patterns. Calcareous oozes have an important application in ocean studies: their state of preservation is used to determine depths of the CCD in different sites on the ocean floor. Observers record the degree to which calcareous shells of micro-organisms lying on the seabed are corroded in order to determine whether the CCD intersects the seabed at those sites (Figure 5.6). Such work is necessary because the CCD cannot be measured directly from the water column.

Organic particles often form clumps by both organic and inorganic processes, and descend to the seafloor as aggregates; aggregation is an important feature which enhances export to the seabed. Deposited organic matter feeds substantial benthic communities, which are also tiered, and form sub-communities at different levels in the sediment to use the substrate to its maximum. Benthic tiering occurs because of competition for space on the seabed, especially in shallow shelf settings. This leads to a subclassification into surface dwellers (epifauna) and subsurface dwellers (infauna), and there is another category: animals feeding and living in the water but near the seafloor, the nektobenthos. Within surface and subsurface dwellers, the communities are further tiered; infaunal organisms live at different depths in the substrate, and leave traces of their depth relationships of competition avoidance. Because the organisms live in the sediment, the potential for preservation of sedimentary structures left by their movement (trace fossils) is high. The study of trace fossils is a major aspect of palaeoceanography, and aids the identification of, for example, soft substrate and hard substrate communities, sedimentation rates, the state of oxygenation of the sediment, and the efficiency of transport of oxygen to the seafloor. Shallow waters within the photic zone are oxygenated by water turbulence, but deeper ocean settings rely on the thermohaline circulation to ventilate the seabed.

Substrate microbia are a key part of the system, and both autotrophic and heterotrophic bacteria exist. In sediments 90% of the nitrogen is from organic matter. Remineralisation of nitrogen-bearing organic compounds (amino acids, proteins) by both aerobic and anaerobic bacteria generates ammonium ions (NH_4^+), making this essential nutrient available for upwelling. Upwelling is not just a physical process; diatom mats have been shown capable of transporting nitrates to the surface waters in the north Pacific Ocean (Villareal *et al.*, 1999). Also of great importance is reduction of SO_4^{2-} by heterotrophic sulphate-reducing bacteria, discussed in Chapter 5. Up to 115 mg of H_2S per litre of water each day may be generated in the upper waters of the Black Sea (Hallberg, 1992), and sulphate-reduction is the most important reaction in the sulphur cycle.

A special sort of marine environment is the deep-sea vent system, where a mixture of volcanic outgassing into the water, with recycled ocean water, is erupted from the seabed (Figure 5.9). The remarkable nature of the communities of bacteria, worms and fishes which cluster tightly around the vents was discovered relatively recently. The entire vent ecology is based on chemosynthetic processes.

6.3.2 FATE OF ORGANIC CARBON

The removal of a mixture of organic carbon (C_{ORG}) and mineralised carbon (as $CaCO_3$ in shells, C_{CARB}) into long-term burial storage represents loss of carbon from ecosystems for, usually, millions of years. C_{CARB} is only about 20% of the flux to the seafloor today. However, sea levels are lower than in the geological past, and therefore the area of shallow epeiric seas is suppressed, so that carbonate platforms (which include reefs) are uncommon at present (Figure 6.5). In fact, about 75% of the carbon reservoir in the Earth's crust is C_{CARB} as allochemical limestone. Furthermore, because of the chemical formula of carbon dioxide (CO_2), as noted in Chapter 5, for every O_2 molecule liberated by photosynthesis, a rough

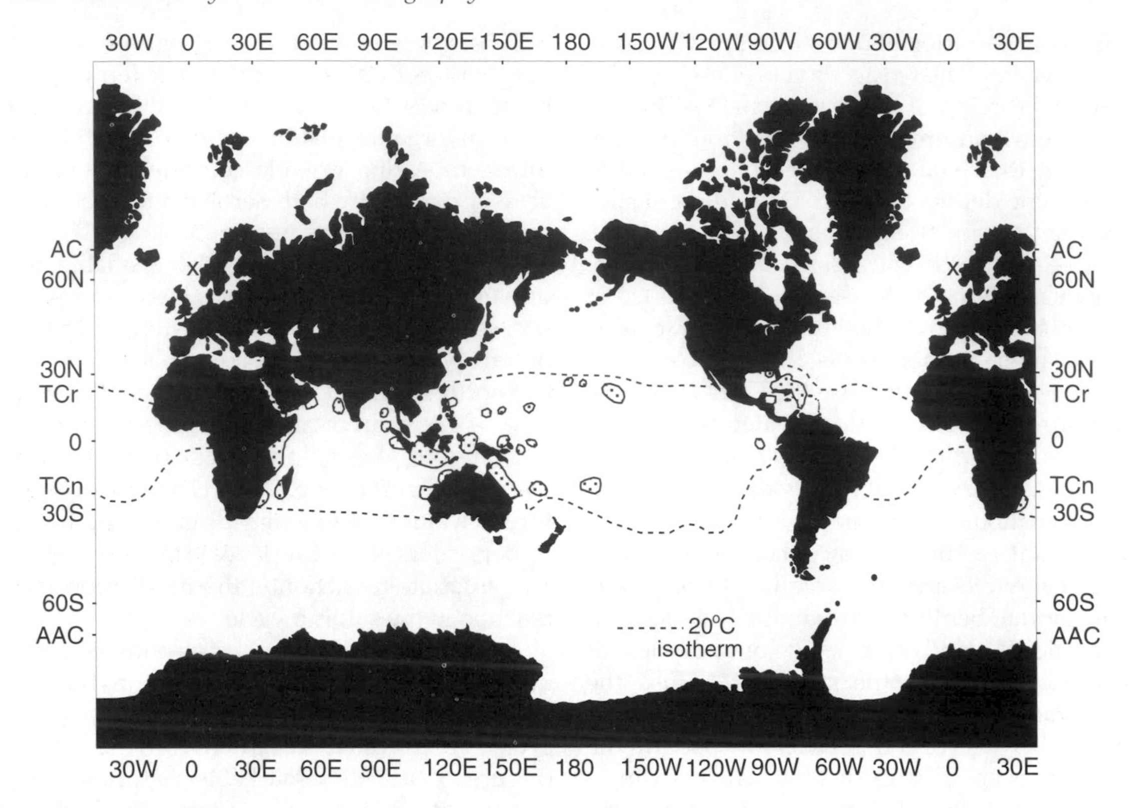

Figure 6.5 Reef distribution (dotted areas) is largely related to equable temperatures of the tropics, but note that the large coral reef complex (marked by **x**) discovered off Norway, shows that reefs are not *restricted* to tropical regions. The difference between cold and warm water modern reefs is that the cold water types are not involved in symbiotic processes with algae.

approximation is that there is one atom of carbon buried in the Earth's sedimentary deposits. Because it is clear that the Earth's early atmosphere and ocean lacked oxygen, then the oxygen must have been created by separation of O_2 from CO_2 present in the Earth's surface systems, principally by photosynthesis as described in Chapter 8.

Furthermore, not only are the continental shelves of the world the largest carbon store, but over geological time, the transfer of CO_2 and O_2 between oceans, atmosphere and sedimentary rocks means that the oceans are really only a staging post in these processes, with any atom of carbon spending an average of 100 000 years residence time in the oceans. The development of rifted basins creates isolated deep-water areas which can collect and bury carbon, such as the North Sea; carbon spends an average 100 Ma locked in the crust before being released by erosion. Plate tectonics thus influences carbon burial and therefore ocean–atmosphere gas composition balance. Hence the wider perspective revealed by these points is that the marine ecosystems we see in operation today are merely a snapshot of a longer-term control process, and this is why the oceans' ecosystems have a key place in the history and support of life. A good discussion of these ideas is given by Francis and Dise (1997), and we return to this theme in Chapter 8.

The action of the global carbon cycle is instructive in this debate. Consideration of the entire carbon cycle is not appropriate to this text, but the following data relating to the oceans from a recent review (Holmén, 1992) illustrate the vastness of the ocean store in relation to other Earth surface reservoirs, but it is insignificant against the carbon stored in the lithosphere (Figure 6.6A).

Because 10% of the production reaches the seafloor as organic detritus, these production figures indicate that 5×10^{15} g of carbon/a are lost from the surface waters, so to balance the carbon cycle for the photic zone, that much must be replaced by ocean circulation from deep water (Holmén, 1992). This serves to emphasise the close interrelationship between ocean water movement and the activity of the ocean ecosystem. Other estimates of primary production suggest that ocean productivity is only about one third of the global total (Hallberg, 1992), but would still imply a large transfer of carbon.

The patchiness of ocean productivity is shown by data from another review (Murray, 1992) (see Figure 6.6B). Note that the total is only about 40% of the estimate given by Holmén. It is important to emphasise that in the collection of data for a system as complex as the carbon cycle, modelling is applied, and models are rarely perfect analogues for real systems. For our purposes, the crucial nature of these various figures is to illustrate the relative importance of the different portions of the oceans for productivity. In addition there is an important aspect of lateral movement between regions of ocean ecosystems; one study of the North Atlantic and North Pacific Oceans

A

Earth surface carbon reservoirs:	
Atmosphere, nearly all CO_2 (as of 1988)	747×10^{15} g = 1.86%
Ocean, dissolved organic carbon	1000×10^{15} g
– particulate organic carbon	30×10^{15} g
– dissolved inorganic carbon	$37\,900 \times 10^{15}$ g
Ocean total	$38\,930 \times 10^{15}$ g = 96.75%
Land	560×10^{15} g = 1.39%
Grand total	$40\,237 \times 10^{15}$ g =100.00%

Lithospheric carbon store:

Sum of carbon in all igneous, sedimentary and metamorphic rocks	20 million $\times 10^{15}$ g

(surface store is therefore 0.002% of the lithospheric total)

Holmén also recorded the organic productivity of carbon.

Net primary production of carbon:

Ocean	50×10^{15} g/a = 45.45%
Land	60×10^{15} g/a = 54.55%

B

Province	% of ocean	Mean productivity gC/m^2/a	Total productivity 10^{15}gC/a
Open ocean	90.0	50	16.3
Coastal zone	9.9	100	3.6
Upwelling zones	0.1	300	0.1
Total			20.0

Figure 6.6 Comparisons of production data for Earth's surface and lithosphere. (A) from Holmén (1992); (B) from Murray (1992).

showed that ocean margins produce more organic carbon than is respired, and that the export of organic matter to the open ocean provides possibly as much as ten times the amount produced directly in the open ocean (Bauer and Druffel, 1998). Modelling of global continental margins also shows a net production of carbon, reducing the ability of the continental shelves to act as a sink for anthropogenic carbon (MacKenzie *et al.*, 1998). Finally, there is a problem with the carbon cycle, because there is an amount of carbon which has not been accounted for in the various reservoirs, referred to as the 'missing sink' problem. What part the oceans play in this is undetermined; the currently favoured location of the missing carbon, temperate forests, has recently been shown to be invalid (Nadelhoffer *et al.*, 1999).

6.4 OCEAN ECOSYSTEM TIMESCALES

The rate of processes is important to appreciating the full scale of ocean ecosystem operation, and it is clear that the ecosystems work at a variety of timescales, simply divided here into two. Sections 6.4.3–6.4.5 discuss their effects.

6.4.1 DAILY TO YEARLY

Daily rhythms have been identified in modern and ancient seas, as tidal processes influence deposition (House, 1995). In modern oceans there is also an annual effect of considerable influence on the physical state of ocean water, related to seasons. The classic picture is that the thermocline in warm seas is transitory (Figure 6.7). It builds in the early summer as water is heated. Productivity is high because of plentiful nutrients accumulated during winter storms (upwelled and runoff). Summer productivity drops as nutrients are depleted above the thermocline (no mixing with nutrient-rich deeper water). Weather deterioration in the autumn leads to an ocean overturn, where upper waters mix with lower, enriching upper layer nutrients and giving a second burst of productivity before the winter decline. This simple model is modified in polar settings where no thermocline forms and productivity is low, except in the better-lit middle part of the year, and in the Antarctic where the unusual conditions cause upwelling (Chapter 3) and high productivity from the release of P and N. In shallow hot coastal areas, the thermocline is permanent and the upper layer eutrophicates, especially where nutrient input is high from the land. Annual shifts in ocean state are recognisable in the recent past in the sedimentary record, but earlier parts of Earth's history do not preserve the signals so well. Nevertheless, annual cycles are recorded in fossils as annual growth bands (e.g. corals, shells).

6.4.2 LONGER TIMESCALES

Ecosystems modify under various pressures, and under long timescales such changes may be regarded as evolutionary forces. Cyclic effects on the oceans over tens to hundreds of thousands of years are considered part of the Milankovitch band. The longest timescale of geological time resulted in the developed diversity fluctuations of ocean environments and biota, discussed in Part B of this book. Separation of continents influenced evolution of life on land, on continents and islands; the consequences of ocean development and the movement of continents are far-reaching, and even affect human evolution, because of their influence on migration pathways.

6.4.3 UNCONFORMITIES, SEDIMENTATION RATES AND THEIR SIGNIFICANCE IN MODERN AND ANCIENT OCEAN ECOSYSTEMS

The geological record is not a complete sequence of deposition; for a variety of reasons sedimentary deposits have a number of breaks of deposition. Breaks can be easily recognised if they represent long time periods (thousands or millions of years) because the erosion of

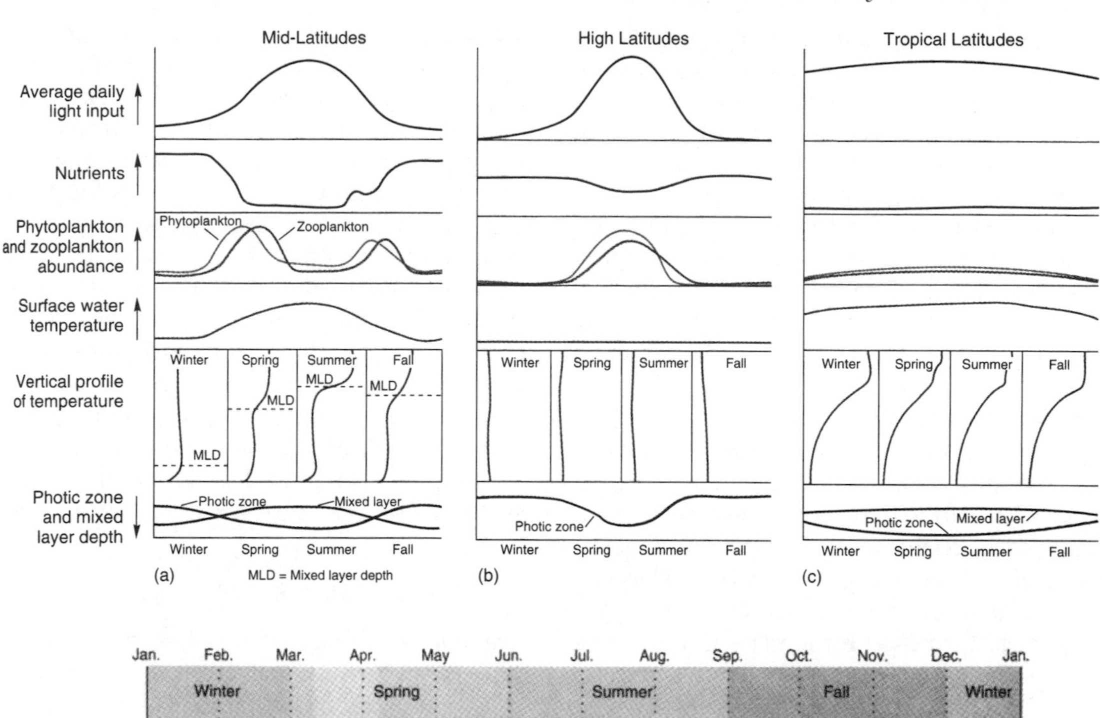

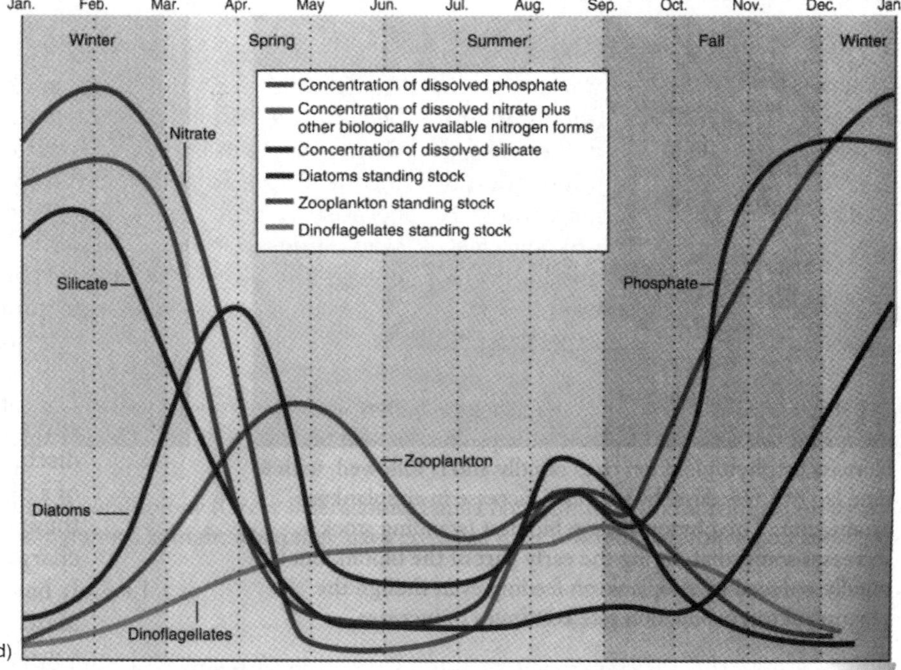

Figure 6.7 Changes of productivity over a year in the oceans (see text for explanation). Reproduced with permission from Segar, D., *Introduction to Ocean Sciences* (with Infotrac), 1st edn (1998), Brooks/Cole Publishing, Thomson Learning, CA, USA.

material leaves irregularities in the sedimentary rocks. On shorter timescales, the breaks are usually not so obvious, but they have important implications for the way certain ecosystem components (especially the benthos) use the seabed. A sediment surface on which little or no new sediment collects for a long time, but which remains covered by sea, becomes colonised by several generations of communities; the longer the depositional break, the more complex the overlayering of communities. In the geological record in sites where input of sediment is very low, such processes leave thin layers of deposits which may represent considerable periods of time (in contrast to other areas where deposition may have been continuous to generate much thicker layers). These thin layers are called condensed sequences. They are important to appreciate, because any area of seafloor left exposed for long periods becomes repeatedly colonised by successive communities; it is a fact of life which palaeontologists recognise in order to interpret ancient ocean conditions at small scale. The study of modern environments plays a large part in the analysis of ancient ocean floors. A good illustration of the effects of sequence condensation is the Early Cambrian record (Droser, 1991; Skoog *et al.*, 1994), in the pipe rock, a sandstone indurated with dense accumulations of vertical burrows. Modern analogues of this intertidal facies in Virginia, USA, show that worms creating similar burrows produce much less dense concentrations nowadays, indicating that the ancient Pipe Rock community represents several phases of worm colonisation on the same sediment surface, which therefore is a small-scale condensed sequence. In a larger scale example from the British Jurassic System, the Inferior Oolite at Burton Bradstock, Dorset, is only 1 m thick and full of fossils, represents several million years, and is correlated to *c.*90 m of strata in Yorkshire, where deposition was more continuous.

A second major feature of ancient sediments which can reveal the nature of environments is time averaging of communities. This refers to settings where the environment has not changed a great deal over a period of millions of years, and the fossil assemblages do not illustrate much modification through the vertical interval of rock representing all that time. In some cases it is convenient for geologists to regard the entire time interval as essentially one environment, and therefore express all the community ecology as a single feature. Thus the community is averaged over a time interval. In the vast majority of geological studies, time averaging is used to a greater or lesser degree, depending on the degree of vertical variability of the deposit studied, for the simple reason that differentiation of small time units (days to years) is often not possible.

6.4.4 BIOGEOGRAPHIC PROVINCES

Organisms are not evenly distributed across the globe, because of physical and physiological barriers. Polar bears live only in the Arctic, and penguins only in the Antarctic, because the tropics prevent migration due to the higher temperatures; coral reef faunas in the Pacific are quite different from the Atlantic because Central America is in the way and stops migration; the Antarctic circumpolar current limits migration into or out of the Antarctic seas. Therefore an important interplay exists between the distribution of continents, climates and ocean circulation on the one hand, and the organisation of biotic communities on the other. The present distribution has a historical background because it depends on climate change and plate tectonics, so the geological past strongly influenced the configuration of present ocean ecosystems. One of the best-known examples in the geological record contains evidence of an opening and closing ocean, called the Iapetus Ocean. This is preserved as the Caledonian–Appalachian mountain fold belt from Scandinavia through Britain and Ireland, to eastern USA and Canada, in which the oceanic record of sediments and fossils plays a key role in understanding the history of the fold belt. For example, as the Iapetus Ocean narrowed during its closing phase, formerly distinct communities began to mix, as recorded in the fossils and sediments from this period.

6.4.5 TAPHONOMY

We complete this survey of ocean ecosystems with mention of a major process affecting both modern and ancient ecosystems. Taphonomy (Greek *taphos* = a tomb) is the study of post-death processes in organisms, and in recent years there has been an explosion of studies of modern and ancient fossil assemblages to characterise taphonomic processes. Taphonomy consists of two subareas:

1. the disruptive processes affecting a biological community, whereby the various organisms are moved or damaged/destroyed (by storm activity, for example), called biostratinomy;
2. the processes of destruction after the dead bodies of the organisms are buried, so that shells may be dissolved and lost from the record, one of many processes collectively called diagenesis. In some cases, however, diagenesis takes place before the death of organisms, and for that reason, diagenesis is considered by some workers to be separate from taphonomy.

For fossil ecosystems, the problems of interpretation that these processes cause can be enormous, and they therefore play an important part in palaeoceanography; but taphonomic processes can also be very useful because, if certain components of the ecosystem are moved to another place, then that can *assist* interpretation. Taphonomy is a pervading feature in Part B, and warrants serious consideration.

6.5 CONCLUSION

Ocean community ecology is a highly complex network of relationships within and between living and non-living components of the ocean system. The grand-scale three-dimensional nature of oceans emphasises the separation of different forms of ocean biology, especially with respect to the contrasts between ocean surface and deep-ocean vent communities. There are two main points that arise from its study:

1. Biological oceanography is not just the study of marine ecology, but is intricately linked to physical and chemical ocean operation, and in a geological perspective has been involved with the whole Earth processes throughout geological time.
2. An understanding of the nature of this complexity is critical to the analysis of ancient ocean ecoystems.

6.6 SUMMARY

1. Because ecology includes a study not only of the interaction between organisms, but also between organisms and their environments, marine ecosystems involve such a wide range of aspects that they constitute biological oceanography.
2. Marine habitats are focused on the photic zone because eukaryotic planktonic green algae are the most efficient primary producers. However, the ocean floor is inhabited everywhere (probably), and key communities are located in ocean vents on mid-ocean ridges, and benthic settings on continental shelves.
3. Marine ecosystems comprise the same general processes of ecology as terrestrial settings, but are typified by rapid recycling because no wood forms, so there is effectively no short-term storage within the ecosystems.
4. Long-term storage (on geological timescales) of carbon in ocean floor sediments occurs because not all dead matter is recycled within the photic zone. Falling dead plankton rain escaping the decomposer organisms in the water column and seabed communities are effectively lost from ocean

ecosystems and become buried. That they do so has two benefits. Firstly, because photosynthesis converts CO_2 and H_2O into carbohydrate and free O_2, every molecule of oxygen free in the ocean–atmosphere–land systems is represented (approximately) by one atom of carbon buried in sediment; if this did not occur, then free O_2 would be converted back to CO_2 and aerobic organisms would not exist. Secondly, carbon burial leads to petroleum generation, which drives human economy!

5. Ocean productivity is closely tied to physical oceanographic factors. In particular: (a) Ekman transport generates coastal upwelling on west-facing coasts; (b) equatorial currents cause upwelling in the oceanic tropics; and (c) polar upwelling occurs because of temperature and salinity differences of water masses, especially the polar convergence zone between sinking cold Antarctic water and rising water in the South Atlantic. In each case, productivity is hugely increased because of updraught of nutrient-rich waters. Phosphorus, nitrogen, iron and silicon are commonly regarded as key nutrients, and phosphorus is the most important because it is in critical short supply.
6. The current configuration of marine ecosystems is the end-result of millions of years of both biological and geological evolution, and shows that the true significance of marine ecology goes a long way beyond common perceptions of its meaning.

PART B

EVOLUTION OF THE OCEANS IN GEOLOGICAL TIME

One of us is old enough to remember a visiting lecturer's opening sentence at a Manchester University student geological society lecture in 1974: 'Everybody knows that geology is not the study of rocks . . .' (*pause; audience is puzzled*) '. . . it is the study of **processes**, preserved in rocks' (*audience enlightened; much relief all round*). (For the record, it was J.R.L. Allen, former sedimentology professor at the University of Reading, UK.)

If you apply that logic to **oceanography** instead of **geology** in this sentence, then you can:

- see that oceanography, past and present, is a study of processes; and
- appreciate the very close interrelationship between oceanography and geology, because the evidence of past ocean processes lies in rocks.

STUDYING THE OCEANS IN GEOLOGICAL TIME

7

7.1 INTRODUCTION

Part B provides an overview of key aspects of ocean process and change since the first record of oceans at *c.*3900 Ma, summarised in the table below. Part B is intended to demonstrate the range of controls on oceans, using landmark examples, and so is not exhaustive – several volumes are needed to describe all ocean change over geological time! The aim is to draw attention to the issues and discuss interpretations, to provide a balanced view of oceans in time as a backdrop to appreciating the nature of possible future change.

It is important to realise that Precambrian oceanography is more difficult to assess than the Phanerozoic Eon, because the evidence has been affected by later tectonic processes, and much is covered by blankets of younger sediments. The critical contrast is that the Precambrian ocean change has been profoundly unidirectional; the Phanerozoic Eon has weaker unidirectional changes, but is characterised more by cycles on varying scales. In Chapter 2 are noted minor changes of ocean composition occurring today in coastal settings, and more profound changes in the Black Sea in the recent past. However, such sites are small marginal areas, only partly connected to the oceans; *global* ocean composition change is far more important, the focus of this part of the book.

There is referenced a wider range of literature in Part B than in the rest of the book, because of its broad scope and often controversial content, to allow readers to follow up aspects of interest.

Summary of ocean change in geological time (Precambrian is unidirectional, Phanerozoic is cyclic)

PRECAMBRIAN (4600–530 Ma)

FEATURES: 1) unidirectional change; **2)** *proven* major compositional changes in CO_2, O_2, Fe, S; **3)** *proposed* major compositional changes in salinity and pH; **4)** the most important date in the history of Earth is probably 1800 Ma, when ocean and atmosphere began to develop surplus oxygen; this has important implications for O_2, Fe and S in the oceans.

SEQUENCE OF EVENTS:

4600 Ma – formation of Earth from gas cloud
4200 Ma – oldest crustal material (detrital mineral grains in sediment)
4000 Ma – oldest rock (metamorphic gneiss) as part of the early continents (called cratons)
3900 Ma – oldest greenstone (sandstone and shale deposited in epicratonic sea)
3800 Ma – oldest Banded Iron Formation (BIF); oldest evidence of photosynthesis (isotopes).
It is generally accepted that the O_2 generated from early photosynthesis was instantly taken up by the dissolved Fe in the sea; this Fe is in the reduced state, but when oxidised it creates insoluble Fe oxide, which is deposited on the seabed, to form the BIF. This process went on until *c.*1800 Ma, when all the Fe supply in the sea was exhausted.
3500 Ma – first fossil remains (bacteria and stromatolites – cyanobacteria), demonstrating that photosynthesis had evolved by then.

– Also first abundant precipitates of $CaCO_3$ crystal crusts on the seafloor; this tells us that there was plentiful CO_2 in the ocean, but why wasn't there $CaCO_3$ precipitation at earlier times? (CO_2 comes from volcanic outgassing, and because the Earth was hot and volcanic, CO_2 is expected to be an important part of the ocean and atmosphere from the beginning of the Earth.) The idea of a high pH (10.5) soda ocean *may* solve this puzzle, because high pH (alkaline) stops $CaCO_3$ from being precipitated on the seafloor. So, why did the soda ocean stop at 3500 Ma? Possibly because subduction and ocean-ridge weathering drew sodium (Na) out of the water, lowering the pH and allowing $CaCO_3$ to precipitate; after that the ocean was a halite (salt) ocean, and continued to the present day.
1800 Ma – as well as the last BIF, this date marks the start of abundant gypsum ($CaSO_4$), and shows that the ocean now had a sulphate (SO_4) component, which it previously lacked.
Late Precambrian – glaciation, followed by the Cambrian evolution of shelly animals; shells made of phosphate and $CaCO_3$, and show that organisms had become highly evolved, which in turn is evidence that ocean and atmosphere O_2 levels had risen a lot since 1800 Ma. Note that evidence of continental postions is poor in the earlier Precambrian, but in the last few hundred million years of Precambrian time, the continents were fused as a supercontinent.

PHANEROZOIC (530–0 Ma)

FEATURES: 1) constant salinity, shown by marine fossils; **2)** constant SO_4, shown by marine evaporite sequences; **3)** cyclic change of mineralogy of ooids (between calcite and aragonite) (Sandberg curve), related to the Mg/Ca ratio: when Mg is high, aragonite forms; when Mg is low, calcite forms; **4)** cyclic change of potassium (K) in evaporites in relation to 3); **5)** cyclic sea-level change controlled by plate tectonics, and corresponding to changes in 3) and 4) (Wilson cycle).

SEQUENCE OF EVENTS:

530 Ma – (= base of Cambrian Period) see end of Precambrian section.
Lower Palaeozoic Era – continents rifting in first half of a Wilson cycle; here mid-ocean ridges (MOR) are active and sea level is high. Weathering of fresh ocean crust draws Mg out of the water, so the Mg/Ca ratio goes down, and calcite forms in ooids (= calcite sea). At the same time, K is weathered out of the MOR volcanics so that in shallow waters, K-rich evaporites form
Later Palaeozoic Era – the continents become fused to form Pangaea, so MOR activity was less and the sea level fell. Mg is carried into the ocean from rivers, so the Mg/Ca ratio goes up, because it is not being removed at MOR; thus aragonite ooids form (= aragonite sea). At the same time, K is not being removed from the MOR volcanics, so K-poor evaporites form in shallow water. The fusion of continents back together again is the end of the Wilson cycle.
Mesozoic – calcite sea again, as Pangaea breaks up at the start of the second Wilson cycle.
Tertiary – MOR activity lower, and aragonite sea develops, continuing today.

7.2 SOURCES OF INFORMATION FOR ANCIENT OCEAN COMPOSITION STUDY

Oceanographers for many years used the widely accepted notion of constant composition through the majority of geological time (e.g. Holland, 1984), but this has recently come under scrutiny (e.g. Grotzinger and Kasting, 1993; Grotzinger and Knoll, 1995). In the geological timescale, there is no empirical reason to expect compositional stability; change is a feature of geological systems, and even James Hutton's concept of uniformitarianism is not wholly applicable for many settings.

How can ocean composition through time can be assessed? The evidence lies in sedimentary rocks of the Earth's surface (the 'sedimentary shell'), and falls into two main areas.

1. **Marine organisms** now preserved as fossils (the majority of which are calcareous), lived in shelf seas. Recycling of ocean crust by plate tectonics leaves the oldest ocean floor at *c.*200 Ma (Early Jurassic), except for ocean floor fragments called ophiolites which are tectonically emplaced on the continental crust. Thus, although much has been learned from ocean floor deposits in the oceano-

graphic study of Jurassic and younger rocks, older evidence is available almost exclusively from marine *shelf* settings and *epicontinental* seas, because these lie on continental crust (which is not subducted), and much is now exposed on land. Applying uniformitarian principles, marine fossils are assumed to have the same tolerance to salinity variations as do their modern counterparts. Common examples are modern brachiopods, corals and echinoderms, which are tolerant to only a narrow range of salinity variation (stenohaline), so geologists deduce that fossil equivalents had similar responses. So it seems that ocean *salinity* was constant (or at least varied only within the very narrow tolerance of these organisms) through the part of geological history when these creatures occur, i.e. Cambrian to recent, but earlier salinity states cannot be assessed using fossils because the fossils do not exist in the rocks. Therefore Precambrian, and also much of the Phanerozoic, ocean composition study relies principally on the next line of evidence.

2. **Sedimentary materials** in modern shallow seas have abundant ancient counterparts. Older Precambrian detrital sediments (that is, eroded from land) contain evidence of low oxygen levels in the ocean and atmosphere, because the sediments contain fragments of reduced material. However, chemically precipitated minerals on the seafloor reflect the composition of the (sea) water they originated from. Although seafloor chemical processes (called early marine diagenesis) appear to be broadly the same in modern and ancient settings, temporal fluctuations of forms of $CaCO_3$, iron and sulphur compounds in particular, have palaeoceanographic significance. They show that ocean composition was not constant in respect of these, and their study has led to major changes in thinking about ancient ocean composition. In particular: (a) iron in the ocean shows profound change; (b) $CaCO_3$ mineralogy has fluctuated in time; and (c) evaporite minerals do not all show the predictable sequence of crystallisation that they should, if seawater composition has been constant; these are explained in the following chapters.

7.3 WHERE GEOLOGY AND OCEANOGRAPHY MEET

The modern oceans are only a snapshot, giving the present state of nearly 4000 Ma of ocean evolution, and the applications of the concept of 'the present is the key to the past', used so much in geology, are relevant in palaeoceanography. Very important in ancient ocean study is the effect of changing sea level. Sea-level change occurs either as regional change, caused by the local tectonic setting, or as global, caused by global tectonics, and accompanying continental drift as the continents are reconfigured through time. Glacial growth and decay have also been important at several points in Earth's history. Global sea-level change is also called eustatic (Greek *eu* = all/full/well; *statikos*=standing), referring to change in sea level without change in crustal elevation. A little reflection reveals that sea-level change must also involve crustal movement by isostatic adjustment, by loading and unloading the crust with water as sea level rises and falls, respectively. Therefore, global sea-level change is not truly eustatic, and it might be argued that the concept of eustasy has no real meaning! Recognition of sea-level change depends on knowledge of the sedimentary record, the analysis of which relates closely to modern studies. However, the organisms which create the modern deep sea oozes evolved only relatively recently, and the modern Earth's climate is atypical, giving challenging problems of reconstructing past oceans' nature.

A successful study also depends on the recognition of low or high, and rising or falling sea level. The most obvious features to help this recognition are unconformities overlying marine sediments, which demonstrate that the area in question has been uplifted and eroded; these are obvious in some sites, but there are problems in their use. In some places, tectonic uplift outpaces sea-level rise, and the consequent unconformity may give the false impression of sea-level fall. Another problem is lateral variation of rocks and ancient surfaces. For example, Mount Snowdon in North Wales is made of Ordovician sediments and volcanics, *c.*450 Ma, and modern streams flowing down its sides deposit modern sediment (recycled from the mountain), so there is an unconformity representing 450 million years underneath the sediment in the stream bed. If you imagine you could walk from the top of Snowdon down to sea level, then continue downwards onto the floor of the Celtic Sea, around Ireland, across the continental shelf and down the continental slope onto the floor of the Atlantic, all that time you are still on a surface which is forming *today*. On the Atlantic seabed, sediment accumulates, and the unconformity so easily recognised on land is equivalent to just another surface in a continuously accumulating sediment pile, in which the unconformity would not be recognisable. So, when you study outcrops of oceanic sediments which are now on land, if all you could see was finely laminated sediment with no breaks, you would not be able to recognise which lamination is time-equivalent to a major unconformity on a land surface in another place, unless it is possible to date the sediments.

Therefore, one of the principal characters in the study of past ocean sediments is the double-edged problem of (a) finding the age of a surface in any one place (called dating a horizon), and (b) correlating the surface between sites. Any geology student would recognise these simple ideas as fundamental, and here geology and oceanography become indistiguishable. The recognition of time-equivalent surfaces, and the processes of sea-level change leading to the accumulation of sediment between them, is *stratigraphy*, a subject as old as geology itself. More recently, the study has been revised in the search for petroleum, and the science of recognition of time-equivalent surfaces and the sedimentary units between them has taken on a life of its own, as *sequence stratigraphy* (Emery and Myers, 1996). The ultimate aim is to correlate a surface to all localities, although where there are no preserved criteria for their recognition everywhere, this is impossible. Nevertheless, sequence stratigraphy provides a framework of thinking in the search for a truly accurate global sea-level curve, and its terms have become part of Earth science language; in particular, *lowstands* (make big unconformities because erosion is increased when sea level is low), *transgressive* and *regressive*, and *highstand* settings leave their results in sediments. On a large scale, the amount of sediment covering the older rocks is important; for example, the recognition that the Upper Cretaceous was a period of highstand comes partly from the fact that the sediment (marine coccolith limestone, chalk) covers much of the underlying crust in western Europe. However, it is not our aim to discuss reconstruction techniques, but to consider the oceanographic implications of such deposits. The huge dataset allows us to divide and recognise several scales of ocean change, considered in the next three chapters.

7.4 'WOULD HAVE' AND 'MAY HAVE': THE THEORISTS' PETS

A large number of geological papers of all disciplines contain these expressions, used by authors to predict what would be expected to happen given the circumstances they describe. We are likewise guilty of using these in previous publications, and this book, but we warn the reader that 'would have' really means 'it could have, but we and everybody else don't know, because we weren't there at the time'; and 'may have' is even more woolly. When you see these written, let the warning bells ring, be aware of the constraints of operating on too-logical lines, and be sceptical of what 'experts' say!

PRECAMBRIAN OCEAN CHANGE 8

8.1 INTRODUCTION: WHY STUDY THE PRECAMBRIAN OCEAN?

Precambrian time represents *c.*90% of geological history, and the profound ocean evolution during that time has determined the composition and structure of the later oceans. Examination of evidence and theories relating to Precambrian oceans helps develop an understanding of ocean–atmosphere–land interactions, and emphasises the importance of change in ocean composition, which seems to involve the whole Earth, not just its surface.

Amongst a vast literature and variety of interpretations, theories of Precambrian ocean composition revolve principally around two major aspects: (a) low oxygen levels in at least the first half of the Precambrian, and (b) variation of CO_2 levels in the atmosphere–ocean system throughout the Precambrian. Low O_2 is reflected in the degree of oxidation of Earth surface materials. CO_2 fluctuations are recognised partly by variations of the type, setting and abundance of limestones and dolomites, a valuable discussion of which is given by Tucker and Wright (1990), and partly by the distribution of carbon isotopes in sedimentary rocks. Because atmospheric and dissolved CO_2 is not preserved directly in sediments, $CaCO_3$ and carbon isotopes in both limestones and organic residues are used as *proxies* to indicate its behaviour in the environment; the application of proxies is widespread in the Earth sciences.

Precambrian oceanography is a study of major contrasts of viewpoint which stimulate thought, and lead to investigations to test ideas and evidence; so this chapter does not present a consensus viewpoint, because that does not exist. The aim instead is to demonstrate the importance of oceanography for Precambrian studies, and starts (bravely) with the origin of the oceans.

8.2 GEOLOGICAL SETTING OF THE EARLY OCEAN

The Earth is believed to have formed rapidly, along with the Sun and other planets, by the condensation of a dust cloud through gravitational attraction, and therefore collision, between particles. Collisions release heat, so the result was a hot early Earth, with magma oceans, and probably no crust. Heat also came from radioactive decay of unstable elements. The initial atmosphere is generally assumed to have been made mostly of hydrogen and helium, and stripped away by early solar windstorm events, but the earliest *stable* atmosphere was probably largely steam, with its greenhouse properties playing a key role in maintenance of surface temperatures of a suggested 1200°C (Kasting and Chang, 1992). That the Earth was hot in the first part of its history is shown by abundant peculiar Mg-rich volcanic rocks (komatiites) which form at high temperatures, rare in rocks after the Archaean Eon (Windley, 1995, pp. 401 ff.). Only when the planet's condensation was complete, or nearly so, could temperatures drop sufficiently for the crust, and then oceans, to form permanently. The oceans may have developed by volcanic outgassing (i.e. eruptions) of volatile components, principally water and CO_2, as in modern eruptions, although just how much water was brought to Earth on comets and meteorites is unknown. It has been suggested that the rate of extraterrestrial input of C, H, O and N is so high that all the Earth's near-surface water and carbon could be accounted for by cosmic influx (Deming, 1999), although this is not supported by much evidence. The depth and extent of the earliest oceans is not easy to determine, but an enlightening discussion (Nisbet, 1987, p. 330) pictures oceans deep enough to cover the ocean ridge systems, and of large volume by at least late Archaean times. The ocean ridges themselves may have come into existence as early as 3800 Ma (de Wit and Hynes, 1995).

The oldest materials on the Earth's crust (*c*.4200 Ma) are zircon grains eroded from the early crust and deposited in a marine clastic sediment, which itself was deposited at a later time of 3800 Ma (Compston and Pidgeon, 1986). The oldest *rock*, in Canada (4000 Ma; McCall, 1996), is crustal gneiss, and the oldest *sedimentary* rock is slightly younger at *c*.3900 Ma in Antarctica (sandstones, shales and volcanics, now metamorphosed and called greenstones). Precambrian greenstones (Windley, 1995, pp. 345 ff.) are found on all continents and contain widespread water-lain sediments at this early stage of Earth history. They indicate a humid Earth surface with weathering and erosion, liquid water by *c*.3900 Ma, and maybe earlier (Figure 8.1). It is difficult to know whether the early water bodies were simply lakes on the cratons (early continents), or whether the Earth's oceans had developed by then. Most people's prejudice is that oceans had formed, and continents may have occupied only 10% of the Earth's crust up to 2600 Ma, with most Earth surface processes occurring in island-arc and oceanic environments

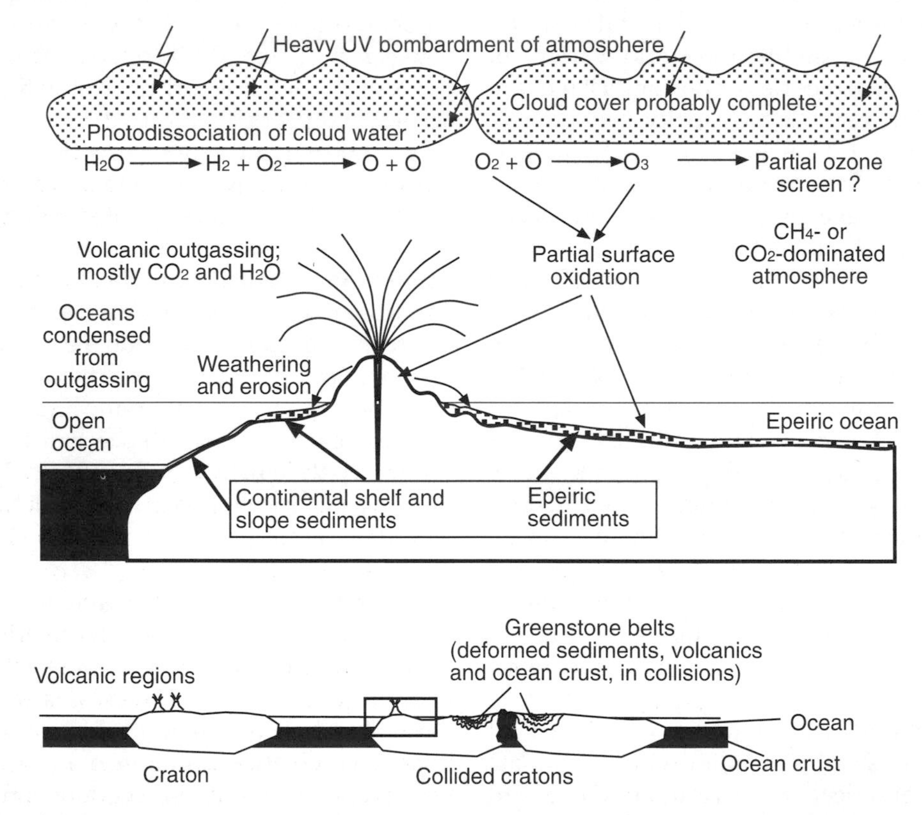

Figure 8.1 Cross-sectional view of early cratons, showing the outgassing of atmosphere and oceans, setting of sedimentary deposits and the notion of an early ozone screen.

(Eriksson, 1995). The earliest eroded landscape is recognised by *c*.3500 Ma (Buick *et al*., 1995), and terrestrial weathering in the later Precambrian may have been assisted by microbial communities, suggesting life on land as early as 1200 Ma (Horodyski and Knauth, 1994). Early continents were small size, but grew larger over millions of years by addition of material through further cooling and continental collisions.

The Earth's interior was unstable during much of the Precambrian, because of major changes in the mantle. Around the time of the Archaean–Proterozoic boundary (2500 Ma), there is evidence of extensive loss of gases and vapour from the mantle, and a consequent large increase in the volume of oceans. Such a change presumably increased areas of continental shelf seas as sea level rose, with likely climatic consequences, and may well relate to the earliest glaciation, the Huronian, *c*.2300 Ma, although just *how* is not clear.

Only in the Late Precambrian has a broad picture of continental positions emerged. Active continental margins were scarce through the Precambrian, although intracratonic basins were common, suggesting a persistent supercontinent for much of the Proterozoic. Most continents seem to have become relatively stable, and many were fused to form a supercontinent (called Rodinia), but with Siberia, and east Russia as apparently separate entities (see Figure 8.7) (McKerrow *et al*., 1992). Earlier, there is no clear information about continent distributions. It is also worth noting that the similarity of Precambrian sedimentary deposits across all continents indicates global ocean processes operated. A summary of the features of Precambrian times is given in Figure 8.2.

8.3 IRON, OXYGEN AND LIFE IN THE OCEAN WAVES AND ELSEWHERE

A key feature of the early atmosphere and oceans is that they were low in oxygen, proved by the common presence in detrital sediments of eroded minerals which *cannot* form particles in oxygenated conditions (e.g. detrital pyrite and uraninite, which breakdown and disperse only when oxidised), and the low abundance of materials which *only* form in oxygenated conditions (iron oxides and gypsum); see Figure 8.2.

8.3.1 THEORIES OF LIFE'S ORIGINS AND THE ROLE OF OCEANS

Within the low-oxygen setting, life developed. Theories on the origin of life are diverse (Figure 8.3). The early Earth was presumed too hot for life to originate here, so the Swedish chemist Arrhenius suggested that life had an extraterrestrial origin, now considered unlikely because cosmic radiation outside the atmosphere is expected to have killed the unshielded biota that we all are. A harrowing account of the rapidity and extent of UV-induced skin damage (Fiennes, 1993) in the present Antarctic ozone hole emphasises the fragility of living matter under UV.

The most popular traditional idea for the origin of life is the Oparin–Haldane theory of chemical evolution, whereby high energies of perhaps UV (in an atmosphere lacking an ozone screen, see later) or lightning created organic matter (amino acids), which then developed into living cells by a presumably convoluted process which has not been satisfactorily demonstrated. Miller's famous experiments with CO_2, ammonia and methane produced amino acids, but it is a very large step from there to even the simplest bacteria. The Oparin–Haldane concept is most commonly considered to have an oceanic setting, with surface waters receiving energy for the vitalising reactions. New discoveries in the deep modern oceans reveal complex ecosystems associated with hydrothermal vents receiving mantle energy (de Ronde and Ebbeson, 1996); and energy may be derived from reactions involving seawater, submarine water and iron sulphides (Russell and Hall, 1997), so that the ecosystems can run without

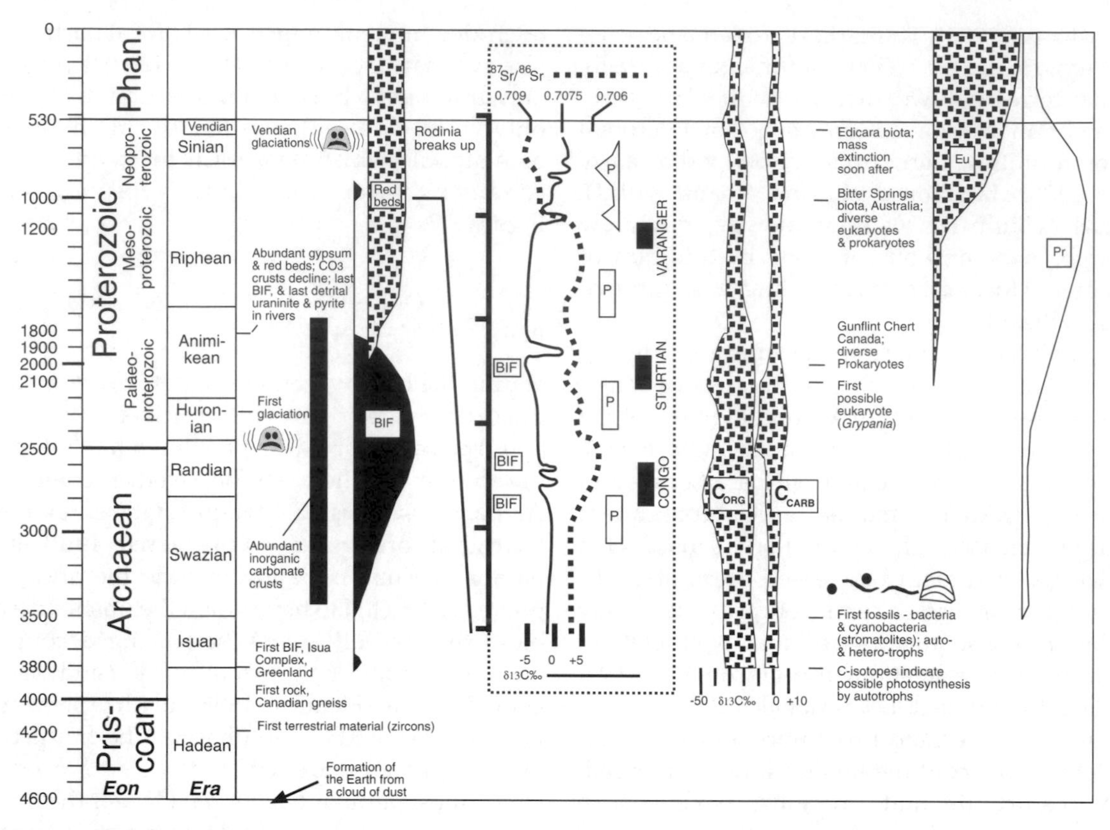

Figure 8.2. Table of events in the Precambrian, showing the main features. (Sources of information include Grotzinger and Kasting (1993), Grotzinger and Knoll (1995), Knoll (1994) and Schopf (1992a, 1994)). The various aspects shown are described in the text. The stratigraphic chart emphasises the relatively great importance of the Precambrian, in terms of its vast timescale. BIF = Banded Iron Formation; P = phosphorite deposits; Eu = eukaryote diversity; Pr = prokaryote diversity; Phan. = Phanerozoic Eon. Congo, Sturtian and Varanger are the three late Precambrian glaciations.

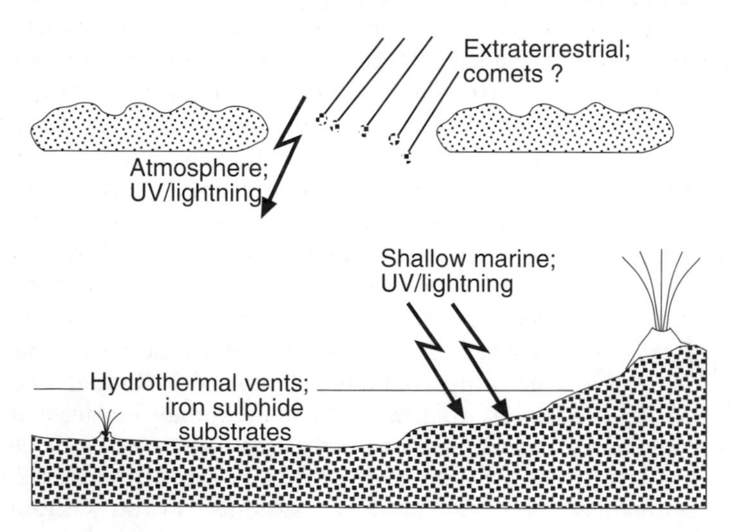

Figure 8.3 Rival settings for the origin of life. The traditional view of life originating in shallow seas contrasts with extraterrestrial, atmospheric and ocean floor settings, each of which has its supporting logical arguments; see text for discussion.

sunlight. It is important to realise that some modern bacteria can live in high temperatures (above 100°C: thermophilic bacteria), so that there is no empirical objection for life to survive the higher surface temperatures, and Russell and Hall (1997) proposed the origin of life at 4200 Ma, in a Hadean ocean of 90°C.

The Oparin–Haldane model interpreted the life-generating reactions as having occurred between compounds floating in the sea, but more recent prominence has been given to the need for reaction surfaces (called substrates) to promote the processes. This is because molecules floating in water are considered by some to be too diluted to promote the necessary reactions, whereas suitable substrates allow reactions to occur with the components in concentrated form. Iron sulphide substrates related to hydrothermal vents could have done the job (Russell *et al.*, 1990). Evidence of major lunar cratering at *c.*3800 Ma (the age of early rocks and a time when life may have existed) suggested that, if large meteorite impacts were still affecting the Earth at that time, then the heat release may have been sufficient to boil off much of the surface ocean water, and therefore that the safest place for life could have been the ocean floor (Holland and Kasting, 1992). Clays (referred to as replicating clays) may have provided suitable substrates (Cairns-Smith, 1985; Libes, 1992).

A novel idea is that life arose in the atmosphere, on dust particles, especially iron sulphide (Woese and Wächterhäuser, 1990), where the reactions could be concentrated on such surfaces, and used energy generated by inorganic surface reactions between iron and hydrogen sulphide gas, a volcanic product. Whether this idea can be reconciled to a UV-bathed atmosphere is debatable. Another aspect of the debate is whether the first organisms were autotrophic (self-feeding, see Chapter 6) or heterotrophic (not self-feeding, like humans are). Evidence of early evolution of photosynthesis comes from $\delta^{13}C$ in 3800 Ma sediments (Mojzsis *et al.*, 1996; Schidlowski, 1988), and ideas involving substrates promote the notion of an early evolution of autotrophs. Most authors, however, believe the first organisms to have been heterotrophs, because of the perceived greater complexity involved in developing autotrophic processes, especially photosynthesis.

Irrespective of the conflicting theories of life's origins, biotas certainly became major components of the oceans at an early stage. Despite suggestions of first life at 4200–4000 Ma (Kasting and Chang, 1992; Russell and Hall, 1997), the oldest fossils are 3500 Ma, with carbon isotopic evidence pushing this back possibly to 3800 Ma. Ocean biology was geared to anaerobic processes, and substrates such as iron sulphide are likely to have been widespread in such times. Bacteria and cyanobacteria (spheres and filaments) and layered domal structures called stromatolites (likely but not proven) to have been constructed by bacterial mats) dominated these early shallow waters; cyanobacteria photosynthesise and produce O_2. Supporters of the Gaia hypothesis suggest that the long timescale of interaction of organic and inorganic parameters of the Earth could lead to a cooperative system (Nisbet, 1995).

8.3.2 IRON AND LIFE

Sedimentary evidence shows that the early oceans were rich in dissolved iron. Modern anaerobic processes have low biological productivity rates, and recent work on modern ocean heterotrophic bacteria demonstrated not only that concentrations of iron in seawater affect productivity rates in bacteria, but this also then influences the primary productivity processes of phytoplankton in the water. This is because bacteria are substantial stores of oceanic carbon (Tortell *et al.*, 1996) and play a large part in oceanic carbon cycling. An interesting possibility, therefore, is that Precambrian oceans, with their high iron content, may well have had much higher productivity rates than their anaerobic photosynthesisers would indicate, and thus ocean productivity may have been substantial. Iron probably entered the Precambrian ocean by hydrothermal input (called exhalation) from the seafloor (Simonson, 1985), because of the association of iron formations with sediments of moderate water depth. Therefore, vast amounts of iron were apparently being pumped into the very place where it may have done the most good for life. Support for the notion of a highly productive biosphere comes from the fact that sedimentary rocks *throughout* geological time contain an average of 0.5% organic carbon, suggesting that even the primitive Precambrian ecosystems were a well-developed feature of the oceans (Francis and Dise, 1997, p. 34). Furthermore, the carbon isotope data, also *throughout* geological time, indicate a considerable degree of burial of organic carbon in the Earth's sedimentary shell from the time of the earliest organisms (see Figure 8.2).

The most popular opinion is that, over millions of years, the O_2 derived from photosynthesis pervaded the sea. Some O_2 presumably reached the atmosphere, although there is no certainty that the oceans and atmosphere allowed the same degree of interchange as they do now (see later discussion of the soda ocean theory). Oceanic O_2 converted reduced material into oxygenated versions. By far the most important reduced matter is iron (Fe), which is soluble in reducing conditions, but when oxidised is precipitated on the seafloor (mostly as Fe oxides in Banded Iron Formation (BIF) from 3800–*c.*1800 Ma); these ideas were promoted by Preston Cloud, a pioneer of such theories. At a critical point in Earth history (*c.*1900–1800 Ma), the seas' vast content of reduced material was exhausted, principally because the BIF stopped forming, which is a proxy to indicate that, from then on, the sea accumulated appreciable amounts of oxygen (which also diffused to the atmosphere). BIF returned briefly in the late Precambrian, which demonstrates the ability of the crust to inject reduced material into the oceans to soak up some O_2 at that time.

It could be argued that the main phase of BIF (2200–1800 Ma) was due to a major episode (lasting 400 million years!) of iron exhalation, although it remains unclear as to whether or not the Fe was pumped into the oceans throughout *all* this time: there may have been such a large initial reservoir of Fe in the oceans that a long time of O_2 production was required to deposit all the Fe. Nevertheless, from 1800 Ma on, the effect of increasing O_2 on ocean biology was profound, with rapid change from the essentially anaerobic bacterial ecosystems to aerobically dominated diverse assemblages (aerobic organisms are 30–50× metabolically more efficient than anaerobes, and have sexual reproduction, accelerating biotic evolution and diversity). It seems there was a mass extinction event (probably the first) associated with the switch from anaerobic to aerobic media, because certain key fossils disappeared at that time.

8.3.3 OXYGEN, OZONE AND THE OZONE SHIELD

At some point in Archaean history (the exact timing is contentious), the atmospheric ozone screen formed, improving the suitability of Earth's surface for life. The pioneering work of Berkner and Marshall demonstrated experimentally that UV will split water by photodissociation into hydrogen and oxygen, then will combine oxygen into ozone. Furthermore, these experiments demonstrated negative feedback: the ozone formed helped block the incoming UV and therefore limited photodissociation. So if we consider that the earliest atmosphere was rich in water vapour, then it follows that at least a partial shield existed then. The key issue is how much was needed for protection of early cells against UV. Well, that also depends on where life was! If life developed in water deeper than *c.*1 m, where most UV is absorbed, then the presence of an ozone layer would not be a prerequisite. If, however, the atmosphere-origin of life operated, then it makes sense that an effective UV barrier existed from near the beginning of life.

An oxygen level of *c*.1–3% of its Present Atmospheric Level (PAL) is generally considered necessary before the ozone screen can form, and that may not have been present until the earliest cells at *c*.3500 Ma (or 3800 Ma if the $\delta^{13}C$ data correctly identify photosynthesis then), but many workers consider this level was not achieved until late Precambrian times. Oxygen concentrations throughout the Precambrian are generally thought to be low, but recent $\delta^{13}C$ data suggest a strong rise in O_2 between 2200–2000 Ma (Kahru and Holland, 1996); the $\delta^{13}C$ information indicates that organic activity accelerated at this time, improving the O_2 output, of significance for the history of BIF.

8.3.4 COMPLICATIONS WITH PRECAMBRIAN IRON FORMATIONS, OXYGEN AND LIFE

The term Banded Iron Formation, found in all texts dealing with Precambrian sediments, actually refers to only one of two types of these iron-rich deposits. BIF is the finer-grained (mud-silt) and more common, with banding on mm to m thick scales in deposits which vary in thickness from 6–370 m. BIF is associated with the other type; that is, the coarser, presumed sandier, unbanded version, called Granular Iron Formation (GIF) (Simonson, 1996). In order to acknowledge this variation, we use the broader term Precambrian Iron Formation (PIF) where appropriate. Sedimentary sequence analysis demonstrates that PIF was deposited in transgressive settings, and sedimentological data from different continents indicate that PIFs were not shallow water deposits; they are not associated with shallow water sandstones, limestones, stromatolites or conglomerates. This implies an ocean stratification of iron-poor upper waters, with partial oxidation of surface waters due to photosynthesis, and iron-rich anoxic deeper waters. PIF was essentially a mid-depth phenomenon, because in some basins the stratigraphic record shows the time when PIFs ceased to deposit was during regressions, whereas in others it was during transgressions, but not in shallow water. Thus the setting of PIF is complex; the general conditions of PIF formation are shown in Figure 8.4 but in detail the controls are not clearly understood. Geochemical data have also been used to suggest that, while the majority of BIF is marine, small deposits of late Precambrian BIF formed in fresh water, in isolated basins (Dasgupta *et al.*, 1999).

Estimates of PIF deposition rate made by Simonson (1996) of 3 m/Ma produced PIF with a periodicity of *c*.20–125 Ma (consistent

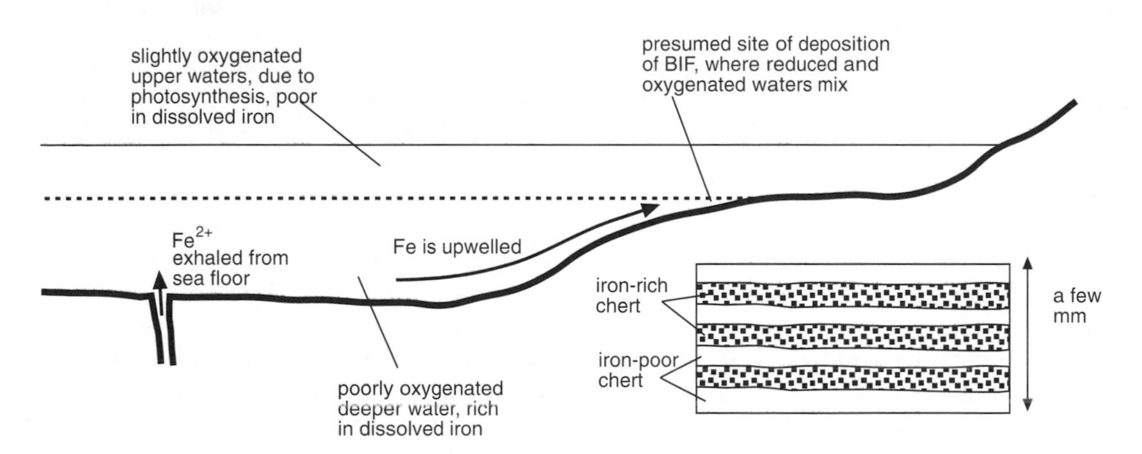

Figure 8.4 Generalised setting for the deposition of BIF. Note that the controls on formation of individual BIFs are complex, and the reader is directed to other texts for a more detailed treatment.

with second-order cyclicity, discussed in Chapter 9). Also, PIF occurs in phases which do not coincide with known glacial episodes (but see Section 8.5). Some small-scale BIF banding is traceable over large areas, suggesting cyclic oceanic processes, which could be forced by Milankovitch processes (if they existed then) or volcanic eruptions; iron is principally from submarine volcanic eruption and its rate of supply presumably critical. Nevertheless, there are several overall controls on PIF deposition noted as possible by Simonson (1996), including shifts in continental size and position, variation of input of Fe to oceans, change of ocean and atmosphere composition, and sea-level change, with all *its* controls. By no means can PIF formation be considered simple.

The main mass of PIF deposition was mid-Proterozoic (*c.*2200–1800 Ma), coinciding with the time of substantial rise in O_2 between 2200–2000 Ma, postulated above. The temptation to conclude that a large oxygen input accelerated removal of dissolved Fe content of the ocean, and exhausted its supply, leads to the possibility that, if the reduced Fe store of the oceans had not been there, then oxygen levels may have risen at an earlier time.

In fact, the history of oxygen is hard to determine. As noted earlier, the last major PIF at *c.*1800 Ma suggests that free oxygen was able to accumulate in the ocean–atmosphere system *only* from that time, but new discoveries draw this into question. Han and Runnegar (1992) Han and Runnegar (1992) recorded the abundant presence of the oldest possible eukaryote (filaments called *Grypania*) at 2100 Ma in Michigan, pushing back the evolution of eukaryotes *into* the time of abundant BIF. Worse still, newer data indicate an even earlier origin of eukaryotes (Brocks *et al.*, 1999). Modern eukaryotes need 1% PAL of oxygen for survival (the Pasteur limit), and their occurrence at 2100 Ma implies at least 1% by then, perhaps as much as 10% (Han and Runnegar, 1992). Before this discovery, altered steranes found at *c.*1800 Ma (Han and Runnegar, 1992) were chemicofossil evidence of the first eukaryotes, with eukaryote body fossils at 1750 Ma, although possible examples have also been reported as early as 2000 Ma (Schopf, 1992b). Riding (1992) stimulated thought by noting that, if such high levels of oxygen were present at that time, then the standard model of the PIF representing a major oxygen sink as photosynthesis accelerated may not be right. According to the chemical processes we describe, PIF should have formed in a low-oxygen environment, and the presence of eukaryotes at the same time potentially shakes the foundations of this long-standing idea. The view that PIF developed in a mildly oxidising atmosphere needs to be tested. Most problematic is that the Michigan *Grypania* is found *within* laminae of BIF, so that it is *not* possible to claim that the *Grypania* grew in shallow water as BIF formed in deeper oceans. This is a highly contentious issue, given the attraction of a model of stratification into shallow partially oxidised and deeper anoxic portions of the Archaean oceans; see Grotzinger and Knoll (1995) for a recent discussion. We must remember, however, that these new problems hinge on the accuracy of interpretation of *Grypania* as a eukaryote, well argued but not proved by Han and Runnegar (1992).

Late Precambrian animals of the Ediacara fauna (e.g. *Dickinsonia*; Runnegar, 1982) may have needed an O_2 level of 10% of present amounts, and there are suggestions that late Precambrian levels were as high as 18% (Canfield and Teste, 1996). If the 2100 Ma-old *Grypania* lived in 1% PAL O_2, then late Precambrian levels were presumably higher, but again, much depends on the interpretation of the nature of these late Precambrian creatures, whose biology may have been unique (Seilacher, 1992). Chapter 9 shows that oxygen also probably fluctuated within the Phanerozoic.

Oxygen is present because of photosynthesis, which draws CO_2 from the environment, thereby creating carbohydrates, a consequence of which is the separation of the

C and O components of CO_2. Therefore, an important feature of the development of an oxygenated Earth surface is that, in order for oxygen to exist freely in the oceans and atmosphere, there must be some mechanism to prevent it from being recombined with carbon; a valuable discussion of this is given by Francis and Dise (1997) and Skelton *et al.* (1997). The C portion must therefore be removed from the surface environments, achieved by burial, largely on continental shelves, where the most concentrated organic sediments form, but also in fluvial and deltaic systems (for example the Carboniferous coal deposits). Also important is the nature of newly rifted continents, which create basins as sites for accumulation of organic matter, for example the North Sea fracture zone, and other key oil sources. Approximately, for every molecule of oxygen free in the ocean–atmosphere system, there must be an atom of carbon buried in the Earth's crust (Chapters 4 and 6), a process which has been going on since the evolution of photosynthesis, possibly as early as 3800 Ma. Aspects of the history of oxygen are given in Figure 8.5.

A final note on BIF is the source of its component silica and iron. We have seen that evidence draws attention to the likelihood of iron being derived from exhalative processes to form primary precipitates on the seafloor; the same seems to apply to silica (Knoll and Simonson, 1981; Markun *et al.*, 1988; Simonson, 1987). The good preservation of microfossils in BIF cherts may be due to rapid silica cementation on the seafloor as an essentially

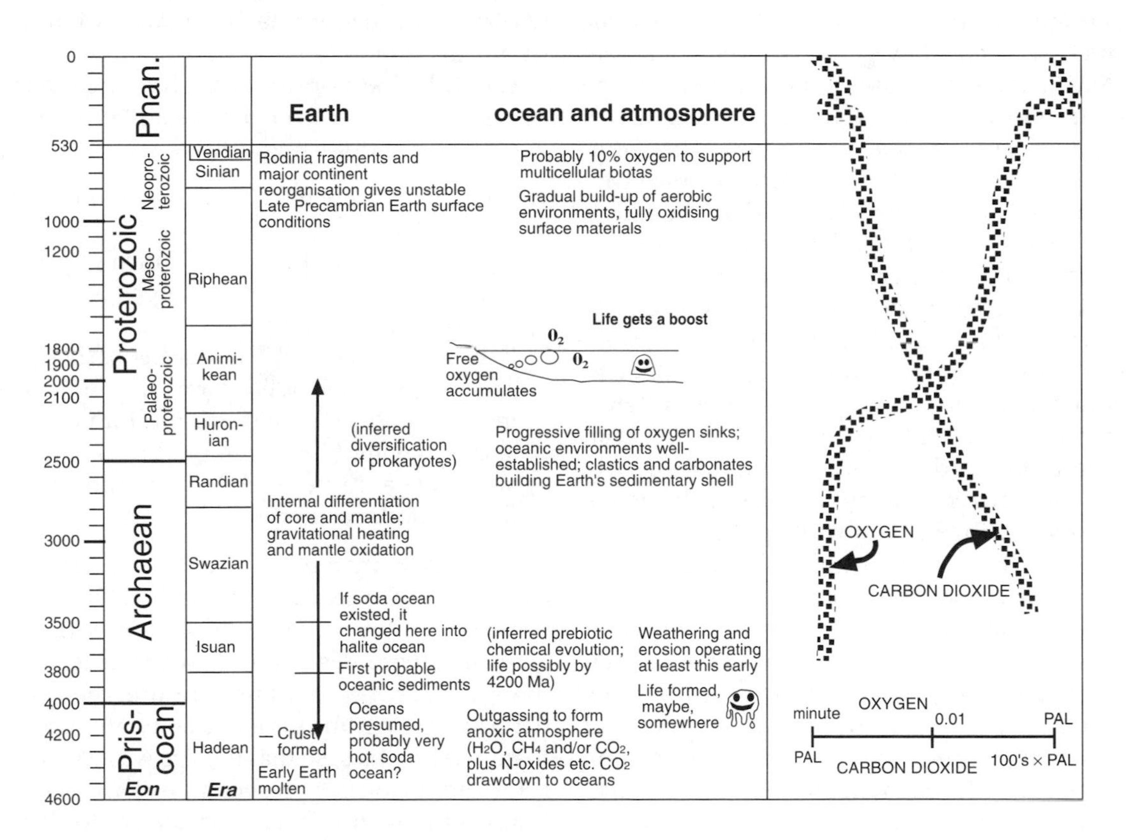

Figure 8.5 Interpretation of the major features of Precambrian ocean-related events, with some Phanerozoic changes for reference.

inorganic process (Simonson and Lanier, 1987). These points emphasise the complexity of BIF formation controls, and it is important to retain an open mind on the simple idea of the BIF as just an oxygen sink.

8.4 SEAWATER CARBONATES, EVAPORITES, CO_2 AND A WARM EARTH

Other features of the Precambrian sedimentary record reveal further complexities of Precambrian oceans, and enhance the potential for ocean studies present in the Precambrian record. (*Brain warning*: this section contains complex arguments and requires careful reading.)

8.4.1 THE FAINT YOUNG SUN PARADOX

Young stars begin as relatively cool objects, and heat up as they get older. General agreement that the Sun and planets formed at the same time implies that the young Earth received less heat than it does now. This led to the 'faint young Sun' paradox, reviewed by Crowley and North (1991, p. 238); the oldest sediments on Earth (3900 Ma) show its surface was humid, and therefore warm, but it should have been frozen under a young star. The early Earth's surface should have been 30°C cooler than present (Eyles and Young, 1994), with solar luminosity only 70% of its present level at 4600 Ma, rising to 82% by the end-Archaean (Grotzinger and Kasting, 1993), but it clearly was not frozen; something was keeping the surface warm, and unravelling this enigma is one of the most intriguing problems in Precambrian ocean studies.

The Earth's internal heat comes from radioactive decay and gravitational forces. Gravitational heating comes about by movement of material within the Earth, and a process of particular importance was the differentiation of the interior into mantle and core. This led to release of a lot of heat, but the timing is unclear, any time between the formation of the Earth and *c.*2000 Ma (Kasting *et al.*, 1993), and may well have been associated with a process of oxidation of the mantle, because the mantle constituents are rich in combined oxygen. Evidence from strontium isotopes indicates differentiation at *c.* 4000 Ma (see Section A.3.4). The amount of heat released also depends on the rate of differentiation, and there is so much uncertainty about this process that other mechanisms of maintenance of the Earth's heat are potentially equally, or more, important. Because there is general agreement that atmospheric and oceanic CO_2 levels were elevated because of outgassing, the ocean–atmosphere control mechanisms are likely to have played a part in keeping the Earth warm. The CO_2 content of these reservoirs is distributed between carbonate (the C_{CARB} reservoir) and organic carbon (the C_{ORG} reservoir) components of sedimentary systems, the only two types of matter which contain carbon in significant quantities.

The early Precambrian ocean has three characters which differ from almost all later times (Grotzinger and Kasting, 1993):

1. There is not much $CaCO_3$ deposition prior to *c.*3500 Ma.
2. Up to *c.*1800 Ma there is not much gypsum deposition in evaporite deposits of a shallow marine origin.
3. Between *c.*3500–1800 Ma marine carbonate precipitation included large amounts of inorganic crusts of high magnesium calcite and aragonite (now preserved as dolomite and calcite respectively), and after that, their abundance declined.

Remember that both $CaCO_3$ and organic carbon abundances are proxies for CO_2, so up to 3500 Ma, when little $CaCO_3$ was formed, *apparently* there was little CO_2 available to keep the early Earth warm. Furthermore, the C_{ORG} deposits during this period are not any larger than in later times, so that if there was a significant amount of CO_2 present in the Earth's surface environments, it was not drawn down to the organic sedimentary deposits, but instead was stored in the ocean–atmosphere system.

The key point here is whether CO_2 outgassing was delayed until *c*.3500 Ma (extremely unlikely), or whether some feature of ocean chemistry prevented $CaCO_3$ deposition until *c*.3500 Ma. So, if gravitational heat came later (as late as 2000 Ma as noted above), there was insufficient heat to keep the early Earth warm in its early history, and so oceanic controls may solve the faint young Sun problem. Also needed are explanations for the limited amount of $CaSO_4$ deposition up to *c*.1800 Ma. Such observations have thrown up three principal theories for the control of early Earth surface environments ranging from moderate to extreme, compared below.

8.4.2 NO CHANGE IN OCEAN COMPOSITION THROUGH TIME?

As outlined earlier, one view (Holland, 1984; Holland and Kasting, 1992) is that ocean composition was constant through time, based on the sequence of minerals precipitated from seawater. Any sedimentology textbook provides details of the proportions of the various minerals and their sequence of precipitation, as a given amount of seawater dries up, and the sequence of precipitation is: firstly $CaCO_3$, then $CaSO_4$ (as gypsum, $CaSO_4.2H_2O$, and anhydrite, $CaSO_4$) then NaCl, then more complex evaporites. Therefore older evaporite sediments should show evidence of this sequence. It so happens that $CaSO_4$ is missing in the earlier Precambrian record of evaporites, which supports alternative views that seawater composition has not been constant through *all* geological time. The next scenario (Section 8.4.3) is an extreme departure from the constancy notion, the third (Section 8.4.4) is more moderate. Section 8.4.5 adds information from land surfaces.

8.4.3 THE SODA OCEAN – AN EXTREME CHEMICAL STATE OF THE OCEANS

Although there are few definites in Precambrian oceanography, it is hard to accept that outgassing/extraterrestrial input of both water and CO_2 were delayed until 3500 Ma, because we are fairly sure from sedimentary evidence that the oceans had formed by 3900 Ma (probably earlier), and because CO_2 is such an important component of volcanic eruptions. It is also reasonable to presume that weathering processes were well developed, and that the oceans were accumulating elements such as Ca, Na, Mg, etc. If the paucity of early Archaean carbonate rocks was because the *already-outgassed* CO_2 was *not* precipitated as $CaCO_3$, then this strongly suggests that CO_2 was somehow stored entirely within the ocean–atmosphere system. This scenario supports the popular hypothesis that the Earth was kept warm by enhanced CO_2 levels in a greenhouse climate; otherwise a glacial record would be expected; the earliest glacial sediments are actually much later, *c*.2300 Ma. The alternative is that the Earth's *cloud* cover was greater, because water vapour is a greenhouse 'gas'. Nevertheless, if oceanic CO_2 was more abundant, with plenty of Ca (presumably weathered out of the hot crust), why is the *early* Archaean sedimentary record (3900–3500 Ma) not rich in carbonate rocks?

A novel solution to this conundrum has been offered recently by Kempe and Kazmierczak (1994) following earlier work by Kempe and Degens (1985), suggesting that the Precambrian ocean had a hugely different composition from the Phanerozoic and was highly alkaline, pH up to 10.5 in early Precambrian. Modern oceans are almost neutral, pH *c*.8.1, and remember that pH is a logarithmic scale, so the Precambrian ocean could have been more than 100 times as alkaline as modern oceans.

Today, oceans are dominated by NaCl, and have very little CO_2: most CO_2 is currently stored as limestones, C_{CARB}, the biggest carbon reservoir, with about 5× the mass of the C_{ORG} reservoir. Kempe and Kazmierczak suggested that the higher temperatures of the early crust caused rapid submarine weathering and transfer of elements to the oceans. Na is a very soluble element, so they propose that the oceans built up a very high Na content by weathering of komatiite crust, which made them alkaline, and under these conditions, the oceans became massively supersaturated with the components of $CaCO_3$. Both CO_2 and Cl derive from mantle outgassing (volcanic eruptions), but CO_2 is much greater in concentration than Cl in modern, and presumably ancient, eruptions. Furthermore, CO_2 is very soluble in alkaline conditions, and rapidly passes along the CO_2 buffer ($CO_2 \rightarrow H_2CO_3 \rightarrow HCO_3^- \rightarrow CO_3^{2-}$) described in Chapter 5; Kempe and Kazmierczak proposed that almost the entire oceans' carbon store could have existed in solution in such conditions. If the soda ocean theory is correct, then the Na_2CO_3 (sodium carbonate, soda) content of the Precambrian sea must have been higher than either Ca or Cl. The oceans may have been so alkaline that Ca could not exist in solution, so was excluded from the oceans; thus, no $CaCO_3$ could be precipitated

For this theory, Kempe and Kazmierczak draw on evidence from modern soda lakes in volcanic rifting areas (e.g. Indonesia) that contain stromatolites, which biogenically precipitate some $CaCO_3$; organisms that build the stromatolites are the only organisms that can live in the water. Kempe and Kazmierczak (1994) suggested that soda lakes represent a microcosm of the world's Precambrian oceans, and the soda ocean state existed until seafloor weathering processes (at mid-ocean ridges and subduction zones) were sufficiently developed to cycle substantial quantities of water between oceans and mantle. Na-rich waters were thus drawn down into the crust and top mantle, and Na was lost into minerals there, thereby reducing seawater pH. That permitted the production of limestones, because the alkalinity lessened sufficiently to start to allow CO_2 to be released from solution, by reaction of CO_3^{2-} with the Ca released from crustal rocks as pH fell. The soda ocean therefore declined as both Na and CO_2 were steadily removed; Na was drawn into the crust/upper mantle to albitise minerals (albite is a Na-rich feldspar), and CO_2 into limestone (via HCO_3^-). Note that Na is still a major part of the ocean today, because there is so much of it and it is very soluble. Thus, seawater was progressively changed to a combination of Na and Cl, switching to a NaCl (halite) ocean at a time they postulated as *c.*1000 Ma (but this date is unlikely, see later). The soda ocean idea indicates that subduction was minor or non-existent in the early Archaean, and therefore that plate tectonics may not have existed. Controversy exists regarding the nature, or even existence, of early Archaean plates (Nisbet, 1987, pp. 193–6).

The ideas of Kempe and Kazmierczak also have an impact on other aspects of early ocean character. Firstly, they provide an alternative explanation to the faint young Sun paradox. Because CO_2 is very soluble in alkaline conditions, the *atmosphere* may have had almost *no* CO_2 (it was drawn down to the oceans in the soda ocean theory) and depended on methane (CH_4) instead, produced by anaerobic fermentation processes, to keep the air temperature warm enough. Secondly, a methane-rich atmosphere has implications for the possible environment in which life might have evolved. Miller (1992) noted from experiments that a more strongly reducing atmosphere is a better environment for the generation of prebiotic organic compounds. If the soda ocean theory is correct, then the CO_2 could have been locked up in the oceans because of high alkalinity, with CH_4 as the major greenhouse gas; therefore the atmosphere may have remained strongly reducing as the ocean soaked up CO_2. Whether such a scenario then promotes the idea of an atmospheric origin of life, while the oceans beneath were a little more oxidising (CO_2 is more oxidised than CH_4), is open to conjecture, but certainly it would mean reassessment of the notion that the ocean–atmosphere system was truly interlinked in these early times.

$\delta^{13}C$ data (Figure 8.2) show the C_{ORG} reservoir was dominated through geological time by light isotopes, indicating burial of much organic matter. Light isotopes suggest organic matter burial because in photosynthesis, ^{12}C, the lighter of the two carbon isotopes (^{13}C and ^{12}C), is preferentially taken up into photosynthesising organisms. Between *c*.3000–2000 Ma, such isotopes were much lighter than would be expected from the burial of normal kerogens (kerogen is degraded organic matter found in sediment). Discussion of this feature (Schopf, 1994) led to the suggestion of abundant methanogenic bacteria (generating methane-rich organic matter which was buried) at these times, and emphasised poor oxygenation. Although this idea coincides with the timing of the soda ocean as envisaged by Kempe and Kazmierczak (1994), we show in Section 8.4.4 that the soda ocean must have ceased to exist by 3500 Ma, and the two phenomena are temporally separated.

Nevertheless, methanogenic bacteria may have played a part in maintaining warm surface temperatures, and indeed it has been suggested that these bacteria were a significant oxygen sink; calculations suggest that there was not enough iron in the Archaean to account for all the oxygen drawdown. BIFs contain a maximum of 30% Fe, and the entire Archaean sedimentary pile is only 5% Fe, see Francis and Dise (1997, p. 66) for a review. Figure 8.2 shows the very light $\delta^{13}C$ isotopes disappear at *c*.1800 Ma, the date generally agreed as the beginning of abundant free O_2 accumulation.

A twist to the scenario of highly alkaline oceans is given by Grotzinger and Kasting (1993) and Russell and Hall (1997) who claim the early oceans were *acidic* because of the high CO_2 content producing carbonic acid, with pH around 5.5. Now, this is not necessarily incompatible with the soda ocean theory, because the ocean could have feasibly started acidic (with a high CO_2 content), then shifted to alkaline as Na was weathered into the ocean from the land. In contrast, more recent modelling supports a pH of *c*.5.8 in the Hadean, rising to nearly neutral (6.8) by the late Hadean, with a succeeding carbonate chemistry similar to today (Morse and MacKenzie, 1998); that study also notes the possibility that seawater Ca levels were low until the late Precambrian, which seems to be incompatible with the abundant $CaCO_3$ crusts and stromatolitic carbonate present in earlier Precambrian time (see next section). Much depends on the nature and timing of outgassing, the timing of the formation of permanent oceans and the nature of submarine weathering, and, of course, to what extent the soda ocean theory can explain the observed features preserved in sedimentary rocks.

8.4.4 THE $CaCO_3$ AND $CaSO_4$ PROBLEM AND A COMPROMISE SCENARIO

How can we test the validity of a soda ocean theory, and can we reconcile the various data into a model for the Precambrian ocean? There are two areas of importance, as follows.

8.4.4.1 $CaCO_3$

Kempe and Kazmierczak's soda ocean hypothesis includes high alkalinity until *c*.1000 Ma, before which $CaCO_3$ should not have been precipitated. However, widespread inorganic carbonate crusts occur in shallow marine environments from 3500 Ma onwards (Grotzinger and Kasting, 1993; Grotzinger and Knoll, 1995), originally composed of crystal fans of aragonite and high magnesium calcite (now found recrystallised as calcite and dolomite). Crystal fans are evidence of inorganic precipitation on the seafloor. Thus it seems that we can exclude a global soda ocean from around 3500 Ma onwards, so if it did exist, the soda ocean cannot have dominated the seas after that time, contrary to Kempe and Kazmierczak's explanation. Nevertheless, the soda ocean may have played a part in the earlier Archaean ocean, and conceivably continued in deeper waters, while the oceans' surface was of normal pH and slightly oxidising, depending, of course, on the degree of interchange between shallow and deep waters. Remember that, although today we are very familiar with the ideas of vertical ocean-water mixing in the current cool climates, with thermohaline circulation (Chapter

3), such mixing seems to have been unlikely in the earlier Precambrian oceans, because only the surface ocean waters appear to have been oxygenated (Section 8.3.4). Nevertheless, because of loss by subduction (we have already noted that the oldest ocean crust is *c*.200 Ma), the oceanic record before that date relies on whatever is preserved in epicontinental sediment. The principal of deep ocean anoxia led to the recognition of oceanic anoxic events, and work using carbon and sulphur isotopes (Lambert and Donnelly, 1991, p. 87) suggests that deep ocean waters acted as partially closed systems, at least in the late Proterozoic. So it is *possible* that soda-type ocean conditions lay undetected beneath the halite ocean through much of the Precambrian. More evidence is needed!

8.4.4.2 $CaSO_4$

There is less gypsum in the Precambrian record than would be expected if seawater had been constant through time (Grotzinger and Kasting, 1993). Note that gypsum and halite are very soluble, so they are not normally found as original precipitates, but as replacements by more stable materials (and are therefore called pseudomorphs). Gypsum is rare until *c*.1800 Ma, when it becomes abundant at about the same time as the strong reduction in occurrence of massive carbonate crusts on the marine seafloor. The lack of gypsum in much of Precambrian time indicates that the seawater content of either Ca or SO_4^{2-} was low. Even if the soda ocean idea accounts for the lack of Ca until 3500 Ma, there was plenty of Ca available after that, so what limited SO_4^{2-} until 1800 Ma? There are two main contenders:

a) *Microbial sulphate reduction* is one of the most important features of anoxic ocean waters, and also occurs in the anoxic zone beneath the redox boundary in sediments deposited in aerated waters, including the present times. Sulphate is reduced to sulphide, mostly by a bacterium called *Desulfurovibrio* which uses the energy released (Chapters 4 and 5). The precipitation of pyrite is commonly attributed to combination of the highly reactive sulphide with ferrous iron. This is well known in the Black Sea, with pyrite apparently forming within the water column as well as in the sediment (Wilkin *et al*., 1997). So, is the lack of lower Precambrian gypsum due to sulphate reduction? Here we can draw on information from isotopes. The generation of sulphides leads to a biological separation of sulphur isotopes (called vital fractionation). Monster *et al*. (1979) showed that ^{34}S values from the Isua greenstones of Greenland are the same as in amphibolites and basalts of the area; so no environmental or organic fractionation of sulphur occurred, and therefore the sulphur in the greenstones was derived only from erosion of the amphibolites and basalts. However, there is a shift in sulphur isotopes in sedimentary rocks of 2500–2300 Ma indicating a significant rise in sulphate concentrations in seawater around that time (Lambert and Donnelly, 1992). Unpublished work by other Precambrian workers suggests that sulphate-reduction processes operated by this time, and published work shows that sulphate reduction was developed in the mid-Proterozoic (Brasier, 1995b, p. 147); nevertheless, it is not likely that sulphate reduction caused the lack of *early* Precambrian sulphate in sedimentary deposits.

b) *Low amounts of free oxygen* in the time before 1800 Ma is generally interpreted for the PIF data, and because there is remarkable coincidence between the latest PIF and the first abundant gypsum, the most parsimonious option for low SO_4^{2-} levels is that the ocean oxygenation was too low. However, this is not compatible with the evidence of higher levels of oxygen in earlier times, discussed earlier in relation to PIF (Section 8.3.4), and remains unanswered.

From the above discussion, it seems that a compromise between the various views is possible. Thus the Precambrian ocean composition was not constant; a soda ocean is possible up to 3500 Ma, at which time salinity reached 'modern' levels, and a low sulphate concentration seems likely until 1800 Ma. One possible model for Precambrian ocean change is given in Figure 8.6.

8.4.5 EVIDENCE FROM TERRESTRIAL ENVIRONMENTS

Because weathering of terrestrial surfaces created detrital materials from 3900 Ma, such surfaces may provide information about the atmosphere, and therefore aid ocean studies. Most data on soil profiles (palaeosols) indicate that before *c.*2000 Ma, palaeosols were reduced, and after then, they were oxidised (Schopf, 1992a). Furthermore, palaeosols of 2750–2200 Ma (Rye *et al.*, 1995) contain iron compounds precipitated from terrestrial pore waters flushed through the soil. In several examples from different continents, calculations indicate that CO_2 levels were *c.*100× more than the amount present in the modern atmosphere, but significantly, *c.*5× *less than* that needed to compensate for lower solar luminosity at that time; this indicates that the models of compensation by CO_2 keeping the Earth warm under lowered solar power are probably not correct, and provides a further boost to the idea of CH_4 as a factor in keeping the Earth from freezing, in keeping with the $\delta^{13}C$ data discussed earlier.

Controversy remains, however, with terrestrial studies; recent work using Fe and Ti concentrations in palaeosol deposits suggests that, between 3000–2200 Ma, the atmosphere had as much as 1.5% PAL O_2, i.e. probably not much different from its level soon after the end of BIF (Ohmoto, 1996). Such evidence may be coupled with the discussion about eukaryotes (e.g. *Grypania*) and draws further into question the accepted views of the history of oxygenation of the atmosphere and oceans.

8.5 LATE PRECAMBRIAN EVENTS

A complex sequence of changes has emerged from multidisciplinary work on the Neoproterozoic, 1000–530 Ma (Knoll, 1994). This work involves data on the following (see Figure 8.2):

- diversification of eukaryotes;
- iron (in BIF);
- phosphorus (in phosporite deposits);
- three glaciations (named Congo, Sturtian and Varanger from oldest to youngest);
- strontium isotopes;
- $\delta^{13}C$ in C_{CARB};
- and finally, of key importance are the major tectonic changes associated with accelerated plate activity (starting *c.*850 Ma) which culminated in fragmentation of the late Precambrian supercontinent Rodinia. The tectonics may have driven the environmental changes.

8.5.1 STRONTIUM ISOTOPES AND TECTONIC CHANGE

Strontium has two major isotopes, ^{87}Sr and ^{86}Sr, and is present in sedimentary rocks, derived from two major sources: the mantle via eruptions, and continents via erosion. ^{87}Sr is enriched if the source rock has been in existence for millions of years, because ^{87}Sr is a breakdown product of ^{87}Rb (present in igneous rocks such as granites), and builds up over a period of time. So, if the ratio $^{87}Sr/^{86}Sr$ found in carbonate rocks on the seafloor is high, then a terrestrial source is suspected; if it is low, a hydrothermal source is more likely. Figure 8.2 shows a low $^{87}Sr/^{86}Sr$ ratio in the time before the Varanger ice age, consistent with a range of regional and global observations that plate activity (and therefore more volcanism) was developing in the early stages of the breakup of Rodinia. In contrast, the sharp $^{87}Sr/^{86}Sr$ rise in the Vendian seems to relate to uplift (Knoll, 1994) preceding the break-up of Rodinia (Figure 8.7), although the youngest part of the curve is associated with

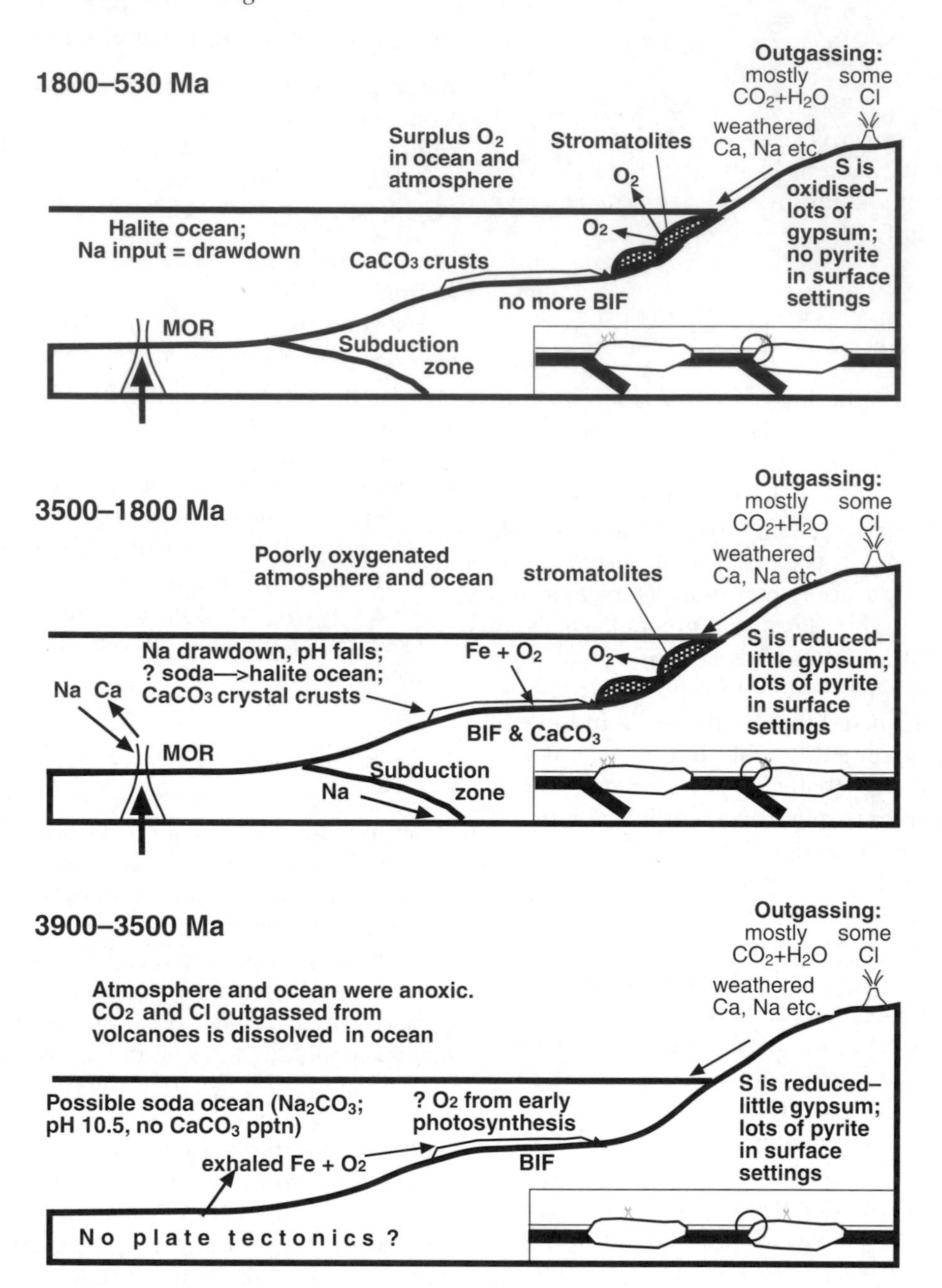

Figure 8.6 An attempt to reconcile the various lines of evidence for ocean states throughout the Precambrian. Lower: the soda ocean theory of Kempe and Kazmierczak (1994), prior to precipitation of major carbonate deposits. Middle: accounts for the formation of abundant carbonate crusts (Grotzinger and Kasting, 1993; Grotzinger and Knoll, 1995), indicating that, if the soda ocean existed, it will have switched to a halite one by then. Upper: highlights the increased SO_4 abundance in later Precambrian times. The arguments relating to these diagrams are discussed in the text. Insets: crustal cross-sections indicating plate tectonic activity; in each case a small circle shows location of the main diagram.

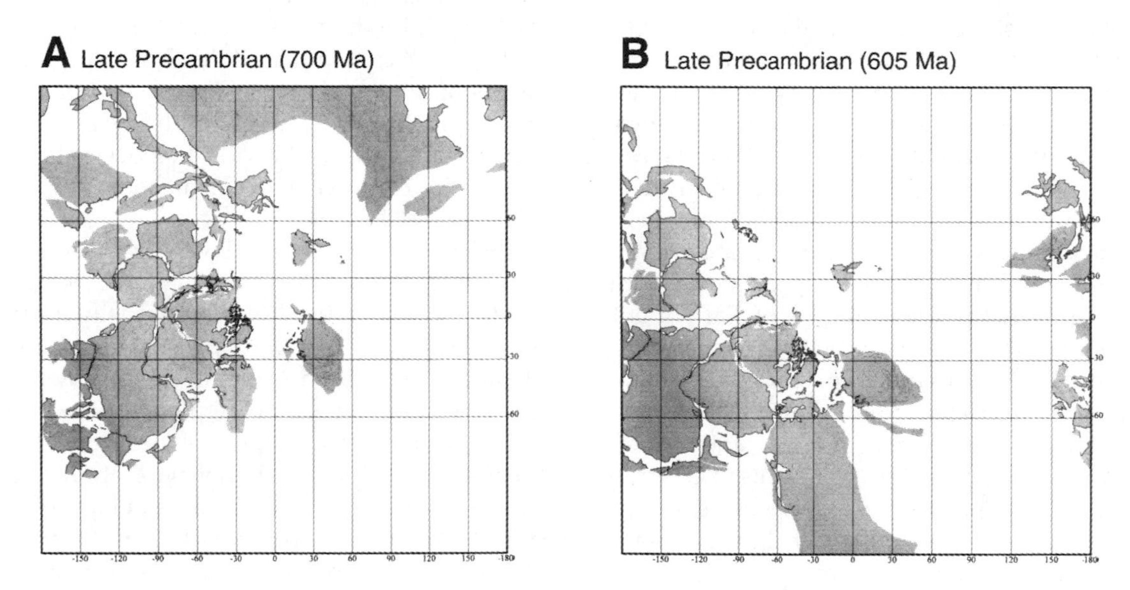

Figure 8.7 Maps showing the breakup of the late Precambrian supercontinent (Rodinia). In (A) note that Siberia (centre) is separate from other continents. Continents located near the poles are enormously exaggerated in these Mercator projections. Reproduced courtesy of Cambridge Paleomaps Ltd.

Early Cambrian rifting and presumably increased volcanism. That part of the curve ought to show that erosion was higher, but active rifting tends to lead to higher sea levels, and so reduced erosion. Therefore the curve there departs from the standard interpretation, and is more difficult to explain. In fact, revision of the Early Cambrian $^{87}Sr/^{86}Sr$ curve (Nicholas, 1996) shows that, although there is a rise in the earliest Cambrian, this is followed by a rapid fall, more in line with increased hydrothermal input of ^{86}Sr into the oceans (Sr ratios are discussed more in Chapter 9). Associated with the developing plate activity is the increase in phosphorite and iron deposits, both from exhalative sources.

Dips in the $^{87}Sr/^{86}Sr$ curve between 850–600 Ma broadly correlate to the Fe and P deposits (from volcanic sources), although the final sharp rise of $^{87}Sr/^{86}Sr$ occurs as the largest phosphorite is produced, which again seems to contradict the trend; the solution to this may be that the phosphorite has a different source, but also indicates that the controls on Sr isotopes are probably more complex than has been traditionally accepted.

8.5.2 BIF AND GLACIATION

Major positive swings in the $\delta^{13}C$ record coincide with the glaciations, and along with the first two (Congo and Sturtian glaciations), there is a resurgence in BIF (Figure 8.2). The early Proterozoic BIF disappeared *c.*1800 Ma, as we have noted, but briefly reappeared in the late Precambrian, *c.*900–750 Ma. What initiated the glaciations is a difficult problem, and is of great interest because the Varanger glaciation may have been the most intense of Earth's entire geological record. Knoll (1994) suggested that the increased erosion from land associated with unstable tectonics drew organic matter into the oceans, where it was buried, and therefore starved the surface environments of CO_2 necessary for maintaining warmer conditions. Hoffman (1999) suggested that the break-up of the late Precambrian supercontinent Rodinia led to increased areas of available continental margin, stimulating burial of organic carbon, which caused the positive swings in $\delta^{13}C$; such large-scale drawdown of CO_2 to the seafloor stimulated the glaciations. The term 'snowball Earth' has been applied to these late Precambrian ice ages (Hofmann *et al.*, 1998).

It is possible that the late Precambrian oceans were stratified. This period may have been dominated by deep-ocean stagnation and stratification related to the build-up of dissolved Fe exhaled from the seafloor and a series of oceanic anoxic events (Grotzinger and Knoll, 1995; Lambert and Donnelly, 1991). Episodes of turnover, oxygenating the deep water, ended phases of BIF, but by the Varanger ice age times it is presumed that the exhalative supply of Fe had also ceased. There is a curious relationship between glacial tills and thick, mainly inorganic, 'cap carbonates' (pure carbonates, made of laminated dolomitic microspar) deposited on top of the tillites. The cap carbonates could have been formed due to inter-glacial warming and/or overturn of a stratified ocean, drawing up alkaline anoxic waters (rich in the CO_2 drawn down, as mentioned above) to the surface, resulting in massive carbonate precipitation. Grotzinger and Knoll (1995) recorded $\delta^{13}C$ data for these beds, in addition to the information provided in Figure 8.2; non-glacial $\delta^{13}C$ C_{CARB} gives values of +8 to +11‰, while cap carbonates are −2 to 26‰. The collapse of the ice age and deposition of cap carbonates has been attributed to large-scale volcanic outgassing (Hofmann *et al.*, 1998).

Enriched (positive) $\delta^{13}C$ levels are normally associated with increased surface ocean productivity, because ^{12}C is more easily taken up (in the form of CO_2) by photosynthesising plankton and buried in the ocean sediments. Conversely, an increase of depleted (negative) $\delta^{13}C$ indicates a collapse of productivity, because the lighter ^{12}C is not being removed from the seawater. The $\delta^{13}C$ curve in Figure 8.2 shows two positive peaks within the Congo glaciation, and short-lived changes right at the end of the Sturtian and Varanger glaciations. The $\delta^{13}C$ fluctuation within the Congo is difficult to explain, but the sharp change at the end of the other two might reflect ocean overturn (Grotzinger and Knoll, 1995, p. 590) with bursts of productivity. What is curious, however, is that the changes occur at the very end of the latter two glaciations, while the $\delta^{13}C$ signal is similar during much of the glacial and non-glacial settings; it might be expected that more productivity would be characteristic of the glacial phases, because ocean stratification is more likely to break down during cooling. These results show the complexity of carbon isotope interpretation, and we emphasise that use of such proxy evidence is always subject to uncertainty.

If the cap carbonates represent vertical ocean-water transport, then thermohaline (T-H) circulation may have existed in the late Precambrian. However, T-H circulation controls in the *modern* oceans are not well understood (discussed in Chapter 3) and the validity of these processes as controls of Precambrian events is unknown territory.

Glaciations do not end with the Varanger phase in the late Precambrian; recent results from west Africa and east South America reveal a glacial phase continuing into the early Cambrian, perhaps not surprising considering the polar location of those areas (Trompette, 1996).

A final curious feature of these late Precambrian times is the ecological change in plankton populations. A group of fossils, called acritarchs (probably algae, which are eukaryotes), exist as rounded structures, and are recognised as early eukaryotes in rocks of *c.*1900 Ma onwards. Although there is an increased diversity of eukaryotes in the late Precambrian to Phanerozoic times, the acritarchs showed a spectacular decline from a peak abundance and diversity at *c.*850 Ma to relative rarity by 675 Ma, and only a few reached the Cambrian (Schopf, 1992b). The reasons why they declined are unknown, but are a further illustration of the profound changes taking place in the later part of Precambrian time.

8.6 OCEAN CHANGE AT THE PRECAMBRIAN/CAMBRIAN BOUNDARY

Major evidence for ocean change during this critical phase of ocean history and organic evolution lies in carbonates, the prime subject of this section. The boundary is one of the most significant of geological time because it is associated with rapid evolution of shelly organisms, commonly referred to as the 'Cambrian Explosion of life', during which all the major groups of animal types (phyla) evolved, and has been under considerable scrutiny in recent years. There are three scenarios which potentially explain the observed changes in both the *physico-chemical* and the *biological* conditions at the Precambrian/Cambrian boundary (Knoll, 1994):

1. they have no interrelation: the evolution of organisms did not influence, nor was influenced by, environmental shifts;
2. evolution was facilitated by environmental controls;
3. sedimentary environments were governed by organic evolution.

It is argued in this section that the sum of the evidence strongly points to an environmentally driven change in biotic evolution, i.e. number 2.

8.6.1 Mg/Ca RATIOS AND THE ARAGONITE–CALCITE SEAS DEBATE

The appearance of phosphatic and calcareous shelly animal fossils, just before the Precambrian/Cambrian boundary, is well documented (Tucker, 1992), with phosphatic shells slightly pre-dating calcareous types. Also, for the first time in geological history, cyanobacteria produced calcareous structures at around the same time (Riding, 1982). Remember that cyanobacteria are photosynthesising bacteria which are presumed to be the key generators of atmospheric oxygen during the Precambrian. The changing nature of $CaCO_3$ deposition across the Precambrian–Cambrian boundary seems to be a key feature of sedimentary environments. Interpretation of this (Tucker, 1992) involves explanation of the large amounts of limestone, and especially dolomite, in marine settings in late Precambrian rocks, including Mg/Ca ratios, and takes its inspiration from the seminal work of Sandberg (1983).

The principle is that, when seawater Mg/Ca ratio is high, calcite is generally considered to be inhibited from forming due to high levels of dissolved Mg, and aragonite forms instead; aragonite and calcite are chemically the same ($CaCO_3$), but aragonite is not inhibited by high Mg levels. Both calcite and aragonite are capable of forming inorganic precipitates on the seafloor, and a common example is ooids (rounded objects 0.5–2.0 mm diameter formed by accretion of $CaCO_3$ layers onto a shell fragment or sand grain as it rolls around the shallow seafloor). Because calcite is stable, and aragonite is not, the presence of recrystallised ooids means that they were originally composed of aragonite, and therefore the seawater from which they formed had a high Mg/Ca ratio. Vendian rocks contain recrystallised ooids, and implies an original aragonitic composition. Also, well-preserved dolomitic ooids of the same age indicate primary dolomite, also proxies for a high Mg/Ca setting. Tucker (1992) and Tucker and Wright (1990) attributed enhanced Mg levels to an increased supply of Mg to oceans by terrestrial weathering, suppressing calcite precipitation in favour of aragonite and dolomite. As part of a larger-timescale theory of cyclic icehouse/greenhouse climates in relation to plate tectonics, which is addressed in Chapter 9, such weathering characteristics and high Mg relate to a state of suppressed plate activity which itself has long been regarded as a result of supercontinent formation. This is because, if plates are very active, Mg, weathered from land to ocean, is drawn out of the seawater by submarine weathering at mid-ocean ridges; when plates are inactive, the Mg builds up in the water. The late Precambrian continental fit, interpreted as nearly supercontinent status, as Rodinia (Knoll, 1994; McKerrow *et al.*, 1992), is an ideal setting for increased terrestrial weathering (Figure 8.7). Indeed the $^{87}Sr/^{86}Sr$ data suggest enhanced weathering at this time (Section 8.5).

This situation contrasts with the Early Cambrian when there was a shift in style of $CaCO_3$ deposition to calcitic characters. Increased plate activity as Rodinia rifted led to increased ocean ridge volcanism, which came into contact with seawater. In the same way that fresh basalt erupted on land begins to weather, so it does in the sea. The result is two-fold: (a) submarine removal of Mg from the oceans as the ocean-floor basalts change chemically; and (b) release of CO_2 from sedimentary rocks as they are consumed in subduction zones, the CO_2 making its way to the surface via magma, then volcanic eruptions. The net result is a reduction in the Mg/Ca ratio, and this is considered to promote calcite (rather than aragonite or dolomite) as a precipitate.

A burst of organic activity across the Precambrian/Cambrian boundary, with rapid appearance in the rock record of abundant fossils in the Lower Cambrian, is accompanied by a large positive $\delta^{13}C$ excursion (i.e. deviation from normal values) in carbonate rocks. We have seen how the late Precambrian $\delta^{13}C$ data are partly anomalous, because of their unusual trends in association with the glaciations and seafloor exhalative processes. In the earliest Cambrian, however, it is easier to apply the widely accepted logic to explain the $\delta^{13}C$ positive trend. We emphasise that such logic works on the principal that positive $\delta^{13}C$ peaks (preserved in the $CaCO_3$ of carbonate rocks) are proxies for high organic activity, and reflect the burst of eukaryotic plankton in the Early Cambrian (Knoll, 1994; Schopf, 1992b) suggesting enhanced organic activity. Also, the biggest injection of phosphate (the most critical of nutrients because it is in such short supply) into the ocean system occurred at that time (Figure 8.2); this was interpreted (Cook, 1992) as caused by redistribution of ocean circulation, involving upwelling on a global scale, and probably not related to the glaciations or exhalative phases which preceded it.

Recent data from Sr isotopes across the Precambrian/Cambrian boundary (Nicholas, 1996) add to the story. As outlined in Section 8.5.1, Sr isotopes (in particular $^{87}Sr/^{86}Sr$) are potentially useful for revealing the extent to which landmasses were being eroded, because the $^{87}Sr/^{86}Sr$ ratio increases with erosion. Although continents rifted in the Cambrian from the late Precambrian supercontinent (Rodinia), the southern continents were rapidly reorganised into a single landmass called Gondwana (Hoffmann, 1999). Gondwana broke up in the Mesozoic Era, several hundred Ma later. So, a decrease in $^{87}Sr/^{86}Sr$ in the earliest Cambrian sedimentary record may relate to a decrease in terrestrial input of Sr into the oceans during the reorganisation phase, but care is required in the interpretation of Sr in sediments, because the changes could have resulted from increased hydrothermal vent activity instead. Brasier (1992) used an increase in light $\delta^{13}C$ values from the Precambrian/Cambrian boundary in China, and a range of trace element and sedimentary data, to describe variations in nutrient enrichment of various settings in the Early Cambrian. The overall conclusion of warm climates, raised CO_2, extra nutrients, and high ocean temperatures, emphasises the backdrop to rapid changes in evolution of life in the Early Cambrian.

8.6.2 CALCIUM AND LIFE

Kempe and Kazmierczak (1994) claimed that the soda ocean idea had a part to play in the Precambrian/Cambrian boundary events. According to this idea, Na in oceans declined by subductional drawdown into the mantle, and the pH fall made conditions more suitable for diverse life to develop. During soda conditions, robust organisms such as bacteria and cyanobacteria might be expected to dominate, as in the modern soda lakes, while eukaryotic life was rudimentary. Therefore pH as well as oxygen was an important stimulus for evolutionary change. In this view, Ca would continue to accumulate dissolved in the ocean as the soda ocean weakened, and work on modern organisms reveal that, when Ca becomes concentrated, it is *toxic* because it disrupts cellular processes. Kempe and Kazmierczak (1994) suggest that the main reason for a sudden appearance in the geological record of abundant calcareous shelly organisms at the base of the Cambrian Period (*c*.530 Ma) is because Ca levels became critically high, and the organisms *had* to get rid of it, the safest way being to lock it up as shell material. However, although we have already seen that the soda ocean could not have existed younger than about 3500 Ma (and therefore changes in Ca concentration must have occurred at that earlier time), Ca toxicity warrants consideration as a feature of oceanic organic processes.

Although most fossil marine organisms have calcareous shells, biomineralisation is a complex process (Simkiss, 1989), and there are significant advantages of having a skeleton, for mobility and protection for example. However, because of the large amount of $CaCO_3$ already present as limestone, dissolved Ca levels were unlikely to have been at critical toxic levels, and therefore the reason for the rapid appearance of calcium carbonates and phosphates as skeletons is not calcium toxicity. Indeed, Grotzinger (1994) suggested that the amount of $CaCO_3$ removed by organisms from the early Cambrian seas was not significant; despite its name, the 'Cambrian Explosion' of life did not suddenly produce great quantities of organically precipitated $CaCO_3$. Also, it must be remembered that the proportion of calcified organisms in modern seas is low compared with unskeletonised creatures (and that as a result the fossil record is usually regarded as being poor). If Ca levels were so high in the early Cambrian seas, why is the precipitation of Ca compounds as skeletons not more common in the fossil record, and why, for that matter, did not calcification occur in prokaryotes of the Palaeoproterozoic and Mesoproterozoic when inorganic $CaCO_3$ was also common? Furthermore, Riding (1982) recorded the curious cessation of calcification of cyanophytes in the Mesozoic, yet there was plenty of Ca available. The control processes of calcification are consequently much more complicated, and unlikely to be solved in simple models.

A final point about the late Precambrian events arises from recent work on the ages of the Ediacara fauna and the overlying shelly Lower Cambrian fossils. Grotzinger *et al.* (1995) noted that improved stratigraphy has revealed that the previously supposed and much-published temporal separation of the Ediacara and shelly fossils is not completely true, and there is some overlap between these two groups. If this is right, then it implies there was no mass extinction event at the end of the Precambrian, and that the sequence simply reflects evolutionary change with no major disruption.

8.7 CONCLUSION

The (sometimes horrendously complex) discussions of this chapter draws attention to questions regarding the notion of constancy of composition of the oceans over geological time. Certainties are minimal, but there seems to have been a *unidirectional* shift in composition with respect to oxygen, which may be extended to include the Fe, Ca, SO_4^{2-}, CO_2 and Na balances of the ocean, but it is unwise to be too firm about the sequence of events. Whatever the geological history of the Precambrian ocean, by the early Cambrian there was a considerable degree of stabilisation of major aspects of ocean chemistry, and subsequent drastic compositional changes are largely limited to relatively sudden influences (generally coinciding with mass extinctions), with a return to stability afterwards. Whether or not the soda ocean theory is fully established (or rejected) after it is considered by geologists over the coming years, some sort of CO_2-related shift seems likely to have taken place. Although the Phanerozoic has aspects of continued unidirectional change, the changes associated with shifts in the Mg/Ca ratio, introduced here, have been interpreted as *cyclic* in the Phanerozoic, and are considered in the next chapter.

Everyone looks for some anchor in their lives, and geoscientists are only human. We all want nice neat theories to explain the past. Unfortunately, with the Precambrian, such anchors, as the 1800Ma change in Earth surface oxidation state are questionable. The sooner you get used to this, the more useful a scientist you will become, and not excessively constrained by traditional wisdom; above all, concentrate on the evidence.

8.8 SUMMARY

1. The Precambrian oceans formed as soon as the Earth cooled sufficiently, probably by 3900 Ma, and life may have formed as early as 4200 Ma in seas as hot as 90°C. The large amount of outgassed CO_2 is presumed to have dissolved in the oceans, although whether that made the oceans acid or alkaline depends on the validity of the soda ocean idea. In that theory, the hot crust yielded Na to the ocean to increase its pH to *c.*10.5, maintained until the plate tectonic drawdown of Na via subduction reduced the pH. Objections to this idea are based on the presence of large-scale inorganic carbonate crusts from 3500 Ma onwards, which would not be possible under the high pH soda ocean. A compromise is possible, that the soda ocean existed until 3500 Ma, when the world's seas became the familiar halite type.
2. Although photosynthesis appears to have evolved early on, the sea remained poorly oxygenated until the last main BIF at *c.*1800 Ma, and was oxygenated from then on. However, a simple coupled BIF–oxygen control is not as clear-cut as has been assumed, because of the evolution of *interpreted* eukaryotes at 2100 Ma, during the main BIF phase.
3. The paucity of gypsum in marine evaporite deposits until 1800 Ma is most appropriately attributed to insufficient oxygen, with the sulphur being converted to pyrite instead. Microbial sulphate reduction, so

pervasive in later ocean processes, seems not to have evolved until after 1800 Ma.

4. Late Precambrian changes were profound, involving the Ediacara fauna, three glaciations and a supposed mass extinction, at a time when the Earth's continents were fused into mostly one mass. All this precedes the Cambrian 'explosion' and significant shifts in the nature of the oceans; these seem to include calcification of organisms, and massive injection of phosphorus into the oceans in the Precambrian/Cambrian boundary interval. Much emphasis has been placed on the role of Mg/Ca ratios in these changes, the ratio supposedly related to ocean processes, plate tectonics and weathering. The fragmentation of Rodinia around that time may be related to these events.

TRENDS IN PHANEROZOIC OCEANS 9

9.1 INTRODUCTION: PROGRESSIVE, CYCLIC AND RHYTHMIC CHANGE GOVERNS THE PHANEROZOIC OCEANS

Chapter 8 examined the profound and unidirectional change in ocean systems throughout the first 89% of the Earth's history. This chapter explores the main features of change in the remaining 11% (from Cambrian onwards, the Phanerozoic Eon), and reveals processes at several scales and due to several causes. The key aspect is that, although unidirectional change remains an element of Phanerozoic oceans, the ocean system was much more stable, with major element concentrations being constant (they have conservative behaviour, see Chapter 4). However, although salinity seems to have remained stable, there are examples of large-scale regional variation, such as in the Messinian salinity crisis of the Miocene Epoch, 6 Ma (Chapter 10). A persistent Phanerozoic feature is cyclic processes, which also affected ocean chemical change. The controlling factors operated at a variety of scales. Large-scale cyclicity relates principally to plate tectonic activity, with global or regional effects, and leaving sedimentary deposits several km thick in places where there were rapidly subsiding basins. At the other end of the spectrum, the smallest scale of cm-thick bedding can be related to local sedimentary dynamics. Paradoxically, however, thin beds deposited over large areas may be controlled by extraterrestrial forcing of variations in planetary motion, affecting climate; so, some of the smallest scale deposits may be caused by the largest scale processes affecting the oceans. Cyclicity scales form a hierarchy, referred to as first order, second order and so on; up to 10 orders have been recognised in some places. Although the concept of cycles has been around for years, the last 20 years have seen great leaps in our understanding of Phanerozoic change, much of which is palaeoceanographic. The information presented and discussed in this chapter therefore draws on understanding of modern ocean processes, and shows the contrasts between the modern and ancient oceans.

9.2 BACKGROUND TO THE PHANEROZOIC OCEANS

9.2.1 NATURE OF THE EVIDENCE

In general, the study of Phanerozoic ocean evolution has fewer obstacles than the Precambrian, because:

1. shelly fossils not only assist palaeogeographic and palaeoceanographic reconstructions to a greater extent than in the Precambrian, but also provide well-developed and fine scale dating and correlations, so that reconstructions of ocean ecosystems and processes are more detailed;
2. the rocks are younger, less destroyed by tectonism;
3. Precambrian continents were smaller, but Phanerozoic deposits are more extensive and better preserved than Precambrian, so there is more material to examine.

9.2.2 PHANEROZOIC CRUSTAL EVOLUTION AND ITS OCEANOGRAPHIC IMPLICATIONS

Phanerozoic geology has been dominated over the last 30 years by plate tectonic theory, and probably the most important oceanographic character is the continuous major continental (and therefore ocean basin) reorganisation. Because plate theory is covered in a wide range of Earth science texts, we stress here only the oceanographic and climatic impacts of plate processes and continent distribution through time. Crustal controls on the oceans are summarised in Figure 9.1. The history of plate movements (Figure 9.2) resulted in major temporal trends in ocean development, summarised as follows.

1. During much of the Proterozoic (Late Precambrian) a supercontinent (Rodinia) existed (Figure 8.7), which broke up in the very late Precambrian to Early Cambrian, part of which then reorganised into the southern continent of Gondwana. Evidence from South China has been used to attribute Rodinia's break-up to a mantle plume rising to the crust (Li *et al.*, 1999). The early Palaeozoic oceans are therefore typified by rifting and sea-level rise, and are generally represented by higher temperatures, except for a glaciation in the Late Ordovician Period (Chapter 10). The early Palaeozoic Era also includes important episodes in evolution of life, including the 'Cambrian Explosion', life on land and several extinction events. Section 8.6.1 describes how active plate tectonics causes enhanced

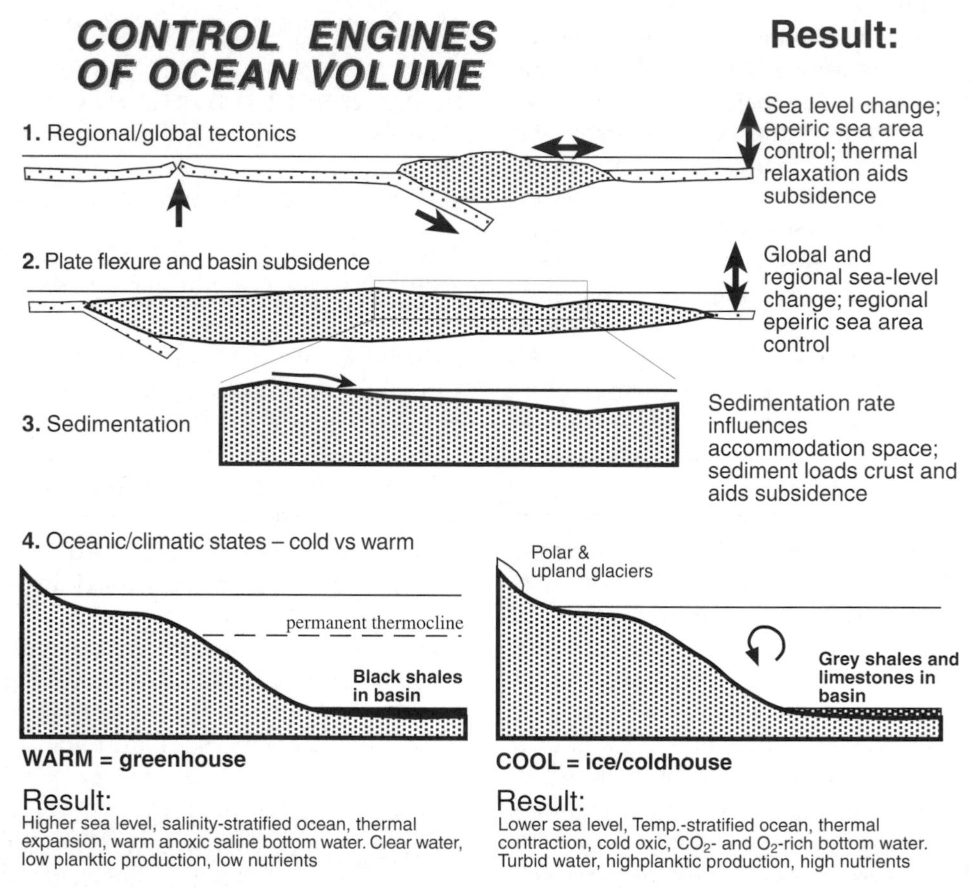

Figure 9.1 Crustal controls on ocean systems showing global/regional tectonics, plate flexure, basin subsidence, sediment loading, greenhouse/icehouse features and interchange of crust and ocean water.

weathering of sea-floor basalt, removing Mg from seawater and encouraging calcite deposition in inorganic precipitates. This early part of the Phanerozoic Eon is typified by such deposits, leading to the term 'calcite seas' described later.

2. In the latter half of the Palaeozoic Era, continents underwent a gradual reassembly, accomplished by continental collision to form fold belts, called orogens (Greek *oros* = mountain, *genesis* = production). The two principal orogens are the Caledonian–Appalachian (late Silurian–Early Devonian), and the Variscan (Devonian–Late Carboniferous). The resulting well-known supercontinent, Pangaea (Greek for all-Earth), throughout the Permian and Triassic, was accompanied by a single ocean, Panthalassa (all-sea), which therefore occupied the rest of the Earth's surface. Pangaea has two recognisable continental masses, Laurasia in the north, and Gondwana in the south (Figure 9.2). A large wedge-shaped sea, Proto-Tethys, developed in Pangaea's eastern margin, and a long-term glaciation occupied much of Gondwana in the Carboniferous and Lower Permian Periods. Sea level fell considerably during this time, not just because of glaciation, but also mid-ocean ridge systems became inactive and the ocean floor subsided. Because part of this period involved an extensive phase of plate inactivity, aragonite predominated over calcite in inorganic precipitates, leading to the term 'aragonite seas'.
3. Pangaea began to break up into the presently defined continental masses in the early Mesozoic, a major feature being Tethys Sea (in Greek mythology, Tethys was the wife of Oceanus; he was the god of the river that encircled the world's land, equivalent in our context to Panthalassa). Pangaea fragmented because of reactivated mid-ocean ridge activity after a period of tectonic quiescence, and led to high sea levels, calcite seas, and warm climates once more, throughout the Mesozoic. The reasons why the plates became active again is unclear, but recent ideas suggest that mantle plumes play a large part in the history of plate movement. Towards the end of the Cretaceous Period, sea level fell, possibly related to the third major orogen, the Alpine, and the ocean returned to an aragonite-sea state, still continuing. Tethys Sea mostly closed at this time, and now exists as a remnant (Mediterranean Sea) because of the collisions between India and Asia, and between the African plate and parts of the European continent. Present sea levels are relatively much lower than in most previous times and, depending on how accurate the current sea-level curve is, are possibly as low as they were during Pangaea times. Earth-surface temperatures are also lower today than in most of Earth history due to the Quaternary glaciation, a feature of considerable importance in understanding ocean circulation processes in present and past times, especially because glaciations are unusual in geological history.

The repeated process of break-up and formation of a supercontinent was recognised by J. T. Wilson in the 1960s, and this notion is in common usage as the Wilson Cycle, accompanied by records of oceanic patterns. It forms the first order of cycles, such that one cycle occupies the entire Palaeozoic Era (from break-up of Rodinia to the end-Palaeozoic continental reassembly as Pangaea), and the Mesozoic and Cainozoic Eras occupy part of another (Figure 9.2). On this scale is a major interpreted fluctuation of ocean chemistry, called the Sandberg curve (Figure 9.3), relating to the mineralogy of inorganically precipitated $CaCO_3$ materials such as ooids. This was introduced in Section 8.6.1 and discussed in this chapter. Second order changes are on the timescale of geological periods, again strongly relating to tectonic change, and emphasise the essential interlink of ocean and crust.

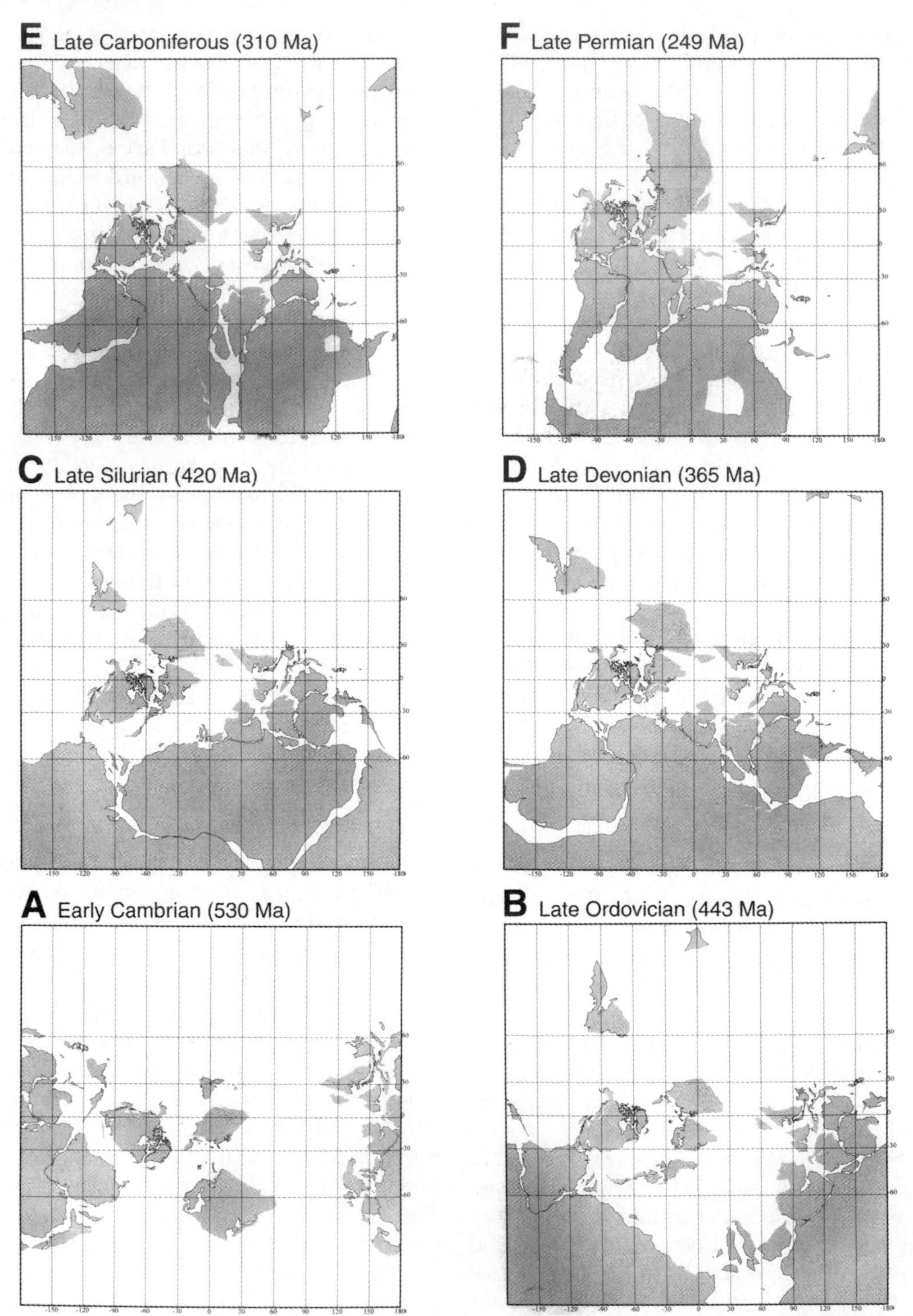

Figure 9.2 Diagram of plate configurations through Phanerozoic time. Reproduced courtesy of Cambridge Paleomaps Ltd.

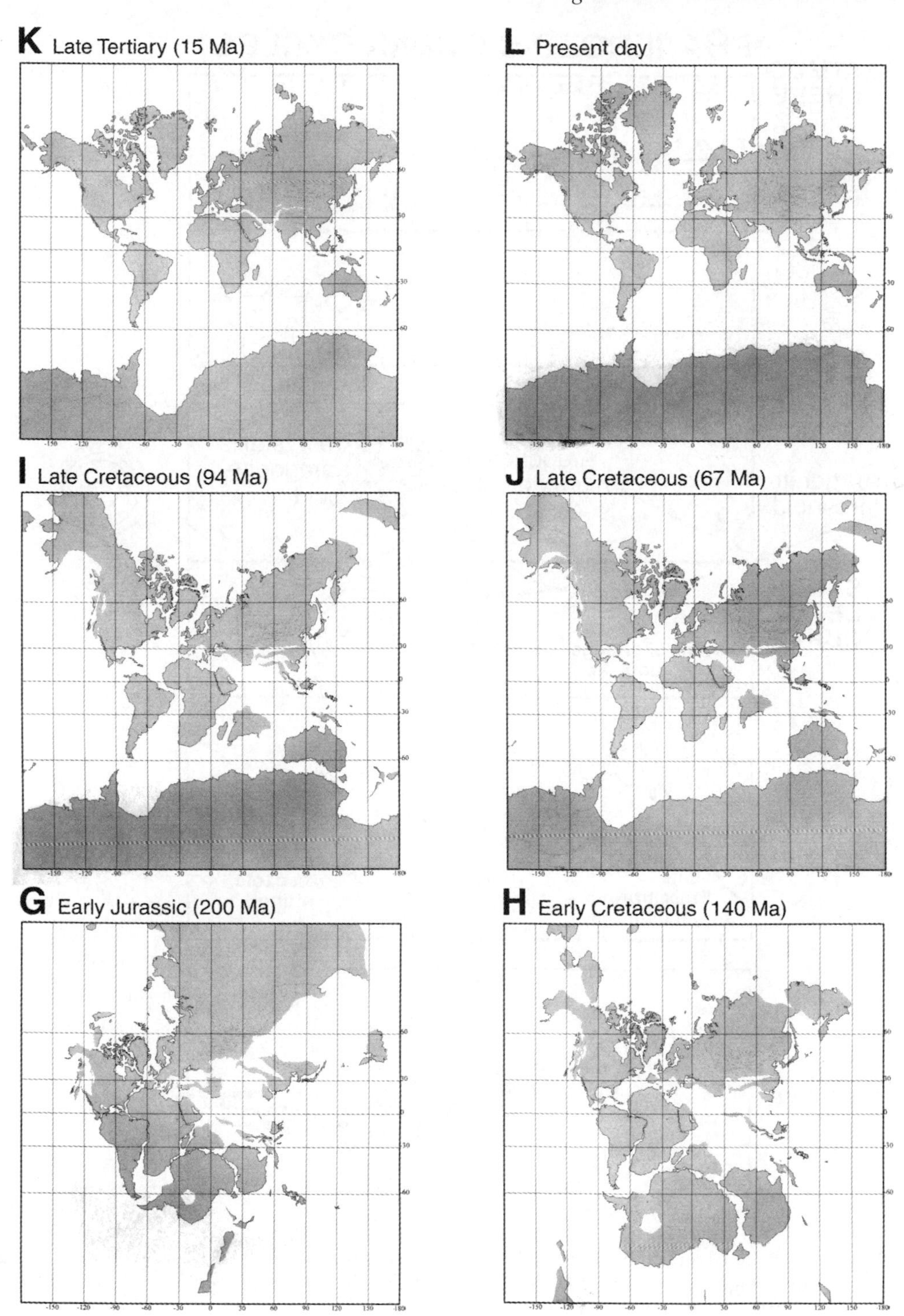

Figure 9.2 *(continued)*

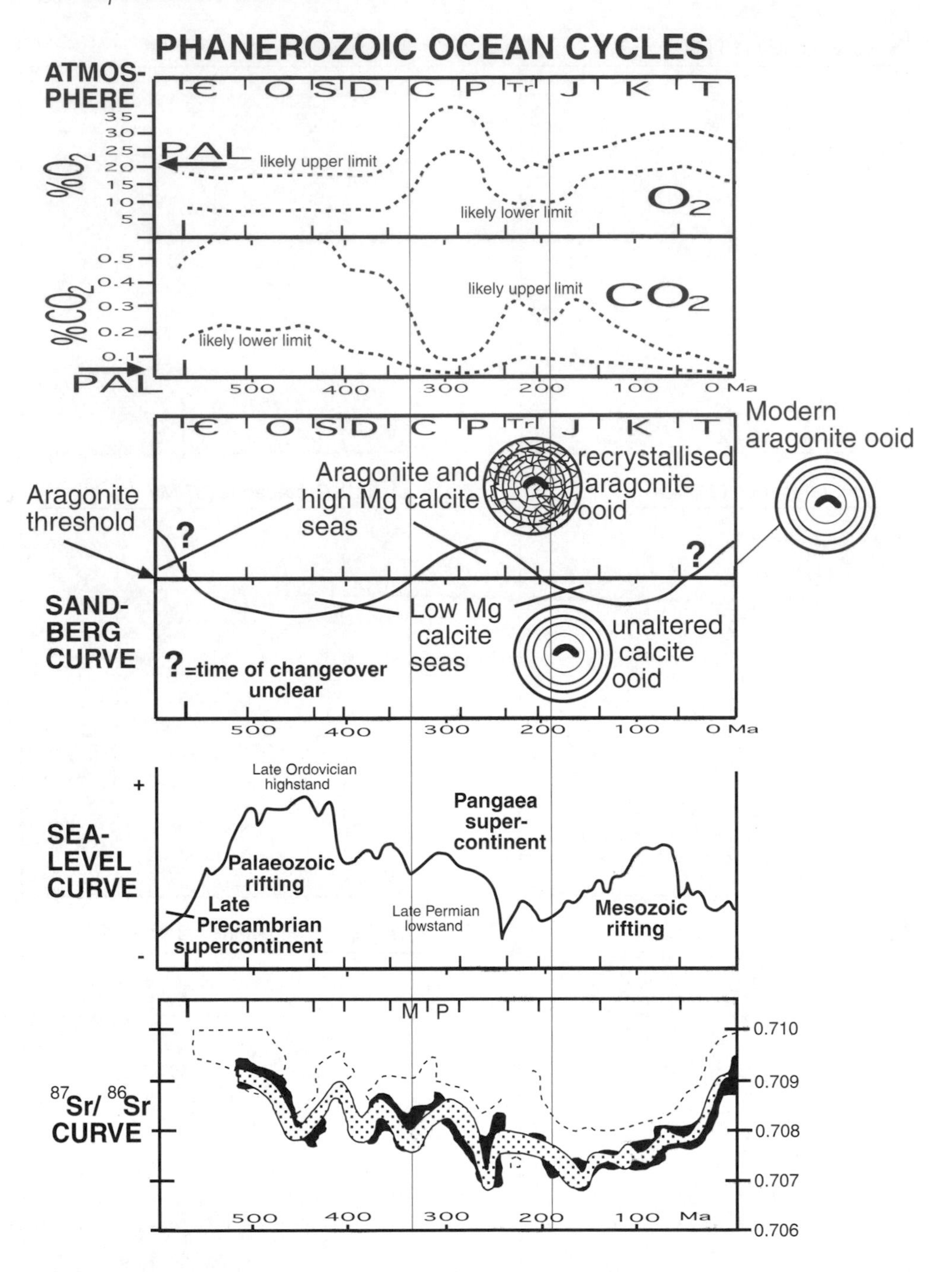

Figure 9.3 Composite diagram illustrating the points of correspondence of several well-known features of Phanerozoic oceanography. It shows Berner's O_2 and CO_2 curves, Sandberg curve, sea-level curve, and strontium isotope curves, all drawn on the same timescale. The fine vertical lines correspond to the boundaries between aragonite and calcite seas in the Sandberg curve. Features of calcite seas are: high plate activity and high sea level; high CO_2, released by reactions at subduction zones; low Mg/Ca ratio because

9.2.3 KEY ELEMENTS OF CHANGE IN PHANEROZOIC OCEANS

Figure 9.3 shows the currently applied sea-level curve for Phanerozoic time, and shows cyclic change in relation to the Wilson cycle concept. However, the *first* Wilson cycle apparently had significantly higher sea level, CO_2 content and (it is estimated) warmer climates, than later times, and demonstrates overprinting of unidirectional change on the cyclic system. Several features relating to continental positions, sea-level change and biological processes play an important part in the development of Phanerozoic oceans, shown in the following list, and detailed in Sections 9.3–9.6 respectively.

1. The ocean circulation system has not been constant through time, seemingly related chiefly to a combination of CO_2 changes and continental configurations.
2. CO_2 and O_2 show secular (= long-term) changes in the ocean–atmosphere system, related mainly to plate tectonics and development of biological communities. Of crucial importance is the history of carbon storage in the various reservoirs.
3. There is a secular change in ocean productivity, revealed by proxy from the geological history of planktonic fossils; this has implications for the nutrient content of oceans through time, a subject of intense debate.
4. Earth history is punctuated by glaciations, and understanding the control of these by ocean–atmosphere–land processes is a long way off. This is worrying because recent work demonstrating the decade-scale speed of change in glaciations shows that Earth surface systems are subject to fluctuations which the current understanding of geological evidence cannot predict.

9.3 CONTINENTAL CONFIGURATIONS THROUGH TIME AND GLOBAL OCEAN CIRCULATION TRENDS

9.3.1 BASIC PATTERNS

Greater understanding of the controls on, and nature of, modern ocean circulation has allowed increasingly detailed reconstructions of past ocean circulation to be made. Modelling ocean circulation is easier for the recent past, because of greater confidence in the knowledge of continental positions and climatic states. General (sometimes called Global) Circulation Models (GCM) seem to be the most applicable, and their recent development and testing on the Mesozoic and Cainozoic record (Lowe and Walker, 1997; Price *et al.*, 1997; Valdes *et al.*, 1995) shows a good match between models and reality. Less obvious are the continental positions before the Permian, which are under continuous revision and therefore more difficult to model, but there is a general pattern.

Throughout the Palaeozoic, continents were concentrated in the Southern Hemisphere, as the long-lived Gondwana landmass (Figure 9.2) which was the focus of glaciation in the Ordovician and Permo-Carboniferous, widely presumed to be related to its position over or near the South pole. The current continental configuration, in contrast, has the bulk of land in the Northern Hemisphere, following

Mg is removed from sea water by submarine weathering of basalts at mid-ocean ridges; K-rich evaporites because K is removed from basalts at mid-ocean ridges, enriching the K content of the sea. Features of aragonite seas are: low plate activity and low sea level; low CO_2, because little subduction activity; high Mg/Ca ratio because of less mid-ocean ridge activity and therefore less weathering; K-poor evaporites because the sea is not enriched in K. Note the incompatibility between the strontium isotope curve and the other curves, indicating a complex control. This curve may relate to erosion rates, and therefore nutrient inputs to the oceans, although the $^{87}Sr/^{86}Sr$ ratio also depends on seafloor volcanic exhalation of mantle material (which is enriched in ^{86}Sr). Stippled band = earlier version (Burke *et al.*, 1982); black band = updated version (Veizer *et al.*, 1997); in which dotted lines show the broad spread of data at certain times. Veizer *et al.*'s curve is a compilation of 3635 measurements.

Pangaea rifting. The changing shape of oceans has important effects on the circulation patterns, e.g. the North Atlantic Gyre formed as America and Europe separated and was well developed by at least *c.*60 Ma, of significance to western European winters today. A subject of great controversy now is the nature and controls of the ocean conveyor system (thermohaline circulation); we have already seen (Chapter 3) how the exact flow patterns and controlling forces of the *modern* conveyor are not fully constrained; this places severe limits on successfully modelling ancient thermohaline circulation patterns (if they existed).

The higher position of global sea level in much of the Palaeozoic Era caused most land to be submerged, probably peaking in the Ordovician Period (Hallam, 1992) (see Figure 9.3). Higher sea level may be presumed to have generated more stable, and warmer, surface temperatures because water has a higher heat capacity than rock, so less heat is radiated to space. The result seems to have been an ocean in which interchange between the ocean floor and surface waters was limited to upwelling events, so the ocean lacked a thermohaline circulation (T-HC) pattern (see Figure 9.3). Remember that T-HC carries oxygenated water down to the sea floor. Warming creates a salinity-controlled stratification with a near-global (?) thermocline and sluggish deep circulation (called halothermal circulation, H-TC), and this promoted seafloor anoxia (the modern Black and Baltic Seas are broad analogues), which leaves a sedimentary record rich in black shales, with a poor or lacking benthic community. There is an increasing recognition of the oceans having existed in one of these two alternative states at different times, and such ideas form the basis of several ocean-related studies (Hoffmann *et al.*, 1991; Jeppsson, 1990; Martin, 1996). Furthermore, higher sea level generated large areas of shallow epicontinental seas; there is much evidence that tectonic history in continental crust of a given region created substantial topographic variations on the surface, revealed as epeiric basins, cut off from the oceans. Epeiric basins are especially important during continental rifting phases which typified the lower Palaeozoic and parts of the Mesozoic, e.g. the familiar Late Jurassic Kimmeridge Clay in the North Sea Graben, of critical value in North Sea oil generation. Consequently there is a prevalence of anoxic basins during past periods of high sea level, in contrast to their rarity today.

9.3.2 CIRCUMGLOBAL CURRENTS

One important oceanographic consequence of continental reconfiguration is that at certain times circumglobal currents may have developed, for example:

1. The present Antarctic Circumpolar Current (ACC) has had an unrestrained eastward flow throughout the mid–late Tertiary and Quaternary, because there is no interrupting land. Deposits at depths as great as 3000 m show large-scale erosion and redeposition of sediment by powerful currents which followed submarine contours; such sediment deposits are therefore called contourites, and these particular ones accumulate at *c.*10 mm/ka (Howe *et al.*, 1997). Furthermore, the circumpolar water flow blocks heat flow to the polar region, so perhaps it is not surprising that Antarctica has 90% of Earth's ice, and 70% of all fresh water. The ACC also acts as a barrier to migration in or out of the Antarctic province, partly resulting in Antarctica's present unique biota. The ACC is quite powerful, with a recorded velocity up to 12 m/s, but even this is considered to be less than that of the last glacial maximum (Pudsey and Howe, 1998).
2. The mid-Cretaceous sea-level rise resulted in flooding of tropical low land, permitting a possible circumglobal connection. At first sight, an equatorial circumglobal current would be expected to develop (see Figure 9.2 for Cenomanian and Maastrichtian times, and Figure 9.4). However, modelling

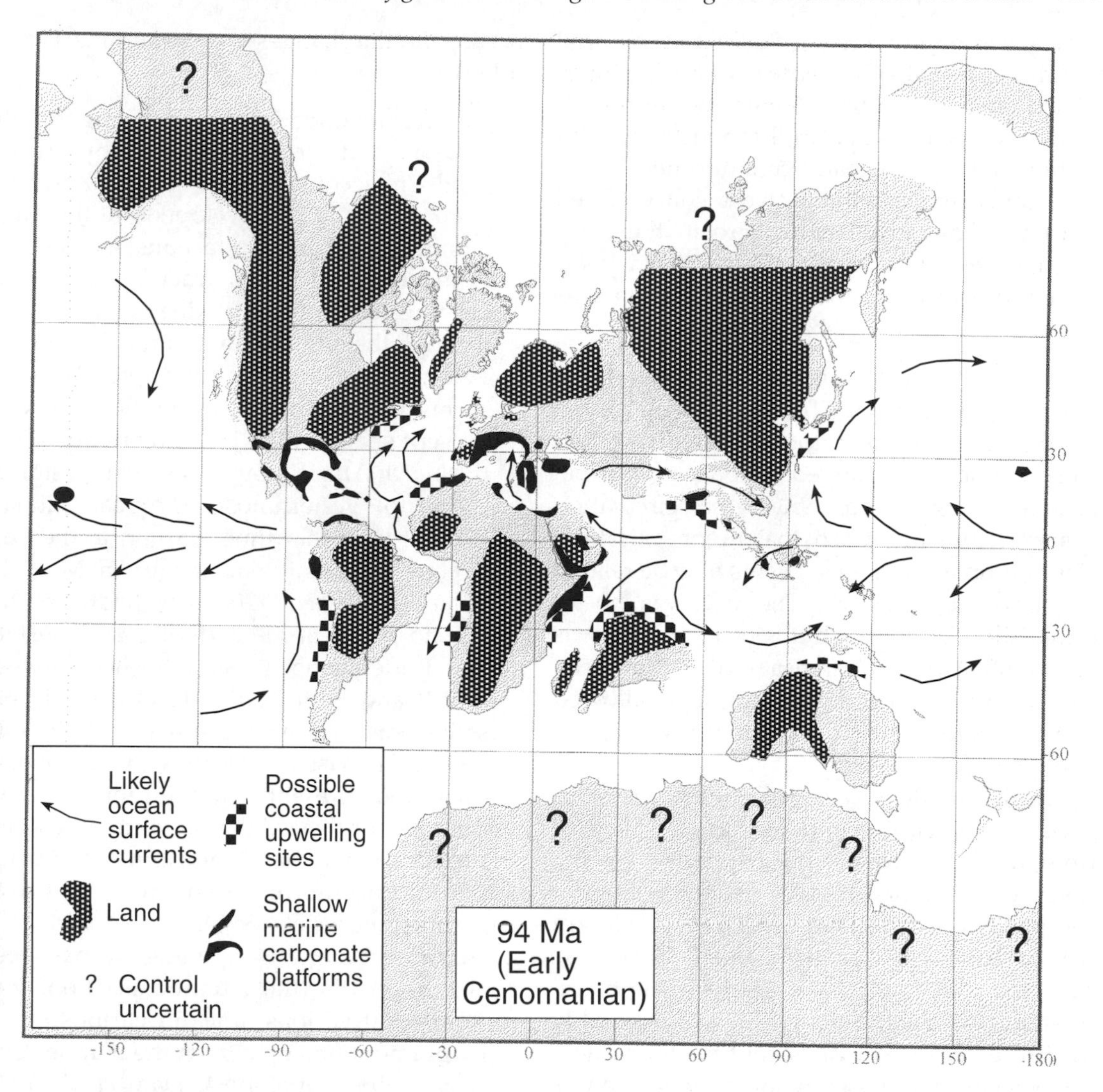

Figure 9.4 Distribution of ocean features in the Cretaceous, with the possibility of a circum-equatorial seaway (about which there is a diversity of opinion). Base map courtesy of Cambridge Paleomaps Ltd.

of the global water flow suggests that a circumequatorial current did *not* develop then, because of the deflection of currents by Laurasia and Gondwana, which caused the Tethyan currents to flow eastwards, while the Panthalassan equatorial currents flowed *westward*. These points were discussed by Crowley and North (1991). More recent models support the view that there was no circumequatorial current then (Poulsen *et al.*, 1998). Further separation of these two continental masses may have created circumequatorial flow for a short time in the Eocene (around 50 Ma), which was only halted when the land bridge between North and South America developed during the Miocene sea-level fall (see Figure 9.2K in the middle Miocene), attributable to continental drift and steady growth of the Antarctic ice sheet from *c*.40 Ma onwards.

The result not only influenced terrestrial faunas, but also prevented equatorial interchange between the Atlantic and the Pacific Oceans, with major implications for evolution of the two oceans' communities.

3. Because the continents were concentrated in the Southern Hemisphere in the early Palaeozoic, *presumably* the northern polar region during the Cambrian–Silurian was subject to circumglobal currents.

The possible early Tertiary circumequatorial current and the mid-Tertiary establishment of the Antarctic circumpolar current are two closely spaced interpreted episodes of unusual oceanic configurations, when ocean circulation systems were in a state of major reorganisation. The culmination of these events may be related to the recent glaciations. The ACC was mostly established by 30 Ma, but South America was still connected to Antarctica until 15 Ma, when full separation of the two continents allowed completion of the ACC, coinciding with establishment of the ice cap.

Finally, models of ocean currents rely on accuracy of continental location at various times, and therefore of great surprise was the discovery of terrestrial deposits on the floor of the present southern Indian Ocean, 2500 km SW of Australia (see *Geoscientist*, 1999, **9**(8), 11), that may once have been on the continental crust. Such a discovery will inevitably lead to a reassessment of continental palaeogeography, and has unknown consequences for ocean circulation patterns.

9.3.3 FEATURES OF OPENING OCEANS: EXAMPLE OF THE ATLANTIC OCEAN

Active ocean spreading and sea-level rise during the Cretaceous is of special interest because of the generation of both deep and surface water circulation patterns in the Atlantic Ocean, critical to the development of the NADW and AABW, and to the surface gyres. The Atlantic is selected as one example to show the changes associated with an opening ocean, and the changes may be summarised as follows.

1. Ocean spreading developed throughout the Jurassic and Cretaceous, and by Aptian (mid-Cretaceous) times (Figures 9.2 and 9.5) deep basins had developed in the mid-Atlantic; by later Cretaceous epochs, the Romanche and Vema fracture zones had become conduits for deep water flow through the mid-Atlantic Ridge (Jones *et al.*, 1995), as they are today (Chapter 3), and it seems the pattern of water flow took its present shape in the late Cretaceous. Thus the North Atlantic Gyre was of great significance to westernmost European climate from at least that time. Earlier, in the Late Cretaceous (Santonian stage, 85 Ma), the Atlantic reached 1200 km width at the equator, and 5000 m depth, and surface water interchange was established between North and South Atlantic Oceans. Dense saline waters flowed into the Atlantic from equatorial basins, but to a greater extent than the present Mediterranean saline outflow into the Atlantic, and its erosive properties may have contributed to formation of unconformities in Cretaceous sediments (Jones *et al.*, 1995).
2. Accompanying the opening, it has been suggested that the Atlantic sea floor was uplifted at various times in various places by plumes of magma from the mantle (Clift *et al.*, 1998; Cope, 1994; Dam *et al.*, 1998; Wilson, 1997; Wold, 1995).
3. Sedimentation and structural patterns in the northeastern Atlantic (UK) area reveal the change from relative stability of the Jurassic crust, followed by increasing activity in the Cretaceous, which culminated in the Cenomanian transgression. Features of note include the interplay between crustal uplift in southwest England and the sediments laid down as sea level rose, creating diachronism of the contact between the Gault clays and Upper Greensand sands as the west of the UK became very unstable in com-

parison with the east (Figure 9.5). Tectonic activity in the shelf west of Britain was ongoing since the Triassic, with thick sedimentary sequences in the Celtic Sea Basins throughout the Jurassic and Cretaceous (McMahon and Turner, 1998) as rifting continued. Examples of complexity include the Bay of Biscay, which opened at about the same time as the deposition of Lower Greensand in UK, when the Iberian Peninsula rotated to its present position.

4. Both Atlantic margins are tectonically passive (no subduction); it has been argued that the North Sea fracture zone with its petroleum sources and associated reservoirs relate to the opening Atlantic, and it may represent a failed rift.
5. The separation of North America and Europe, and South America and Africa, created terrestrial biotic provinces; testament to the tectonic control of land organisms as well as oceans. Note that the North Atlantic opened first, during the Cretaceous, with the South Atlantic development being largely a Tertiary feature (Figure 9.2).

Accompanying these changes were some global events also associated with the Atlantic Ocean:

1. The Antarctic Circumpolar Current, established *c.*30 Ma, isolated Antarctica by 15 Ma.
2. The Cretaceous was a greenhouse 'waterworld', with high sea levels and consequently the total area of exposed land was less than at present. The Cretaceous world is generally thought to have been warmer at the poles than today, although recent isotopic work in the Pacific Ocean disputes that by suggesting that equatorial temperatures were not much different from present times (Price *et al.*, 1998), and therefore implies that there were ice caps after all. Verification of this is difficult, not least because any evidence from Antarctica is hard to obtain.
3. During the later part of the Early Cretaceous (Barremian and Aptian Epochs) there was an event of near-global deep-ocean anoxia of uncertain cause, described in Chapter 10.

9.4 CO_2 AND O_2: MODELS, MEASUREMENTS AND RESULTS

Much Phanerozoic geological and oceanographical interpretation relies on understanding processes involving the two main geologically critical gases, because of the great range of their organic and inorganic controls.

9.4.1 PHANEROZOIC CARBON STORES

The carbon store of Earth is divided between a number of reservoirs. The atmosphere, oceans and land surface form a temporary store for only small amounts of carbon at present. The geological record shows that, of the two long-term carbon stores, carbonate and organic carbon (C_{CARB} and C_{ORG}), the C_{CARB} portion holds most carbon, as limestone. The vast majority of limestone is fossiliferous, formed on continental crust as carbonate platforms, and it seems that the secular decline in ocean–atmosphere CO_2 can be explained in terms of progressive drawdown, principally into limestones, via crustal weathering as described in Chapter 5. Theoretically, this should have led to secular global cooling, but it is important to remember that the Sun has been warming up over geological time, and that is believed to have compensated for the removal of CO_2.

The evolution of shelly biotas from the Cambrian onwards led to an increase in the development of carbonate platforms and associated reef systems, which was largely responsible for the progressive transfer of CO_2 to the oceans, as limestone. Sea-level rise from Early Jurassic times led to widespread epicontinental seas, which continued the CO_2 removal after the interruption during the time of Pangaea (Figure 9.3). Large carbonate platforms developed in the Permian and Triassic, especially in southern Asia. However, although this presents a simple picture, the

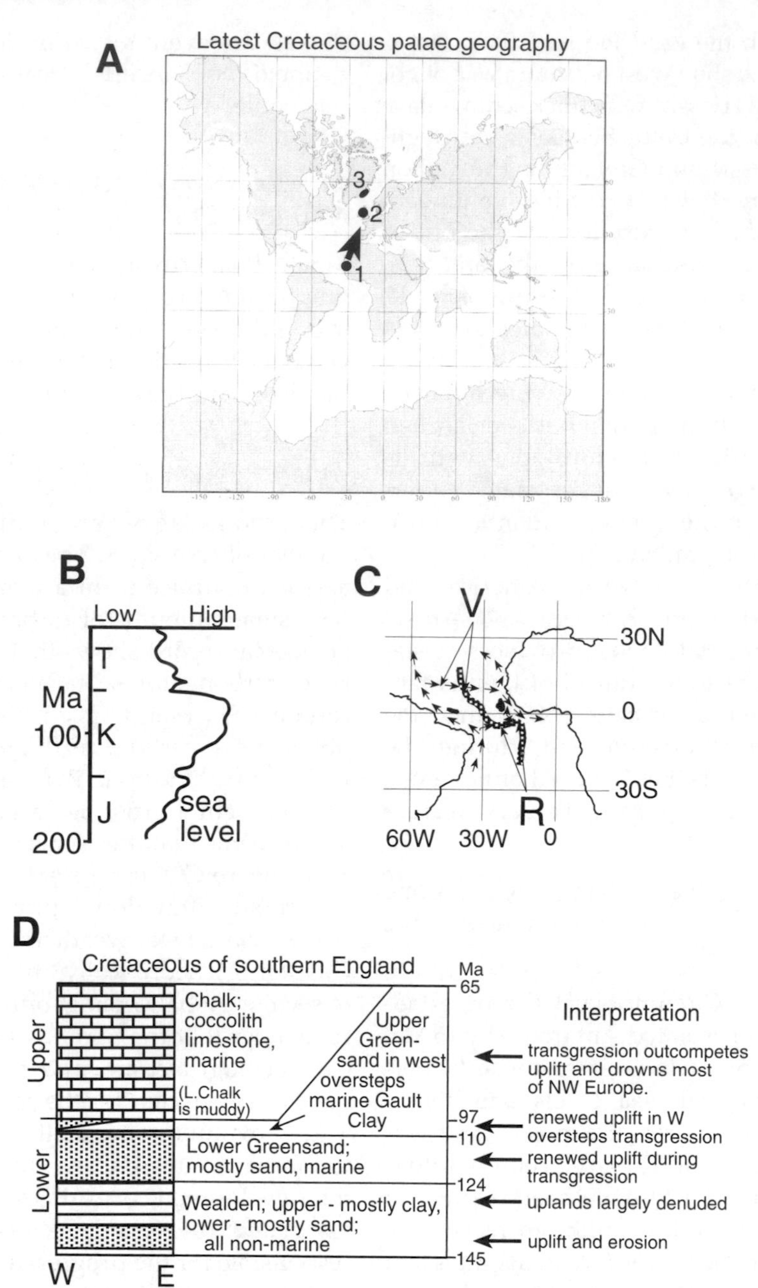

Figure 9.5 Features of opening of the Atlantic in the Mesozoic Era. (A) Palaeogeographic map of Late Cretaceous Period, *c.*94 Ma. 1 = location of Early Jurassic hotspot and NNE drift of mantle plume, a component of the break-up of Pangaea and the separation of Europe and North America (Wilson, 1997); 2 = location of Cretaceous uplift and *c.*2 km of erosion from British Late Mesozoic–Early Tertiary, due to likely hotspot (Cope, 1994); 3 = location of *c.*2.5 km of early Tertiary erosion in East Greenland, resulting from an interpreted hot-spot, which induced uplift from a passing mantle plume (Clift *et al*, 1998). Base map courtesy of Cambridge Paleomaps Ltd. (B) Global sea-level curve, showing the rapid rise of sea level

growth of carbonate platforms also depends on evolution of shelly organisms. The evolution of one particular calcareous algal group (coccoliths) in the mid-Cretaceous led to the formation of a huge blanket of $CaCO_3$ in Europe (Figure 9.5); this also had the effect of shifting the balance of dissolved elements (especially Mg and Ca), because coccoliths are low-magnesium calcite (LMC), and left the seas relatively enriched in Mg. Coccoliths also created a new substrate over large areas of epicontinental seas, which led to evolutionary adaptations amongst the benthos which lived on this calcareous soft sediment, now familiar to us as chalk.

9.4.2 MODELLING THE CHANGES

Measurement of past ocean–atmosphere gas changes cannot be achieved directly, but requires proxies, and a commonly applied technique is *modelling* the changes, based on a series of rules derived from measurements and estimates. For CO_2, construction of the curve (Berner, 1991) uses estimates of the rate of addition (e.g. from volcanic activity, weathering of organic carbon stores) and removal (e.g. weathering of crustal silicate minerals, burial of C_{ORG} and C_{CARB}) of CO_2 to and from the atmosphere respectively. Modelling of global CO_2 indicates a progressive fall through the Phanerozoic. Levels of *c.*16× present atmospheric level (PAL) CO_2 in the early Palaeozoic are suggested, but there is a major drop in the Carboniferous and sharp recovery in the Permian Periods. This model is a development of the earlier so-called BLAG model, taken from the first letters of the authors' names – Berner, Lasaga And Garrels (Berner *et al.*, 1983). A separate model of atmospheric O_2 (Berner and Canfield, 1989) used weathering and burial rates of sediments, associated with the carbon and sulphur cycles. Both these elements are subject to oxidation and reduction; sulphur is represented by the common mineral pyrite, which as we have seen in Chapter 8 is a good indicator of oxygenation. The O_2 curve shows generally opposite changes to CO_2 (Figure 9.3). Higher levels of CO_2 in the Palaeozoic and Cretaceous were attributed by Berner (1991) to the lack of land plants in the former, and increased global volcanic degassing in the latter time period; this emphasises the complex interplay of both organic and inorganic components of the Earth's systems in control of ocean and atmosphere conditions. Of great significance, therefore, is the shift attributed to the rapid widespread development of swamp forests in the middle to late Carboniferous, which caused drawdown of CO_2 from the oceans and atmosphere to the biosphere, and released substantially higher O_2 than PAL, possibly as much as 35% PAL, due to the huge increase in photosynthesis, with obvious implications for life (Graham *et al.*, 1995); for example it has been suggested that the large insects of the Carboniferous could exist only because of the higher O_2 levels, because insects have no lungs and rely on O_2 diffusion into tissues. Also it seems there was an increase in the frequency of forest fires during that time, because of the preservation of abundant charcoal (reviewed by Skelton *et al.*, 1997). Later the CO_2 and O_2 trends reversed in the late Permian because the drier climates (and lower sea level) of the Pangaea supercontinent suppressed the vegetation growth, and also may have reoxidised much of the accumulated organic matter.

through the Jurassic and Cretaceous Periods, an interpreted consequence of accelerated continental rifting as Pangaea broke up. (C) Interpretations of bottom current flow during Oligocene Epoch (*c.*30 Ma), controlled by sea-floor topography of the mid-Atlantic Ridge (grey band), the Vema Fracture Zone (V) and Romanche Fracture Zone (R), much as they are today; black patches are the Ceara Rise (left) and Sierra Leone Rise (right) (based on Jones *et al.*, 1995). (D) Schematic geological section from the Cretaceous of southern England, with interpretations of interplay of tectonically induced uplift and erosion, sedimentation, and sea-level rise (based on Gibbons, 1981).

During this CO_2 rise, the calamitous end-Permian mass extinction occurred, when an estimated 96% of species died out, and CO_2 levels were apparently increased by particularly large-scale volcanic eruptions in Siberia (discussed in Chapter 10). The fall in O_2 can therefore be reasonably attributed to the reoxidation of stored C_{ORG}, and seems to have affected both atmosphere and ocean, because both terrestrial and marine biota suffered extinction.

9.4.3 MEASUREMENTS AND MODELS: LIMESTONE PETROLOGY, GEOCHEMISTRY AND SANDBERG'S CURVE

A different approach from using CO_2 and O_2 modelling in understanding ocean change has developed through the study of limestones, using observations of the nature of types of limestone components under a microscope, plus geochemical measurements. This underlines the importance of analysis using sedimentary petrology, a skill normally found in Earth science university courses, but one which palaeoceanography cannot do without.

We have seen that the Wilson Cycle process is a global feature related to plate tectonics, and that the major sea-level changes are well established. In the Permian and Triassic Periods, for example, the amount of sediment deposited on land (alluvial, lacustrine, fluvial), plus large unconformities, demonstrates the global sea-level lowstand at that time. However, oceanic information can also be drawn from the chemical precipitates in limestones, and leads us to recognise the truly interdisciplinary nature of ocean studies, seen in the Sandberg curve (Sandberg, 1983) which has revolutionised the way geologists view ocean processes.

Sandberg attempted to explain the reasons for preservational differences amongst inorganic carbonate precipitates at certain times in the rock record, particularly related to ooids, 0.5–2 mm objects having $CaCO_3$ inorganically accreted in concentric layers on shell fragments or sand grains (found today in places such as the Bahama Banks). Ooid composition is therefore in equilibrium with seawater, such that the mineralogy of ooids reflects the $CaCO_3$-secreting capacity of water. This is in stark contrast to organic secretion of many $CaCO_3$ shells, which may have vital (= biotic) effects resulting in structures *not* in equilibrium with seawater. Modern ooids are aragonitic; aragonite is an unstable mineral which inverts to calcite during diagenesis, and the original concentric layering is lost. Therefore, if ancient ooids were aragonitic originally, they normally exhibit recrystallisation fabrics. Sandberg's survey of ancient ooids revealed a pattern of change which suggests that, at certain times, ooids are recrystallised and at other times they are not (Figure 9.3); the recrystallised fabric is a proxy for indicating an original aragonitic mineralogy in some ancient ooids, in contrast to unrecrystallised original calcite. *Un*recrystallised ancient ooids are often formed around recrystallised mollusc fragments (a proxy for betraying vital effects of original aragonite in molluscs), thereby providing further evidence for a stable, calcitic mineralogy for *those* ooids.

So what controls ooid mineralogy? The Mg/Ca ratio has long been regarded as critical for the mineralogy of carbonates; the sum of current knowledge indicates that a high Mg/Ca ratio (> 5:1) inhibits calcite precipitation, and aragonite forms instead. When Mg/Ca falls below 5:1 calcite is not inhibited, so it forms a primary material in inorganic components, such as ooids. The Sandberg curve demonstrates long-term change in carbonate precipitation, and implies a large-scale control operated on Phanerozoic (and probably late Precambrian) ocean composition. As discussed earlier for the Precambrian–Cambrian boundary (Chapter 8), the control engine reviewed by Tucker (1992) and Tucker and Wright (1990) is plate tectonics throughout Late Precambrian and all Phanerozoic time, which controlled the Mg/Ca ratios of the oceans. Briefly, when plates are active, seafloor volcanic eruptions are extensive and the interaction between seawater and the new volcanics extracts Mg from the water by submarine weathering, and calcite forms. When plate activities are subdued during supercontinent times, constructive and destructive margins are relatively inactive, and so oceanic Mg is not removed from the sea, and its levels rise with input from terrestrial runoff, so aragonite forms.

More recent work augments these results. Firstly, although the influence of vital effects means that mineralogy of fossils is not necessarily a reliable proxy, calcareous worm tubes change their mineralogy from calcitic in the Lower Carboniferous (Mississippian) to aragonitic in the Upper (Pennsylvanian) (Railsback, 1993), coinciding with the switch from calcite to aragonite seas. Of course this could be a coincidence, but emphasises the complex controls on biomineralisation discussed in Chapter 8. Secondly, Hardie (1996) presented evidence that the content of Mg in potash-rich evaporites through time fluctuates in tune with aragonite and calcite sea episodes. Thus the evaporites contain excess $MgSO_4$ only during aragonite sea times, reflecting the accumulation of Mg in oceans by terrestrial erosion at times when it is not being removed by submarine weathering. Nevertheless, Hardie (1996) noted other studies which reveal different patterns in the Mg/Ca ratio through time from the ones predicted by Sandberg, and this emphasises that these datasets are only *proxies*. Furthermore, there is disagreement between workers on the precise controls of interchange of dissolved elements between ocean and crust; examination of the comments and replies section of journals can give an insight into a topic more than full papers – see *Geology* 1998, **26**, 91–92 for an example relevant to the evaporite controversy.

A synthesis of Phanerozoic change in seawater chemistry (Stanley and Hardie, 1998, 1999) is given in Figure 9.6. Stanley and Hardie's work drew attention to different organisms' ability to control precipitation of $CaCO_3$ to form skeletons, and recognised two categories: those that have (a) weak and (b) strong control over their ability to calcify; the latter category being called hypercalcifiers. Also involved, probably, is temperature; it has been argued that warm conditions are conducive to calcite deposition, while cooler settings favour aragonite (Morse and Wang, 1997). These very interesting results from inorganic and organic carbonates, and evaporites, give a clear indication that seawater chemical composition has not been constant over Phanerozoic time, although no synthesis models are perfect. For example, Stanley and Hardie (1998) placed some emphasis on the Palaeozoic stromatoporoid sponge group being calcitic. Stromatoporoid specialists (including the senior author of this book!) and others have presented evidence that stromatoporoids were more likely to have been aragonitic, but their original mineralogy is not clear-cut, and once again we warn that while broad-scale models provide a framework for viewing changes in ancient oceans, they are still only models.

9.5 MATCHING THE CURVES: GREENHOUSE AND ICEHOUSE WORLDS

Figure 9.3 synthesises the data from CO_2 and O_2 modelling, sea-level change and Sandberg's observations. Compare the changes with Figure 9.2, and note the relationship between the continental configurations and the curves. Comparisons of the datasets have led to recognition of warm and cool phases in Phanerozoic history, referred to, rather simplistically, as greenhouse and icehouse respectively. The concept is easily extended back into the Precambrian, although the *cyclic* change between the two is not recognised there. Tucker and Wright (1990) synthesised a model using plate tectonic activity through time in relation to the assembly and disassembly of late Precambrian and mid-Phanerozoic supercontinents, drawing on evidence from the Mg/Ca fluctuations. Accordingly, when active continental separation occurs, sea levels are higher due to ocean floor expansion, spilling water onto the

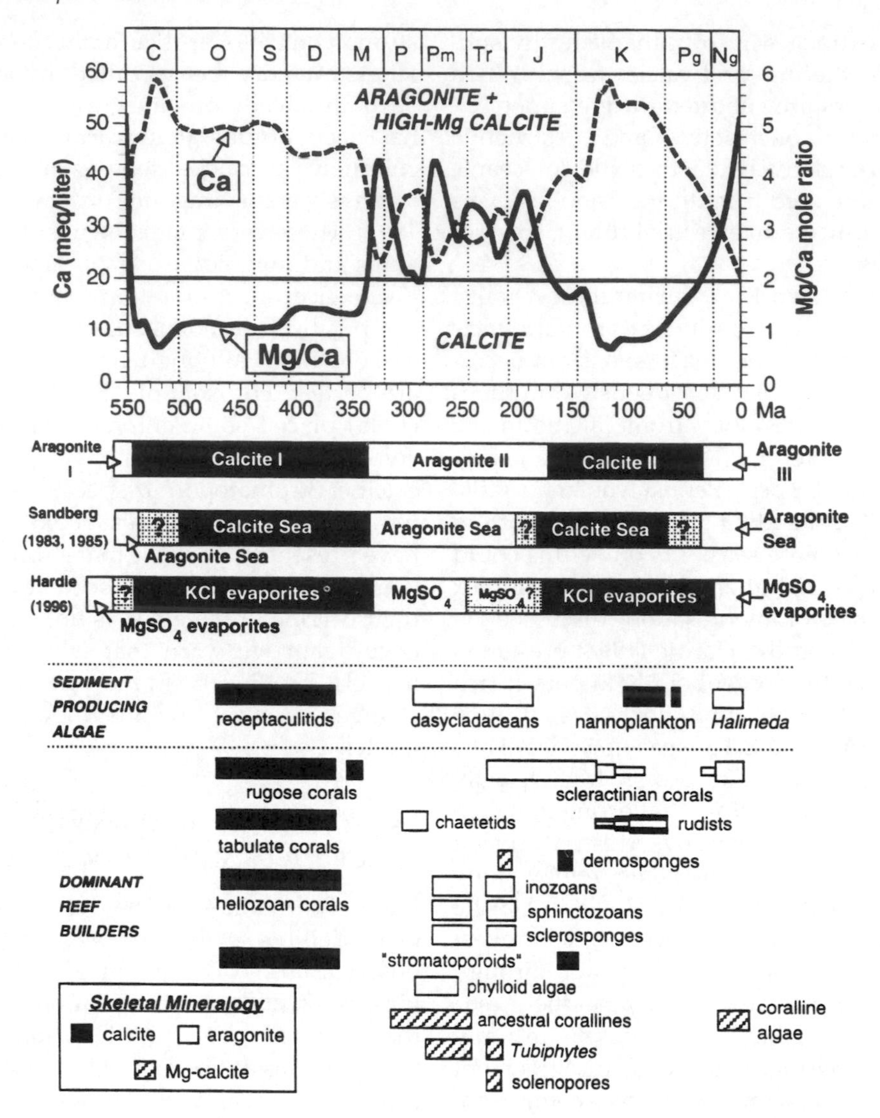

Figure 9.6 This composite figure attempts to show the correspondence between calcite seas and the importance of certain calcifying organisms which had a strong control over their skeletal secretion. This is in contrast to the aragonite sea times, typified by organisms which calcified in response to environmental influence, and less to intrinsic control. The accuracy of these interpretations is yet to be unequivocally demonstrated. Reprinted from *Palaeogeography, Palaeoclimatology, Palaeoecology*, 144, Stanley and Hardie, Secular oscillations in the carbonate mineralogy of reef-building and sediment-producing organisms driven by tectonically forced shifts in seawater chemistry, pp. 3–19, Copyright © (1998), with permission from Elsevier Science.

continents. Also, subduction zone metamorphic reactions lead to breakdown of seafloor carbon stores, releasing CO_2 back into the ocean–atmosphere system and warming the Earth's surface. As we have seen, the broad result is that calcite seas are accompanied by higher sea levels and warmer climates, while aragonite seas have lower sea levels and cooler climates. This in turn leads to the notion of icehouse and greenhouse states of world climate, with aragonite seas representing times of icehouse, and calcite seas times of greenhouse (see Figure 9.3).

However, Sandberg's curve has question marks at critical places (e.g. Early Cambrian and Tertiary), because the data are not sufficient to permit ready recognition of the record at these points. Resolution of this problem is awaited; in the meantime, this is a good *working model* of the processes governing the oceans in the Phanerozoic, and they reinforce the notion of cyclicity.

Discrepancy also lies in another key indicator preserved in limestones. $CaCO_3$ is capable of accepting extra elements into its structure; Mg is the most common, but another is Sr, present in several isotopes (Figure 9.3). Wide fluctuations of the ratio $^{87}Sr/^{86}Sr$ (Veizer, 1997) do not match the Sandberg curve, and are regarded as governed by the interplay between hydrothermal input to the ocean floor (adds ^{86}Sr), and the erosion of terrestrial material (adds ^{87}Sr). Such features demonstrate the continuing influence of the mantle on oceans through geological time and do not necessarily operate on cyclic scales.

9.6 PHANEROZOIC OCEAN BIOTIC CHANGE AND ITS CONTROLLING FACTORS

9.6.1 BASIC PATTERNS AND PROBLEMS

Ocean ecosystem change through time is well known. Following the increase in organic complexity through the Precambrian (possibly constrained by O_2, but also by exhalative processes on the sea floor; see Chapter 8), the diversity and complexity of Phanerozoic ecosystems has apparently continued to rise. This information is based on the fossil record, which, although it cannot solve the problem of preservation of soft-bodied organisms, is the only dataset of organism types. The best-known approach to quantifying the complexity of organisms through time is that of J. Sepkoski, to group organisms into families, and plot their abundance across the Phanerozoic (Figure 9.7). Results for the oceans seem to indicate increasing complexity through the Cambrian and Ordovician, presumably reflecting the exploration of ecological niches following invention of shelly skeletons and the rise of oxygen, then

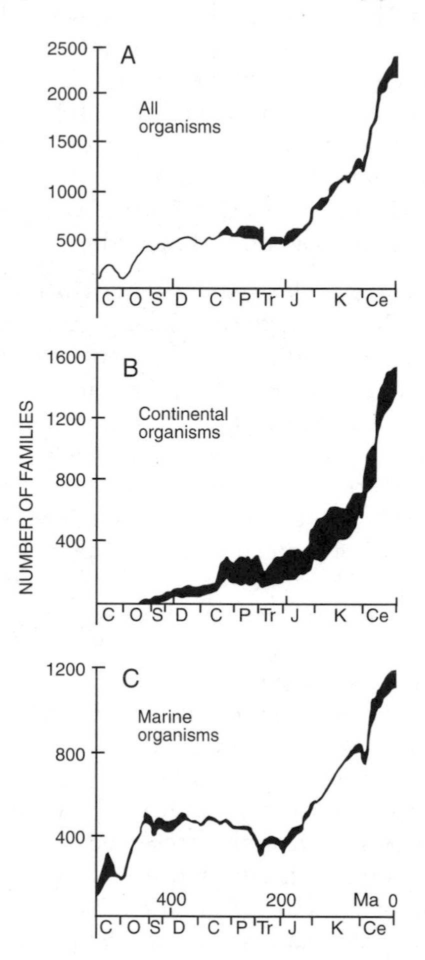

Figure 9.7 Curves of numbers of families of fossil organisms through the Phanerozoic Eon, demonstrating increasing numbers in both terrestrial and marine environments. Note the more rapid development of marine families, with a sharp increase in the Lower Palaeozoic Era. Curves are based on estimates, with upper and lower limits shown by the black regions of each curve. Reprinted with permission from Benton, Diversification and extinction in the history of life, *Science*, **268**, p. 53 (1995), American Association for the Advancement of Science.

maintenance of diversity at a plateau during the rest of the Palaeozoic, possibly due to equalisation between rates of origination of new species and extinction of others (Francis and Dise, 1997, pp. 65 ff.). Then, following the end-Permian mass extinction, diversity recovered and rose further, up to the present-day levels. The trend is interrupted by several other mass extinctions (see Chapter 10), and debate continues about whether or not humans are causing a mass extinction at present.

While the trends match our intuitive view of what ought to have happened to diversity through time, things may not be so simple. Four aspects can be highlighted.

1. There is a long-standing criticism of the family-group diversity approach, that it does not allow for the differences in numbers of species between different families; the point being that some families may have many species, others only a few. Furthermore, at mass extinctions, a family may be wiped out except for a few individuals of one species, yet still appears on the graph. The hope is that these discrepancies even themselves out between families at any one time, but being strict, that still remains a hope.
2. It is impossible to quantify the unquantifiable: there is no satisfactory way of accounting for the (presumably) vast number of soft-bodied and poorly mineralised species which are preserved only under exceptional circumstances (for example, the famous Middle Cambrian Burgess Shale, the Devonian Hünsruck Shale and the Jurassic Solenhofn Limestone all reveal part of what is missing at those times in less favourable conditions).
3. Ecologists measuring diversity of living organisms do not take account of just species *numbers*, they also need the relative *abundance* of each. A population of 100 individual organisms of two species is low diversity if 99 are one species, while 1 is the other; it is a maximum diversity if there are 50 of each. If the two species are in different families, then they would get equal status on the Sepkoski curve irrespective of their number. However, the species *number + abundance* diversity approach does not work for most fossils because of preservation problems (called taphonomy), so that it is unrealistic to obtain abundance data of species in the once-living community; remember that it is difficult enough to measure in the *modern* oceans! Nevertheless, reasonably good estimates of the abundance of species can be obtained for microfossils if they are present in large numbers, and therefore have value in oceanographic models. Microfossils play an important role in Mesozoic and Cainozoic studies (Brasier, 1995a; Murray, 1995).
4. Chapter 8 indicates that the C_{ORG} and C_{CARB} $\delta^{13}C$ data show a fairly uniform pattern from earliest records of life, to modern times (Figure 8.2), and that this may be taken to suggest that the overall productivity has not changed a great deal through time. The possibility that the skeletal fossil records shown in Figure 9.7 are not a true account of diversity change must be considered, and that the Precambrian oceans seem to have teemed with life. Indeed there is good evidence that at least the Proterozoic Era supported organic diversity sufficient to show variations in relation to water depths (Butterfield and Chandler, 1992), and that the late Precambrian Ediacara fauna overlapped in time with Phanerozoic-type fossils (Brasier and McIlroy, 1998; Grotzinger *et al.*, 1995); such newer work shows that the Precambrian–Cambrian boundary is not such a sharp line in Earth history as has been previously assumed, so starting the diversity curve at the Precambrian–Cambrian boundary is probably not realistic.

Despite these misgivings, acceptance of the validity of the diversity curve may be found in the changes associated with mass extinctions; such extinctions are accompanied by changes in sedimentary sequences, so the sudden decline in abundance and diversity of fossils is a real feature of the biotic record. While it must be emphasised that the curve can really only be validated for skeletal organisms, the chances are that it also applies to soft-bodied creatures; it is a reasonable argument that the Phanerozoic skeletal record is a reflection of the organic world as a whole. For the shelly fossil assemblages, increasing organic diversity through time seems to be accompanied by a trend of *decreasing* background extinction through the Phanerozoic; mass extinctions can be related to catastrophic ocean events, such as global anoxia, as possibly at the end-Permian, but background extinction is the continuous loss of species unable to adapt to changing conditions. Biotic recovery from mass extinction led to a large step increase in skeletal organism diversity according to the Sepkoski curve; the concept of stepwise change is commonplace in evolutionary ideas, and so increasing diversity must mean that background extinctions do not keep pace with new species formation.

9.6.2 DIVERSITY CONTROLS

Whether oceanic controls are part of the reasons behind the family abundance trend or not is difficult to assess; there is no obvious linkage. Sepkoski also tried to relate diversity to an *evolutionary* scale, by dividing the Phanerozoic record into three evolutionary faunas; i.e. groups of families which, somehow, seem to be related in a characteristic pattern of appearance and disappearance in the fossil record, and these are also unequally affected by mass extinctions. It is not easy to assess the value of divisions of the fossil record into such evolutionary faunas; much depends on the true biological validity of having only three categories, and whether the groupings chosen really reflect the balance of the communities through time; such discussion warrants a book on its own. Only partly does the changing pattern of Phanerozoic ocean configuration (Figure 9.2) relate to diversity increase, but diversity trends better match the Phanerozoic ocean trends, which themselves are partly related to plate tectonics! Note the following:

a) The Cambrian rise in diversity has been attributed to ecological niche-filling in the early millennia of skeletonisation, which may have been enhanced because it also coincides with the plate fragmentation of Rodinia (creates barriers, aiding speciation), plus the associated sea-level rise and consequent increase of area on continental shelves. This draws attention to the well-known (and suspect) species–area effect theory, that larger modern-day areas for habitation lead to greater diversity of life (Skelton, 1993). However, it may be too simple just to transcribe this idea to the Cambrian, and its effects are unquantifiable, especially in such old rocks.

b) The Sepkoski diversity trend does have some relationship with the CO_2 curve. If the effects on both curves due to Pangaea are ignored, diversity *increases* as CO_2 *decreases* through time. It might be argued that this is a large timescale effect of changing ocean circulation patterns, and may relate to the progressive removal of CO_2 from the ocean–atmosphere system, having nothing to do with evolution. Nevertheless, these trends draw attention to the relationship between geochemical and biotic data (Figure 9.8). Remember the principles of ocean circulation controls, that higher CO_2 levels cause warm climates, leading to ocean stratification, and limited vertical mixing; thus deep ocean stays anoxic, and nutrient supply to surface waters is less due to restricted upwelling. That situation seems to apply more to the Palaeozoic than to later times, because of the higher CO_2 levels. The trend of planktonic diversity has increased through time (Martin, 1995, 1996) (Figure 9.8), and it might be argued that part of the reason was increased vertical mixing.

Phanerozoic $\delta^{13}C$ and $\delta^{34}S$ isotope records in limestones become important in this

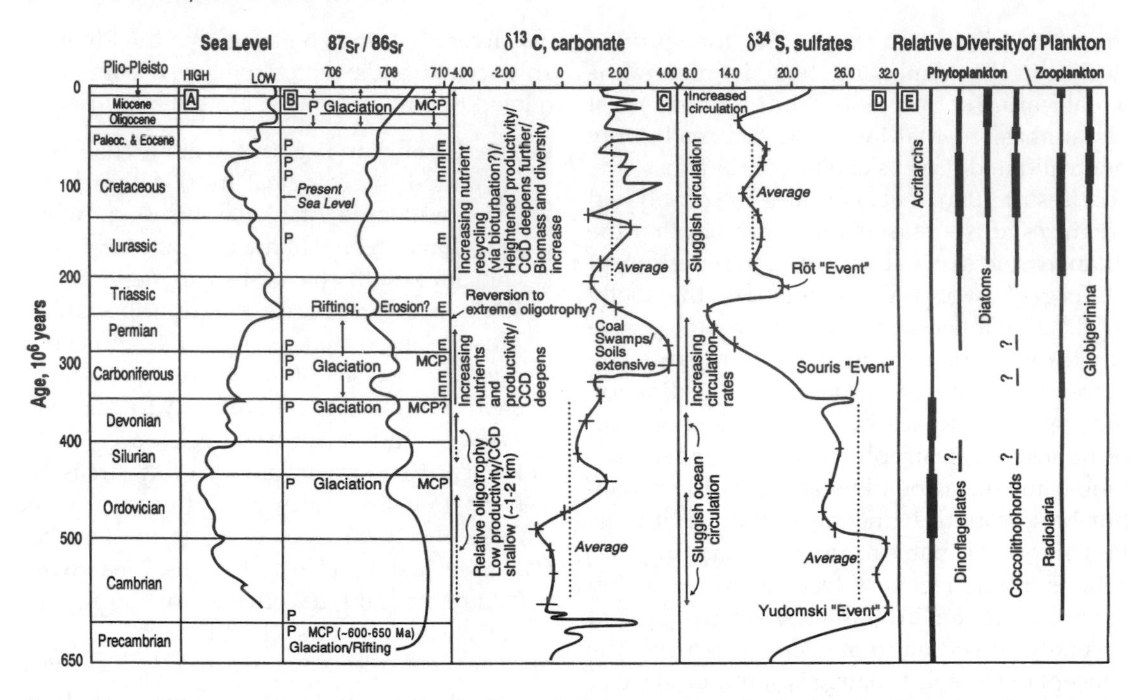

Figure 9.8 Diagram of trends of C, O, Sr and S isotopes, and planktonic diversity through the Phanerozoic. In particular, note the decrease in the sulphur isotope curve with time, corresponding to an increase in plankton diversity; this is attributed to a long-term change from poorly oxygenated deep seas in earlier times (when nutrients were mostly contained in deeper water) to better oxygenated seas in later times (when there was more ocean mixing, and movement of nutrients to the surface). Trends shown have enormous implications for interpretations of deep ocean circulation of the present type, at different times in the geological record. (From Martin, 1996.) MCP = marine carbon/phosphorus episodes, indicating eutrophication; E = possible eutrophication episodes; P = phosphorite peak. Reprinted from *Palaios*, **11**, Ronald E. Martin, Secular increase in nutrient levels, Copyright © (1996), with permission from SEPM (Society for Sedimentary Geology).

discussion (Martin, 1996). If $\delta^{13}C$ increases in limestone precipitates, then ^{12}C must be removed from the environment, and it is generally taken that this is organic assimilation of the lighter carbon isotope, which then accumulates on the seabed as organic matter (C_{ORG}). Note that in Figures 8.2 and 9.8 the C_{ORG} curve indicates a long-term pattern of consistent enrichment of $\delta^{13}C$, indicating ^{12}C removal from the active biosphere, which in turn aided the growth of atmospheric oxygen. Martin (1996) also showed that as $\delta^{13}C$ increased, $\delta^{34}S$ reduced throughout the Phanerozoic (see isotope description in Part A). High $\delta^{34}S$ in sedimentary rocks is taken as a proxy for ocean anoxia, and therefore most of the Palaeozoic suffered a greater degree of this than later times (Figure 9.8). The decrease of $\delta^{34}S$ through time implies that the poorly oxygenated deeper ocean waters were becoming more oxidised, as a result of improved upwelling and ocean overturn. This in turn supplied more nutrients to the surface waters, where organic productivity was higher, leading to the trends in Figure 9.8. In support of this, there is also a number of events of increased phosphorus injection into ocean waters. The buried phosphorus content is expressed in relation to the carbon interred along with it, termed the Marine Carbon-to-Phosphorus (MCP) ratio; this ratio has increased through time in a series of steps, some related to glaciations (Figure 9.8) which are periods of overturn and release of phosphorus to the oceans. There is also a broad relationship between MCP events and times when there seems to have been a global state of eutrophication.

From an ocean-state viewpoint, such ideas make sense; the higher global temperatures of the Palaeozoic greenhouse world lead to the suspicion of a maintained stratified water system where reduced oceanic oxidation would be expected, a scenario supported by the record of thicker and more widespread black shale deposition in the Palaeozoic compared with the Mesozoic (see Martin, 1996, p. 213). Such ideas are supported by the association of anoxic events and greenhouse conditions in the Cretaceous.

However, never forget that isotope data are *proxies*; just to show how difficult it can be to rely on proxy data for an interpretation, the simple story of nutrient trends was questioned by Martin himself. For example, the Cambrian and Ordovician $\delta^{13}C$ data are *negative*; this might reflect the fact that, during those times, there was no terrestrial flora, so there was no removal of ^{12}C on land from the atmosphere. According to this idea, because of the diffusion interlink between ocean and atmosphere, the amount of ^{12}C extraction by photosynthesis in the sea was not sufficient to leave an excess of ^{13}C in the water. It therefore means that the Cambrian–Ordovician part of the $\delta^{13}C$ curve cannot be reliably used to infer deep-ocean anoxia and low productivity. Nevertheless, the sedimentary data of widespread black shales at these times are not proxy data; instead they are fundamental sedimentary features, good indicators of the poorly oxygenated deep waters. Such discussion shows that proxy data can be conflicting, and confusing, and the lesson is that they should never be taken at face value. Discussion of the end-Ordovician extinction event, in Chapter 10, demonstrates these problems further.

9.6.3 COMPLEXITY IN DIVERSITY STUDY

Despite the encouraging matching of patterns discussed in Section 9.6.2, there is a degree of *mismatch* between the biotic diversity curves (Figures 9.7 and 9.8) and the others in Figure 9.3, as follows.

1. The overall sea-level fall since the Late Cretaceous is accompanied by the biggest increase in diversity, counter to the species–area concept.
2. The Carboniferous rise of O_2 is matched by an increased diversity on land (Skelton *et al.*, 1997), and lots of adaptations to take advantage of the higher oxygen (Graham *et al.*, 1995), but in the oceans the diversity fell slightly. The gradually falling sea level through the Carboniferous may have played a part.
3. A detailed discussion of the relationships between diversity and nutrients in two related papers (Brasier, 1995a,b) makes sobering (but exciting) reading. Productivity is clearly higher in times of cooler climate; the main nutrients, P, N, Fe and Si, are controlled by upwelling, which only happens on a large scale if there is a climatic cooling. Remember from Part A that upwelling relies on prominent thermal gradients between the equatorial regions and the poles, so that wind strength influences upwelling. Implicated here is the Ekman spiral feature of wind-induced water flow, and upwelling happens because of the disruption caused by enhanced wind friction on the sea surface. Conversely, when the climate is warmer, the upwelling is suppressed, and water stratification established. Examination of a productivity map of modern oceans (Figure 6.4) shows that productivity is greater in the cooler regions of the oceans. Studies of the Cainozoic record (Brasier, 1995a,b) reveal a significant increase in productivity in the cool phases of the Quaternary, and in the global cooling of the early Tertiary after the end-Cretacous extinction event. So, glacial phases should

be associated with higher diversity. Now examine the Sepkoski curve for the Upper Carboniferous to Lower Permian (Figure 9.7). The sharp increase of terrestrial diversity in the middle Carboniferous may be a feature of increased oxygen, but the small drop in marine diversity does not match the largest and longest-lived glacial period in Phanerozoic history. There is good evidence that thermal gradients were higher, and that winds were stronger; an example is the sizes of Early Permian sand dunes in Britain, which are much larger than those of today, and could only have been created by greater wind strength (Glennie, 1990b), so it is reasonable to assume that upwelling was also enhanced. Why the recorded skeletal-fossil diversity did not increase could be due to at least two possible causes:

a) the diversity curves do not reflect true marine diversity, as discussed above;

b) the oceans lacked the other vital parameter to drive global ocean mixing – a thermohaline circulation.

Thermohaline circulation is needed to transport oxygen and heat and nutrients around the global deep-ocean waters, and we have already seen the uncertainty of what controls it in modern oceans (Chapter 3). During the Permo-Carboniferous times, continental assembly was gathering pace; the continents were bunching together, and lacked the separated configuration they have today. Present continental positioning could be critical to the thermohaline circulation because one of the proposed driving forces is salinity gradients between the modern Pacific and Atlantic oceans. Although we might expect an unrestrained deep-ocean circulation within Panthalassa Ocean, that does not seem to have left a record in the organic diversity data. However, this discussion is wandering into the realms of unsupported speculation, but may provide stimulus for future research, and certainly makes you *think* about ocean controls and their relations to the biosphere.

The application of diversity curves and isotopic proxies is only one line of study in understanding nutrient supply and organic change in the oceans over time. A wealth of data has been amassed in the last twenty years on biotic responses to nutrients in modern oceans. The identification of feeding styles in modern organisms has been applied to understanding these features in fossils, with a spectrum of feeding characteristics ranging from oligotrophy to eutrophy. Oligotrophic communities are those which utilise sparse food resources, and thrive in low-nutrient settings, e.g. coral reefs, where nutrients are retained within the community, and the envrionment itself has little. Eutrophic communities are those which thrive in high-nutrient conditions, e.g. many algal groups which grow exponentially in an unlimited food supply. Several groups of planktonic microorganisms demonstrate this range of feeding style, and the presence of oligotrophic and eutrophic habits are interpreted on the basis of structure in fossils. One such group is the foraminifera, where feeding habit is recognised by size of the fossils, and a range of morphological features, as proxy evidence of these feeding styles (Brasier, 1995a,b).

Oligotrophic foraminifera show *K*-selection features (e.g. large size, slow growth), while eutrophic forms show *r*-selection patterns (e.g. small size, fast growth). (The well-known biological concepts, *r* and *K*, refer to the dynamics of populations; May, 1976.) The Cretaceous–Tertiary extinction (with the likely asteroid impact) is presumed to have led to widespread disruption of ocean systems, and the overturning of ocean water is presumed to have released huge amounts of nutrients into the upper waters (Brasier, 1995b). Productivity collapsed with the widespread demise of ecosystems (Figure 9.8), so the uptake of these nutrients was limited. The continuing warm climate of lowest Tertiary (Palaeocene and most of Eocene) led to re-establishment of stratified ocean and return to nutrient-poor conditions at the ocean surface. In this setting, diversity of foraminifera expanded (Figure 9.9)

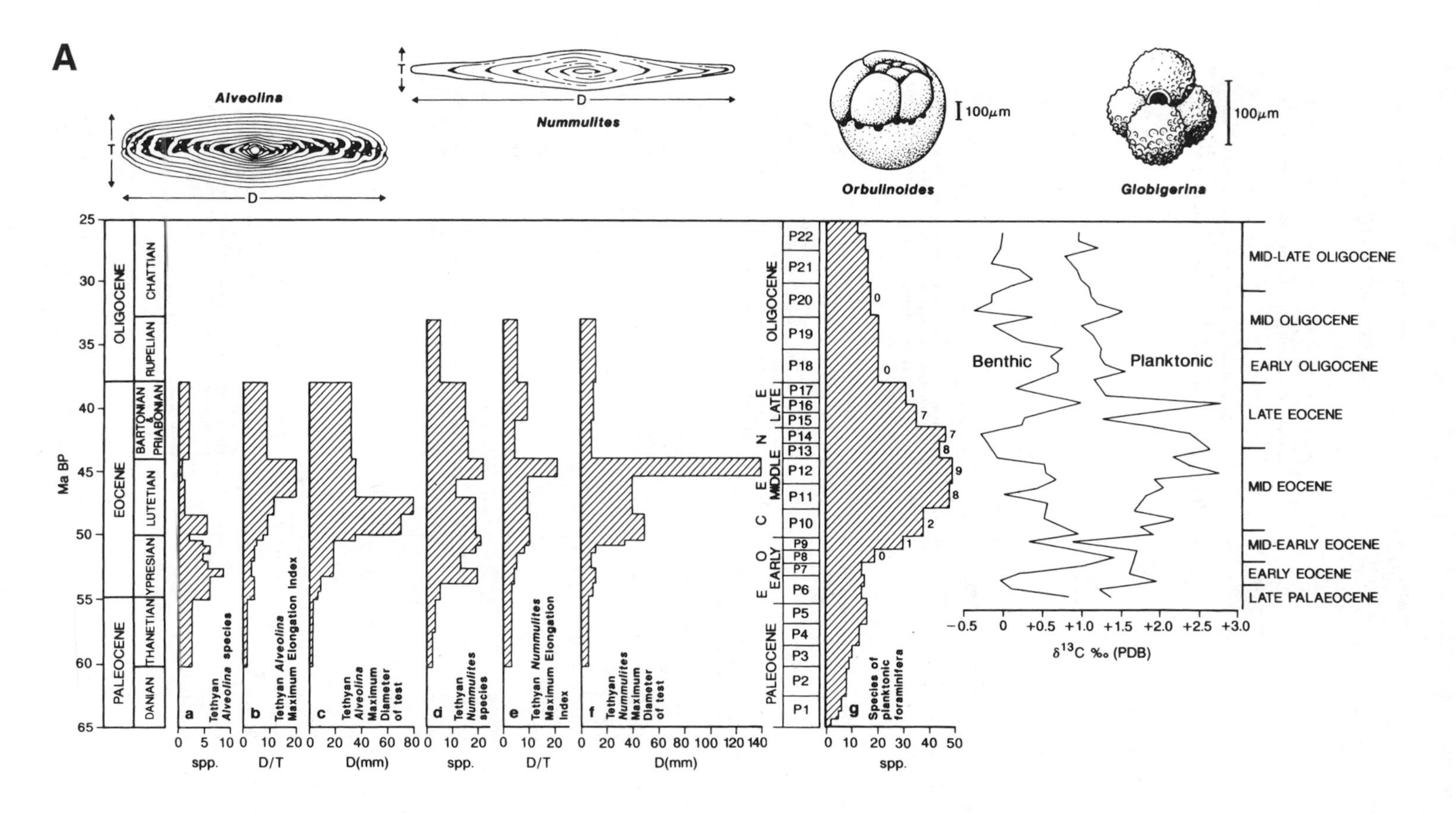

Figure 9.9 Early Tertiary changes in foraminifera. (A) Changes in diversity and skeletal architecture in two genera of large-size foraminifera (*Alveolina* and *Nummulites*) and of typical planktonic foraminifera (*Orbulinoides* and *Globigerina*) during the early Tertiary. Numbers written next to the planktonic foraminifera histogram show the number of species which have additional apertures in their structure. (From Brasier, 1995b.) (continued overleaf.)

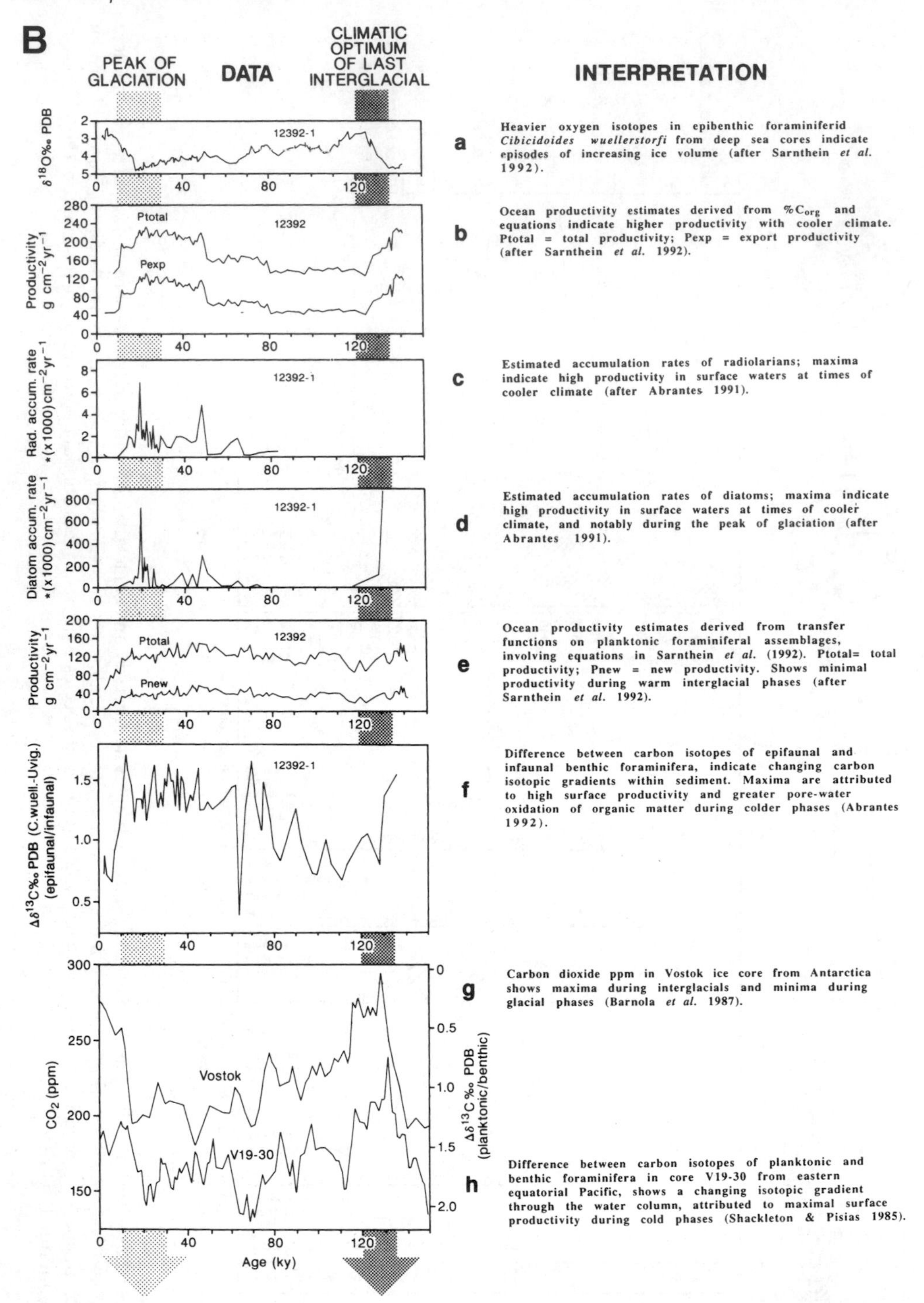

Figure 9.9 (*continued*) (B) Application of palaeobiological data to identify eutrophic episodes of West Africa during Quaternary glacial maxima (at *c.*20 000 and 130 000 years BP); the data are correlated to information from the eastern equatorial Pacific sediments and CO_2 levels from the Vostok ice core in Antarctica; see text for discussion (from Brasier 1995a). © The Geographical Society of London, reproduced with permission.

and developed morphological features consistent with autotroph symbiosis, as would be expected under an oligotrophic regime. $\delta^{13}C$ data showed divergence of curves for benthic and planktonic species, thereby implying a decoupling of these two portions of the system, expected for a stratified ocean. In the late Eocene to early Oligocene, cooling occurred as the Antarctic Circumpolar Current developed, ocean overturn is proposed, and the replacement of oligotrophic plankton communities by eutrophic assemblages taking advantage of the higher nutrient levels brought to the surface by upwelling. The development of a thermohaline circulation in the Tertiary may have been a key feature of that part of Sepkoski's family-diversity curve, and is certainly indicated by the palaeoceanographic changes illustrated in Figure 9.9.

Nutrients and their effects seem to have fluctuated enormously through time. For example, Brasier (1995b) noted that the Jurassic and Cretaceous explosion of evolution of marine microfossil communities (especially coccoliths, foraminifera and diatoms) led to the accumulation of their skeletons on the sea floor, generating the oozes that we now see preserved as fine carbonate and siliceous sediments. These represent a net export of carbon from the ocean system, and storage on the sea floor, which are not only partly responsible for the lowering of atmospheric CO_2, but also (presumably) suppressed nutrients in the water. Such processes might lead to increased diversity but lower the overall productivity of the ecosystems.

The overriding conclusion is that falling CO_2 levels have played a part (maybe the main part) in driving the changes in ocean stability in the Phanerozoic. The apparent stability and stratification of the early Palaeozoic could be attributed to warm, CO_2-rich surface environments, locking nutrients in the ocean sediments, with more frequent release and cycling of nutrients as time progressed. Indeed Figure 9.8 shows a fairly good match between the cooler phases of Phanerozoic history and the occurrence of phosphorites (rapid productivity leads to rapid burial of organic matter, taking phosphorus down to the seabed in concentrated amounts, ultimately forming phosphorites).

A final aspect of the diversity story relates to ocean-floor communities. In times of stratified oceans and anoxic bottom waters, the ocean floor biota was limited; today, diversity of the ocean floor is increasingly recognised as being high. Also, deep-sea vent communities are diverse, with 464 species known (McArthur and Tunnicliffe, 1998), and a fossil record traced back to the Silurian (Little *et al.*, 1998). Ocean floor biotas also include microbial components of course, and iron bacteria, for example (Konhauser, 1998), and it is of interest that the current oxygenated ocean floor allows them to thrive. The wreck of Titanic, lying in *c.*4000 m of water, *c.*500 km south of Newfoundland, is bathed by oxygenated water from the ocean conveyor, and is reported in the media as being gradually devoured by iron bacteria, informally called 'rusticles'; the ship is expected to last only another 90 years or so, before she disintegrates. If the ocean conveyor did not operate, Titanic may have been converted to pyrite in anoxic sediments!

9.6.4 CARBONATE PLATFORMS AND REEFS

The largest carbon store is all around us in epicontinental sedimentary rocks, familiar to any geology student. Deposited in epeiric seas when sea levels were higher, most ocean–atmosphere CO_2 of earlier ages is stranded in limestone, nearly all of which is skeletal material secreted by organisms. Referred to as the carbonate factory, sites of limestone production lie largely on shallow continental shelves, which focus productivity in the photic zone, with $CaCO_3$ stored on the seabed. The shape, size and location of carbonate platforms provides critical regional data on production controls, and draws attention to the application of sedimentary sequence analysis in oceanography. Carbonate platforms are forming today in only a few sites (e.g. the

Caribbean), because sea level is relatively low, and the widespread shallow seas of former times are not a key feature of modern surface environments.

Because carbonates depend on organisms, limestones form best in settings lacking large amounts of clastic sediments (sand and mud, derived ultimately from terrestrial erosion) carried to the depositional area; they also develop best in places where the growth of organisms is faster, i.e. warmer sites, and in general terms provide some information on climate and sea level. The Mesozoic greenhouse in particular led to large-scale carbonate platform development. Carbonate platforms are implicated in much of the discussion of Phanerozoic ocean processes in this book, and details are given where appropriate. Note, however, that while the principle of greater carbonate productivity in warmer waters has a good database, arguments for the deposition of many carbonate deposits in cool, not warm, waters, have been presented, based on modern distributions (James and Clarke, 1997). This has sparked a lively debate, because if many ancient carbonates were deposited in cool, not warm, waters, then this has a huge impact on palaeoclimate models and continental configurations. One example of a response relates to bryozoans, which are abundant in modern cool waters, but less so in warmer water; bryozoans are abundant in the rock record. There is a striking difference between Palaeozoic and post-Palaeozoic bryozoan distributions (Taylor and Allison, 1998), with the former being clearly linked to tropical environments, and fitting well with established palaeogeographic reconstructions. Nevertheless, one has to be careful not to divide up the natural world too strictly; the growth of coral reefs (important carbonate producers) is now recognised as not limited to just warmer waters: a huge living coral bank 13 km long was discovered at 64°N, off the coast of Norway at 300 m depth (Freiwald *et al.*, 1999). Although the coral does not have a symbiosis with algae, well known as an important feature of shallow water coral reefs, it has been shown to grow at a similar fast rate, 2–3 m/ka, so it is important not to make too many assumptions when studying the fossil record.

9.7 CYCLES WITHIN CYCLES: OCEAN CHANGE IS COMPLEX

9.7.1 GLOBAL AND REGIONAL CYCLICITY, AND SEDIMENTARY SEQUENCES

Cyclicity is endemic in Phanerozoic sedimentary sequences, shown in the wide range of stratigraphy and sedimentology texts, and it can be recognised in several forms. While the link between Wilson cycle plate processes and ocean states operates on a large timescale, recognition of important ocean state change on scales of only a few million years has developed in the last twenty years, from the combined study of palaeontology and sedimentology, and forms an excellent illustration of the integrated nature of oceanography within the Earth sciences. Thus, subsumed within the Wilson cycle scale (first order) are cyclicities of varying genesis and timescales, with a vast literature database. Cycles exist because of regular changes in the forces controlling the conditions under which sediment is preserved. Like first order, second order effects are on geologically long scales, of tens of millions of years, and relate to tectonic processes.

The consequence is that ocean basins open and close, and ice sheets grow and decay, all leading to regional and global sea-level changes, which inevitably leave behind large-scale cycles of sediments. The fact that the Earth's crust is not so rigid as was once thought has led to recognition that:

1. the crust flexes due to lateral compressional stress, and
2. it sags under sediment, ice and water loading; for example, a 100 m thick water body depresses the continental crust by 30 m (van Andel, 1994); ice has a slightly

lesser effect, and sediment more, because of their relative densities.

Orogenic belts, flexure and loading may be focused regionally, and have a relationship with oceanography in the sense that the controls on sediment supply influences regional marine depositional processes. Studies and modelling of loading and flexure show the important part they play in the development of sedimentary basins (Lerche *et al.*, 1997; Nadirov *et al.*, 1997). However, many Earth scientists would not regard such aspects as strictly oceanographic; but not only do regional effects appear in sedimentary sequences, but those may extend to global scale if, for example, global sea-level change is stimulated by a strong regional effect. Such discussion emphasises the difficulties of defining the limits of ocean study.

Broadly, globally influencing Earth-bound processes fall into two major groups, recognised by well-established terms: tectono-eustasy and glacio-eustasy, discussed in Section 9.7.2 (but remember our comment in Chapter 3 about whether there *is* such a thing as eustasy). A third key feature is extraterrestrial control of inconstancy of planet motion (Section 9.7.3).

9.7.2 TECTONO- AND GLACIO-EUSTASY

Descriptions of these abound, and they may be interlinked or independent. They are examined here in three examples from the stratigraphic record to demonstrate the interlink between oceanography and cyclic sedimentary processes. A detailed treatment might be regarded as straying too far into the realms of sedimentology, and away from the scope of this book.

9.7.2.1 Silurian reef cycles

In 1990 Lennart Jeppsson published a theory to explain why changes in patterns of evolution and extinction of certain abundant marine microfossils coincide with shifts in sedimentary patterns in both shallow and deep waters. Drawing on observed processes in modern oceans, a recent ocean-state theory (Aldridge *et al.*, 1993; Jeppsson, 1987, 1990; Jeppsson *et al.*, 1995) has received increasing attention from other authors, and shows how an interesting idea stimulates other research.

Most work done so far on this is in the Silurian record, with conodont microfossils (phosphatic tooth structures of a mobile animal in the Palaeozoic), which are important biostratigraphic indicators. Conodonts in open marine shelf settings show a pattern of stepwise extinctions and evolution, associated with extinctions of other groups such as trilobites. The key to developing useful theories to explain the faunal changes is a high resolution stratigraphy. Correlations based on conodonts are well developed. Nevertheless, it is important to realise that stratigraphy is a living science; adjustments to the correlation system continue, and it is likely that amendments will be made to Jeppsson's theory as stratigraphic refinements are made.

At present, this theory is based on changes in patterns of both conodont extinctions and originations (i.e. evolution of new species), in tandem with certain shifts in sedimentation styles in deep and shallow water. These led to application of modern oceanographic principles to the Silurian record, using the differences between salinity- and temperature-dense waters at different times. Water density is controlled by both temperature and salinity, and the relative importance of each depends on the temperature and salinity range of the body of water examined (Chapter 2). The results of integrated study suggest cyclic fluctuations in the patterns of global ocean circulation, driven by changes in atmospheric CO_2. Jeppsson's theory envisages two ocean states on the basis of the faunal and sedimentary data, Primo and Secundo (P and S), which flip from one to the other (Figure 9.10). During the transitional stages, extinction events occur, which may even be related to disruptions in each state (thus P–P or S–S events). The Earth is presently in a P-state.

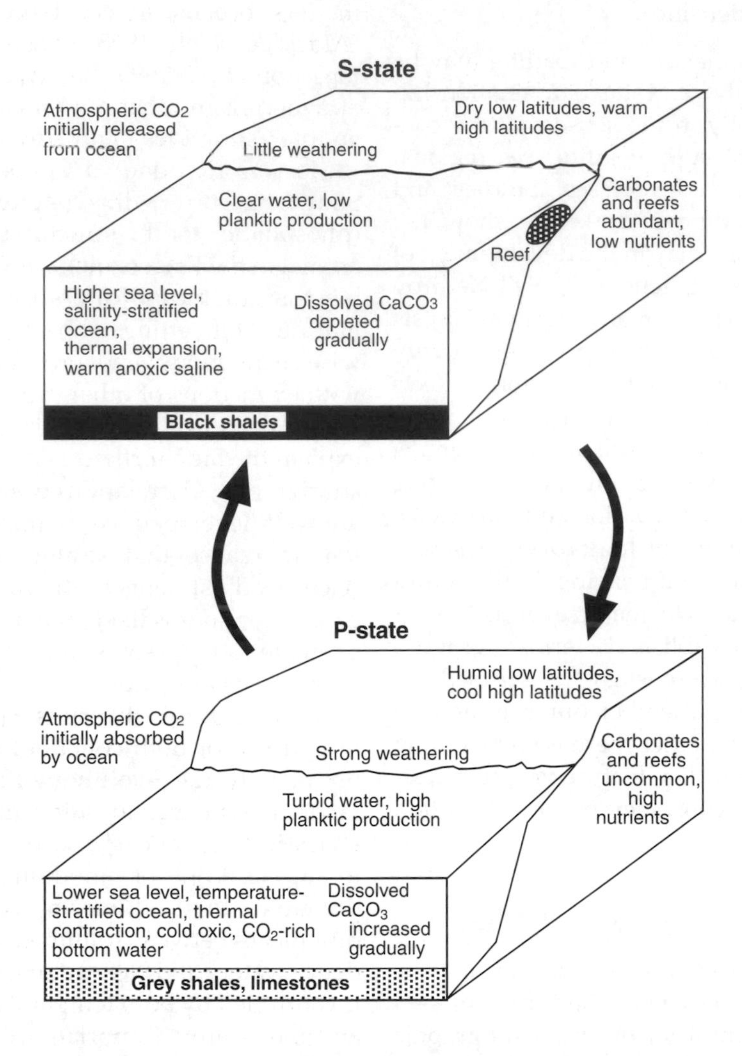

Figure 9.10 Diagram of Jeppsson's (1990) P- and S-state ocean theory. P-states are typified by cooler conditions and active ocean circulation, while S-states are present under warmer conditions and poor ocean circulation. See text for discussion.

During P-states, the oceans are fed with clastic sediments in cool humid climates, such that not only are shallow water facies dominated by clastic sediments rather than carbonates, but also that the deeper water sediments are typified by grey shales and silts. The latter indicate that deep waters were oxygenated, occurring when the water density is controlled principally by temperature, such that cool, oxygenated bottom waters are generated, much as they are in Arctic and Antarctic waters today. This is temperature-controlled stratification. Such water holds more CO_2, so that the climate is cool. Conversely, S-states are typified by warmer climates which lead to drier conditions, and therefore less clastic supply to the oceans because terrestrial weathering is suppressed. Warmer sea causes salinity-controlled stratification, as the upper waters are more evaporated, and there is consequently less CO_2 in the deeper water. This water, being warmer and saltier than in P-states, is not as rich in oxygen. Because salinity stratification results in a severe reduction in deep-ocean circulation, the resulting deep facies are de-oxygenated, leading to deposition of black shales. The cooler waters of P-states force seawater to slightly contract, so that there is a small sea-level fall (*c.* 3m), and the converse happens in S-states, although a critical point is that this is a climatic theory, not a sea-level change theory.

Jeppsson's CO_2-controlled oceanic stratification model has an important aspect of the control of carbonate platforms, which contain reefs, the most diverse of all marine ecosystems. These occur at certain intervals of sea-level highstand and clastic-starved seas during warm periods when seas were salinity-stratified and bottom waters were poorly oxygenated (S-states). Hallock and Schlager (1986), among others, recognised that oligotrophy enhances modern reef growth, and such a feature is consistent with the S-state. Two portions of the Silurian succession on the famous reefs of the Swedish island of Gotland examined by several authors revealed that reefs dominated by a major group of calcified sponges (stromatoporoids) can be related to this oceanic model (Kershaw, 1993); the chronology of these reefs matches both the Silurian global sea-level curve (Johnson *et al.*, 1991) and the P- and S-states. The correlation of S-state reef cycles has been attempted on a global scale to reveal eight episodes of reef growth (Brunton *et al.*, 1997) (Figure 9.11). The first four of these correlate with known interglacial phases in South America; thus a glacio-eustatic control can be attributed with some confidence to the Llandovery Epoch. However, the knowledge of sea-level change and glaciations is less well developed for the Upper Silurian (Wenlock, Ludlow and Pridoli Epochs), and so the causes of the last four reef cycles may or may not be the same. A GCM of the global Wenlock ocean (Moore *et al.*, 1993) suggested that the Wenlock Epoch was a time of high productivity in a generally stratified ocean, without ice caps. The asymmetry of the continents (the Earth's land was concentrated in the Southern Hemisphere), with Gondwana stretching to the equator, was regarded by Moore et al. (1993) as preventing polar glaciation, assisting the maintenance of warm global conditions. Although this model suggests seasonality was an important feature of the Wenlock world, the finer scale of change interpreted as diversification into CO_2-controlled climatic states is not identified by Moore *et al.* Resolution of controls on Upper Silurian reef cycles remains for future research.

An interesting aspect of this study is a reconsideration of the conditions of reef growth. Because reefs grow up to sea level when they are fully developed, there is a widely published preconception that reefs develop in regressive settings. The position of the Silurian reefs in relation to the sea-level curve suggests the opposite: that reefs are more related to sea-level *rise*. This has a critical impact, not only on the understanding of reef controls, but also on the value of reefs as sea-level indicators in the geological past. Modern reefs show features which make it possible to use them to identify past sea levels, especially coral microatolls, which identify the position of the sea level, but such things are rare in the past. Thus palaeoceanographers (= sedimentologists and palaeontologists) need to be aware of such constraints.

9.7.2.2 Carboniferous change

The general pattern of sedimentary deposits in the Carboniferous of northern Europe is as follows. Following the regression represented by the largely Devonian Old Red Sandstone (ORS) continent which was created by the Caledonian Orogeny, renewed global transgression in the Early Carboniferous (Dinantian Epoch) led to widespread deposition of limestones and

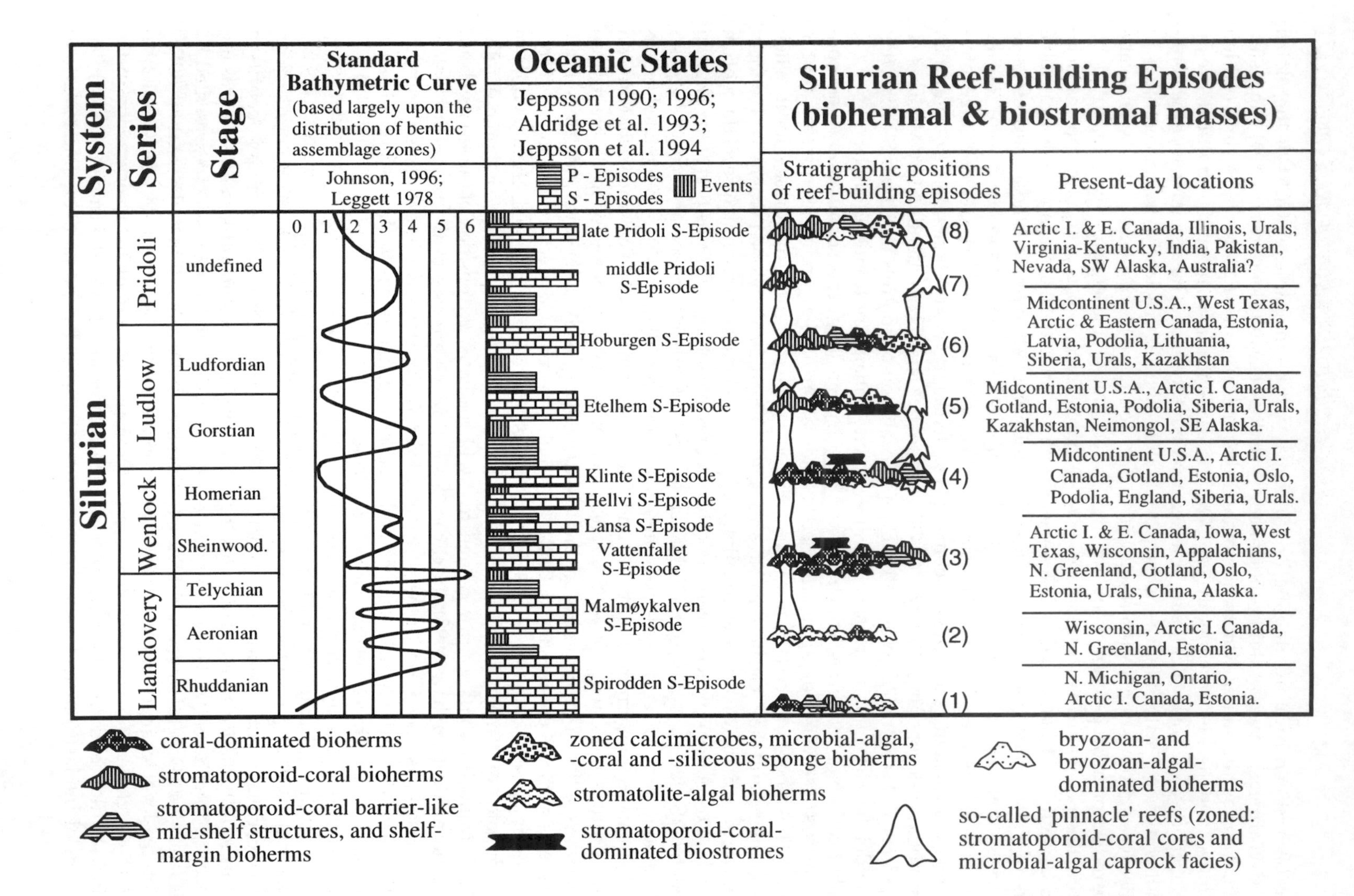

Figure 9.11 Silurian reef cycles, and their relationship with sea-level change, and ocean state. In general, reef systems developed in either rising sea-level settings or at the highstand of transgression, rather than lower sea levels, although the illustrated patterns are complex because of regional overprint of global sea-level change (from Brunton *et al.*, 1998). Reprinted courtesy of the New York State Museum, Albany, NY.

shales, which show cyclicity (Busch and Rollins, 1984; Busch and West, 1987; Ramsbottom, 1973). The middle (Namurian) and later (Westphalian and Stephanian) Carboniferous records (Ramsbottom, 1977) show gradual regression, but with a superimposed cyclicity of sea-level change (Figure 9.12). The cyclicity is recognised as being of such a large lateral extent that correlation of Lower and Middle Carboniferous sequences (Dinantian and Namurian) is based on the cycles (Read, 1991), with less reliance on fossils. The Dinantian transgressions are not all global, and there is a distinction between regional (especially in the tectonically active Western Europe) and global features; the global units are attributed to growth and decay of the Gondwana ice sheet.

The change from limestone-shale dominated sequences of the Dinantian to sandy deltaic facies in the Namurian Epoch is attributed to uplift in Northern Europe, so that the facies were dominated by freshwater deposits. Transgressions in these non-marine sequences generated fast facies shifts of marine sediment, flooding epeiric shelves and creating marine bands. The evolution of the goniatites (nektonic molluscs) found as fossils in these marine bands was so rapid that they make an excellent stratigraphic zonal system.

North American Carboniferous sequences differ from European in one important respect: the North American continental crust was more stable, so that sea-level change led to easily correlatable units of accumulating sediment, spread over much wider areas than are possible in Europe. These are therefore aggradational, interrupted by episodic sea-level falls, and referred to as Punctuated Aggradational Cycles (Busch and West, 1987). Such work also distinguished between sea-level changes driven by external processes (such as glacial expansion and retreat in Gondwana in the Upper Carboniferous) and those by regional effects (such as switching of direction of river systems, or limiting water depth in certain areas); these two controls are called allocyclic and autocyclic respectively.

Thus the Carboniferous sedimentary system as a whole illustrates the interplay of tectonics, sedimentary processes and sea-level change in the development of a mixed facies system, with short-lived epicontinental marine episodes, and a complex set of controls on cyclicity.

9.7.2.3 Permian carbonate–evaporite sequences of the North Sea area

During the first half of the Permian Period, Britain and the North Sea area were under arid desert settings, after the completion of the assembly of Pangaea (Figure 9.2) under sea-level lowstand. The change in climate from the humid Late Carboniferous to arid Early Permian is recorded in the sedimentary successions of Europe and also in cyclic sequences of sediments in North America (West *et al.*, 1997). Dunes, wadis and desert lakes provided the sediments which later became the reservoir for North Sea gas in the Rotliegende sandy facies (Glennie, 1990a), and these developed during maximum glacial growth in Gondwana. In the Rotliegende facies are the huge sand dunes only possible in times of steep global temperature gradients of glacial times, referred to in Section 9.6.3. As time passed into the last half of the Permian, the Gondwana ice sheet collapsed, causing a global sea-level rise. The land-locked North Sea basin appears to have lain below sea level, but was accessed from Panthalassa Ocean along the incipient North Sea fracture zone developed by tension in the Northern hemisphere crust. Rapid flooding of the North Sea basins (both northern and southern basins) covered the Rotliegende sandy deposits with marine shales. These rapidly became anoxic, preserving copper-rich shales (Kupferschiefer), because of the lack of circulation in the deep portions. The basin margins, however, in shallow water, became sites of biotic diversification, with the famous later Permian (called Zechstein) reefs (Figure 9.13). However, the reefs were short-lived because the transgression did not continue, leading to isolation of the North Sea basin, and as it

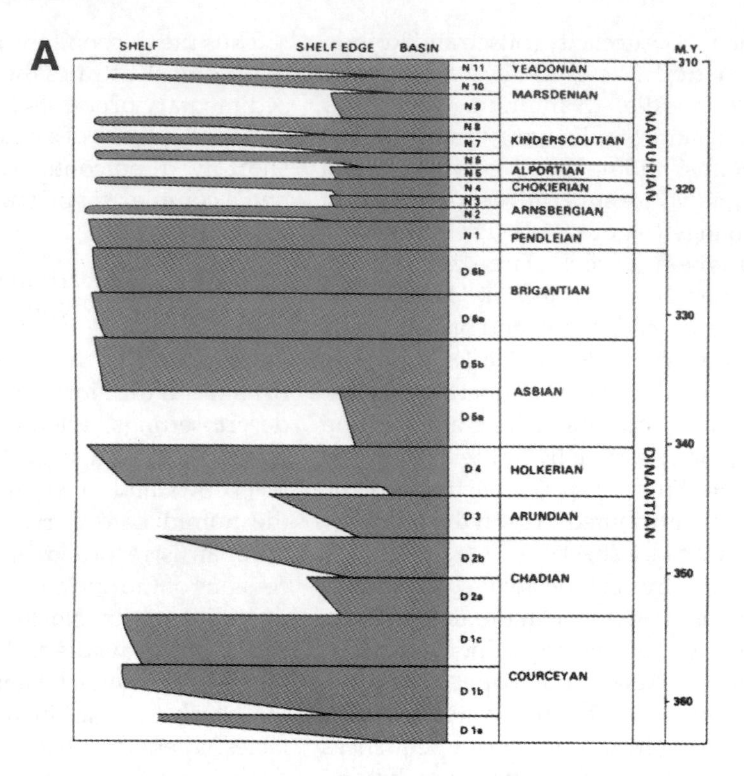
A
SHELF
SHELF EDGE
BASIN
M.Y.
N 11
N 10
N 9
N 8
N 7
N 6
N 5
N 4
N 3
N 2
N 1
YEADONIAN
MARSDENIAN
KINDERSCOUTIAN
ALPORTIAN
CHOKIERIAN
ARNSBERGIAN
PENDLEIAN
NAMURIAN
D 6b
D 6a
BRIGANTIAN
D 5b
D 5a
ASBIAN
D 4
HOLKERIAN
D 3
ARUNDIAN
D 2b
D 2a
CHADIAN
D 1c
D 1b
D 1a
COURCEYAN
DINANTIAN
310
320
330
340
350
360

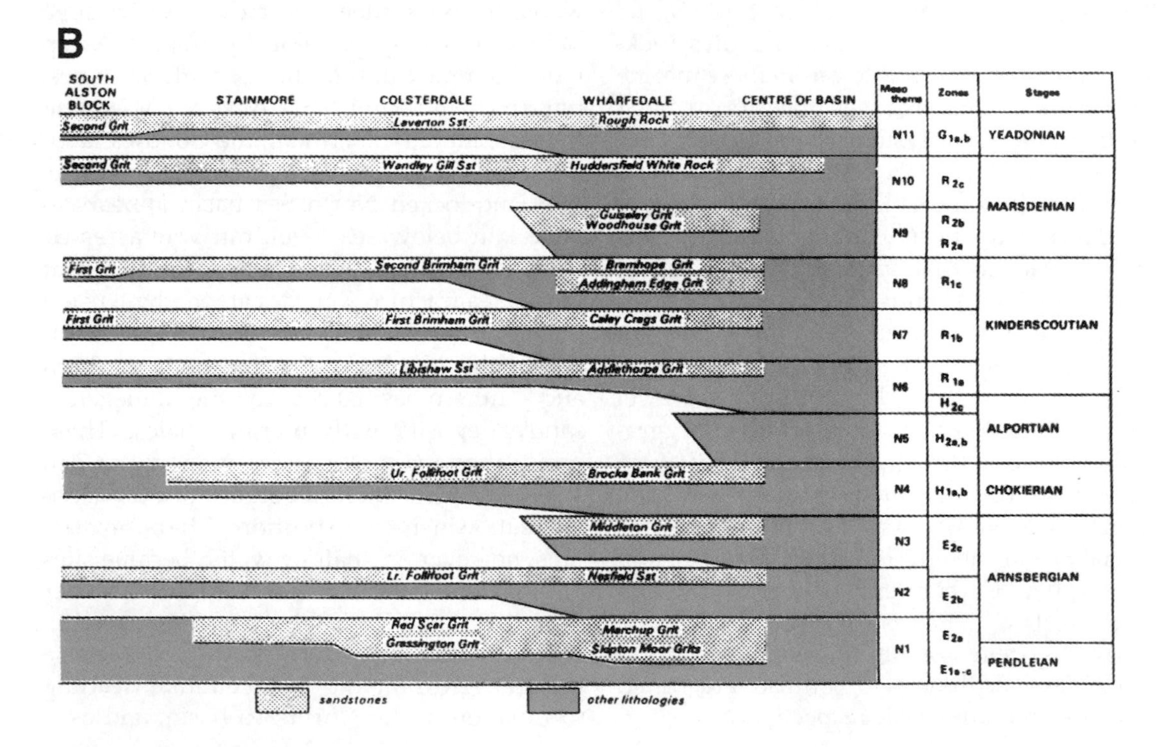
B
SOUTH ALSTON BLOCK
STAINMORE
COLSTERDALE
WHARFEDALE
CENTRE OF BASIN
Meso thems
Zones
Stages
Second Grit
Laverton Sst
Rough Rock
Second Grit
Wandley Gill Sst
Huddersfield White Rock
Guiseley Grit
Woodhouse Grit
First Grit
Second Brimham Grit
Bramhope Grit
Addingham Edge Grit
First Grit
First Brimham Grit
Caley Crags Grit
Libishaw Sst
Addlethorpe Grit
Ur. Follifoot Grit
Brocka Bank Grit
Middleton Grit
Lr. Follifoot Grit
Nesfield Sst
Red Scar Grit
Grassington Grit
Marchup Grit
Skipton Moor Grits
N11
N10
N9
N8
N7
N6
N5
N4
N3
N2
N1
YEADONIAN
MARSDENIAN
KINDERSCOUTIAN
ALPORTIAN
CHOKIERIAN
ARNSBERGIAN
PENDLEIAN
sandstones
other lithologies

evaporated, gypsum, then halite, was deposited in huge quantities, being replenished by small flooding events from the north; thus a substantial thickness of evaporite minerals accumulated (Figure 9.13B).

Repeated major flooding and evaporation generated cycles of carbonates and evaporites leaving the five Zechstein cycles (Figure 9.13C). Thus the oceanographic message here is that glacio-eustasy appears to have influenced the sedimentary regime of a marginal basin system, and, by the way, created unique conditions of stratigraphy; Carboniferous humid coal swamps are overlain by a porous and permeable sandstone, then an impermeable evaporite cap. There are certain places in the southern North Sea basin where all three units are preserved intact, leading to one of the world's largest natural gas fields (gas comes principally from wood decay, while oil is largely a breakdown product of buried phytoplankton).

9.7.3 SPLENDID INSOLATION: ORBITAL FORCING AND OCEAN PROCESS REVEALED BY SMALL-SCALE FEATURES

The ocean record, ancient and recent, is littered with thin-bedded sediments, ranging from the Precambrian banded iron formations (Nisbet, 1987) to modern laminated deposits (Kemp, 1996), in which small-scale bedding cycles abound. Early Precambrian BIF show alternating iron-rich and iron-poor layers on cm scale, correlatable across thousands of km^2, the cause of which has not been established; at the other end of geological history are Quaternary lake varves, which have slight variations of grain size due to seasonal fluctuations in sediment supply. Such small-scale alternations are called rhythmites, and are being recognised at various points throughout the geological record (e.g. Elrick and Hinnov, 1996; House and Gale, 1995). Controls on minor cyclicity may be regional climate, but there is now a wealth of evidence that variations in planetary motion influenced oceanic processes, epitomised as Milankovitch cyclicity. Thus the Earth's distance from the Sun and positions of the poles vary regularly and influence erosion rates through climate change driven by variations in solar influx; this is now formally termed orbital forcing (Figure 9.14). One result of these changes is that the tropics are shrinking: in 1998 the tropics of Cancer and Capricorn came *c.*29 m closer together, transferring *c.*1100 km^2 of Earth's surface from the tropical belt (*Geoscientist*, 1999, Vol. 9). Tidal cycles are unusually preserved, but examples are known in Carboniferous tidal sediments; lunar cycles have been recognised in coral banding (House and Gale, 1995), seasonal growth in oxygen isotope data from Carboniferous brachiopods (Mil and Grossman, 1994), Cretaceous bivalves (Steuber, 1996), and varying types of sedimentary cycles. A record of annual sedimentation in the Cariaco Basin, Venezuela, through the late glacial to Holocene episode, 12.6–9 ka BP (Hughen *et al.*, 1996), indicates the potential of

Figure 9.12 Diagrams of vertical sections through Carboniferous sedimentary cycles related to sea-level change. These diagrams are *chronological* in nature, and are not simply vertical sections through the rock sequences; thus the white portions represent locations where there is no preserved rock through the vertical interval illustrated. In higher sea levels, sediments are deposited on the shelf, while when the sea level is lower, only the basins receive sediment, and the shelf regions are eroded, or receive no sediment. (A) The lower and middle Carboniferous systems comprise a set of major sedimentary units, (mesothems) strongly influenced by large-scale sea-level change; sea-level change was rapid, so the extensions of sediment onto the shelves are almost flat-based, thereby indicating little time elapsed during sea-level rise. (B) Details of the Namurian series, showing that shelf deposits are thinner, and have more erosional breaks than basin deposits, where water depth is less affected by sea-level change, so that sequences are thicker (from Ramsbottom, 1977). Reproduced with permission of the Yorkshire Geological Society, from *Proceedings of the Yorkshire Geological Society*, **41** (3), Ramsbottom, W. H. C., Major cycles of transgressions and regressions (Mesotherms) in the Namurian, pp. 261–291 (1977).

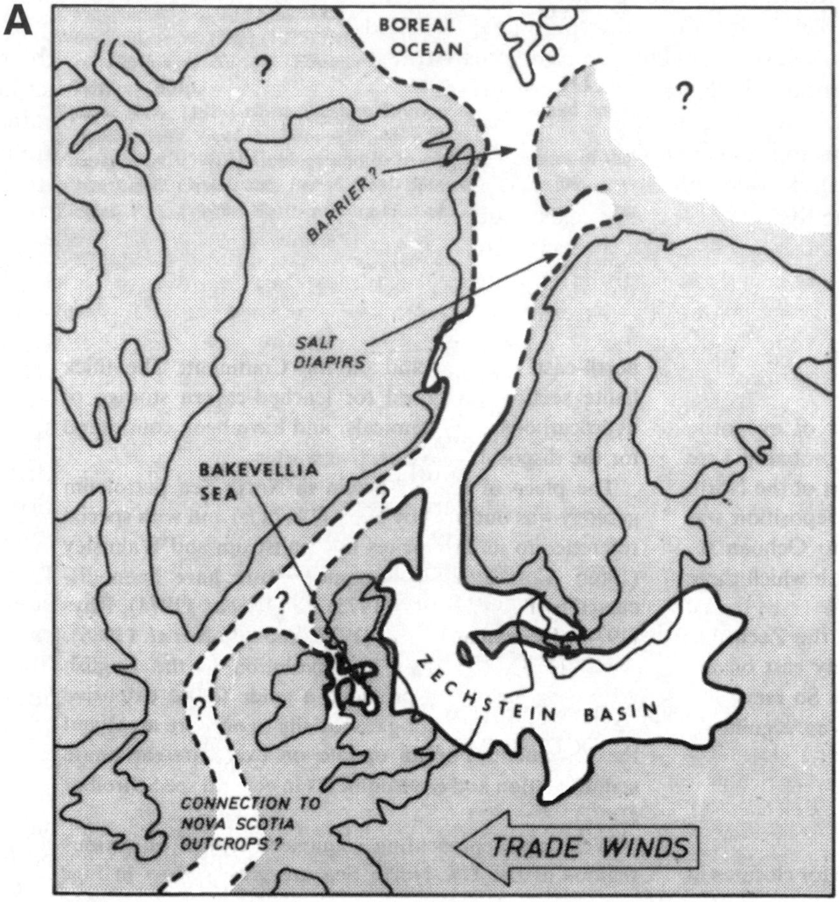

A

Figure 9.13 Zechstein cycles of carbonate-evaporites, in the North Sea region. During this time the Northern European region was part of Pangaea. (A) General setting of the Zechstein Basin, showing its tenuous link with the Boreal Ocean to the north.

the oceans to preserve annual signals. In this case, between 10.8–9.8 ka, increased runoff is shown by thicker dark laminae, while (upwelling-related) productivity increased (thicker diatom-rich laminae). After 9.8 ka, these deposits thinned rapidly, and the changes are coincident with the end of the Younger Dryas cold phase, which had such profound effects in the Atlantic circulation patterns, discussed in Chapter 10.

More focus, however, has been placed on the larger scale cycles, related to the three rotational/orbital variations of the Earth, which are: precession (19–23 ka), obliquity (41 ka) and eccentricity (54, 106 and 410 ka). All these scales are described in a recent review (House, 1995). Orbital forcing most commonly leads to alternation between two sediment types, for example limestone and shale, and organic-rich and organic-poor shales. For limestone and shale, the control is that, during one part of the cycle, the land area adjacent to the basin receives less heat from the Sun, leading to cooling and less erosion; therefore supply of clastic sediment to the basin is reduced, so limestone is deposited. In the

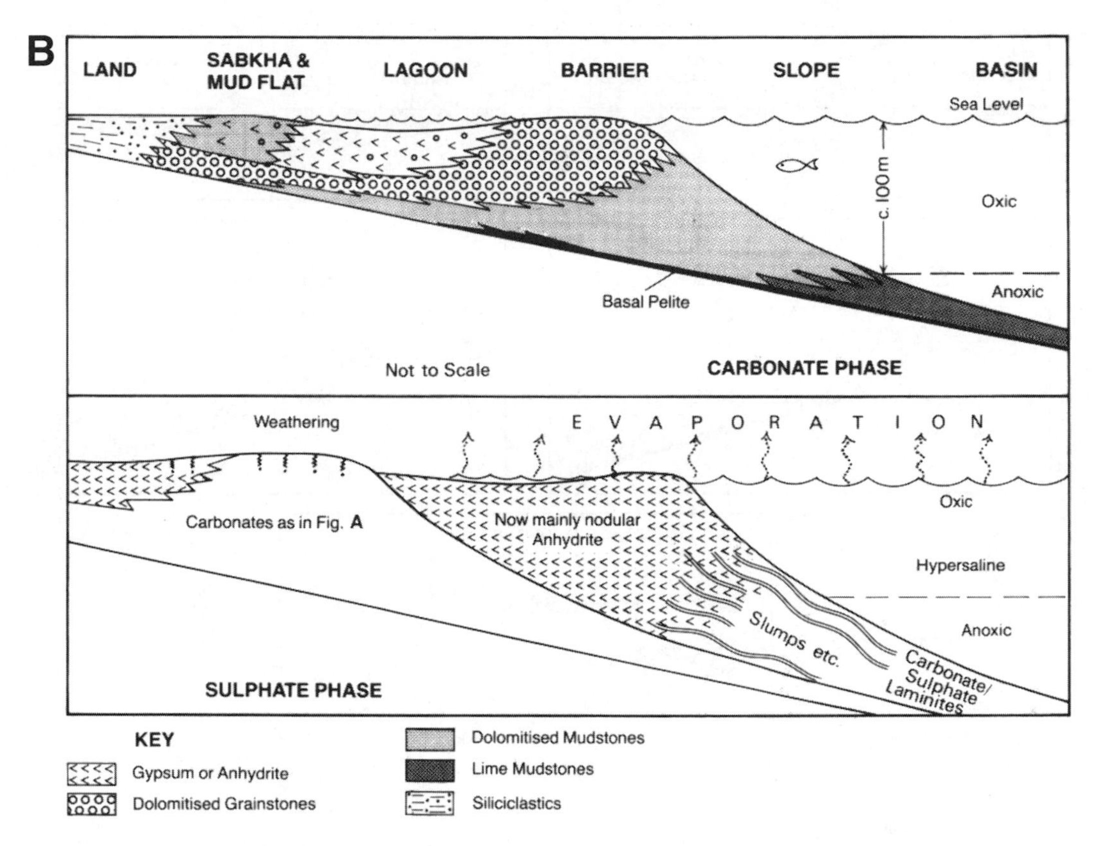

Figure 9.13 (*continued*) (B) Schematic model of the alternate states of the Zechstein Sea; following an influx of seawater from the north, a carbonate phase developed; the basin then became isolated, leading to large-scale evaporation of water, with evaporite deposition of the sulphate phase. Note the anoxic deep water in the restricted circulation of the basin in both settings (continued overleaf).

other part of the cycle, erosion is greater, so that mud is carried into the basin and swamps the limestone formation, depositing mud (as shale) instead, as might apply to the Lower Jurassic of southern Britain. For organic content variations, the sediment supply does not change, but the degree of oxygenation of the water is affected by climatic control on water circulation, with episodes of eutrophication (giving surface algal blooms and therefore organic matter accumulation on the seabed) alternating with times of nutrient reduction giving organic-poor deposits, as might apply to the Kimmeridge clay in Dorset, southern England. Facies interpretations based on microfossil palynomorphs show orbitally forced cycles (Waterhouse, 1995).

In the Precambrian, BIF banding may be orbitally forced (Simonson, 1996), but the mechanism is unclear; presumably it would require a process to modulate oxygen supply, but the alternative is that mantle processes regulated Fe input into the ocean via exhalation from the sea floor. In the Phanerozoic oceans, orbital forcing has been proposed for many Mesozoic sequences, as in the two examples above, and the evidence is growing for its application to Palaeozoic rocks, for example, the huge Devonian Catskill Delta in Eastern USA (Tassell, 1994). Great care is needed in

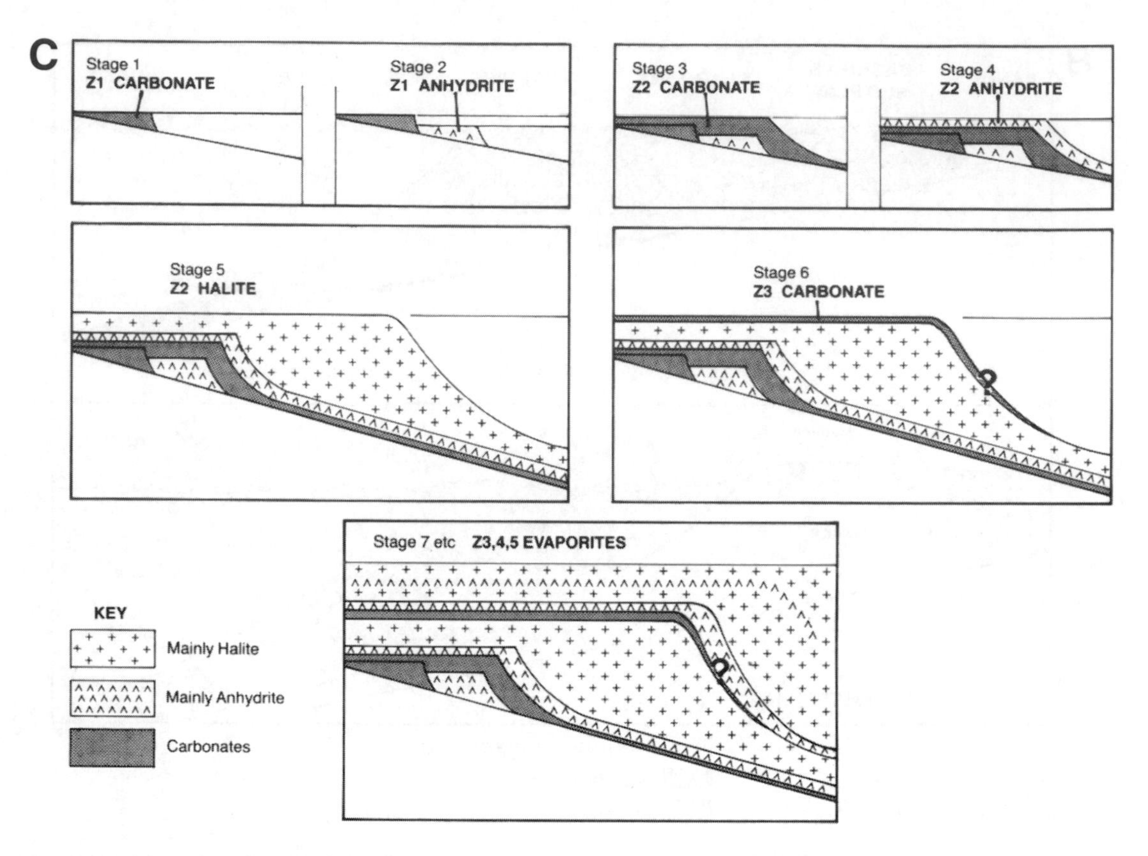

Figure 9.13 (*continued*) (C) Schematic models of the sequence of events throughout the evolution of the Zechstein Basin; note the successive carbonate and sulphate phases, which result from repeated flooding of the basin. Only the first three cycles contain carbonate units, and the third one is very thin; thus the basin progressively receives less replenishment from the Boreal Ocean (from Glennie, 1990b). Reproduced with permission, copyright Blackwell Scientific Publications.

the identification of orbital processes, and environmental setting is critical; for example, in the Pennsylvanian of mid-USA, depositional sites near orogenic belts are more likely to be influenced by the regional tectonics, while in basin sites a climate signal should be more easily discerned (Klein, 1992). Basin sequences may be highly sensitive to orbital processes; the potential for using their results in correlation of strata between sites (cyclostratigraphy) is in its infancy (House and Gale, 1995).

The technique of recognising orbital forcing involves careful logging of sequences, and then relating the regular changes of sediment type (limestone–shale alternations, for example) to a timescale; only when the amount of time for deposition of a particular sequence is known can the variations be equated with one or more of the forcing agents. Problems lie in the superimposition of one orbital forcing scale upon another, and also in the presence of small gaps in deposition. Some beds may be missing from the predicted sequence, because one particular area did not receive sediment due to some regional disturbance of the regular pattern. Some sequences preserve sediments fitting one or other of the orbital features

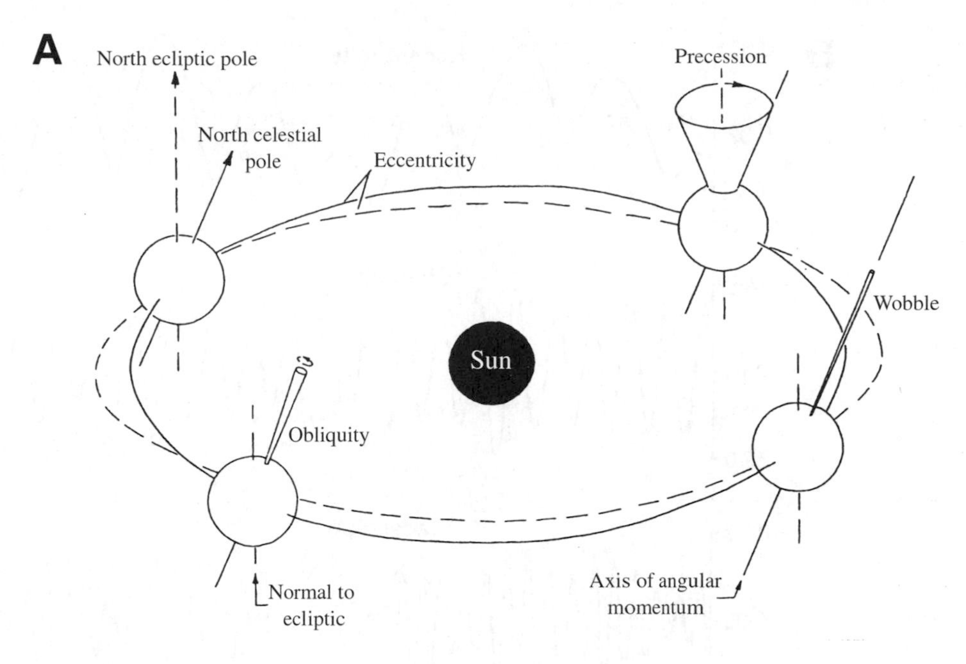

Figure 9.14 (A) Orbital variations of the Earth, which influence the degree of insolation in a cyclical fashion (continued overleaf).

(House and Gale, 1995) (Figure 9.14E). The types of preserved features range from simple sediment supply variations, variations in organic supply, to more complex changes, such as in the Cretaceous Chalk of England (flint cycles, hardground cycles and nodular chalk cycles: Arthur, 1986; Bottjer *et al.*, 1986) (Figure 9.15). Of the three orbital forcing cycles (eccentricity 100 000 a, precession 23 000 a and obliquity 41 000 a) affecting global ice volume of the late Quaternary glaciation, the weakest of the three (eccentricity) seems to provide the strongest influence on the glacial cycles, for reasons unknown (Broecker and Denton, 1990b, p. 44).

Finally, orbital forcing processses are invoked to explain rhythmic changes in pollen, spores and other terrestrial plant material in sediments in which the cyclicity is not detectable in the sediments themselves (Waterhouse, 1999), in examples from the Jurassic of Britain.

9.8 CONCLUSION

The Phanerozoic oceans demonstrate both unidirectional and cyclic change at different scales, with a range of control engines. The opportunities for detailed study are much better than in the Precambrian, because the continents are larger, and more unaltered rock is available to study. As always, the more information, the more complex is the story. Study of sedimentary sequences also emphasises the interdisciplinary nature of palaeoceanography, and the consequent difficulty of defining the limits of its study. Finally, this chapter has focused on trends in ocean change and their controls. The trends are interspersed by events of varying magnitude and effect, case studies of which are given in the next chapter.

9.9 SUMMARY

1. Phanerozoic ocean change is complex, but can be divided broadly into two sets of

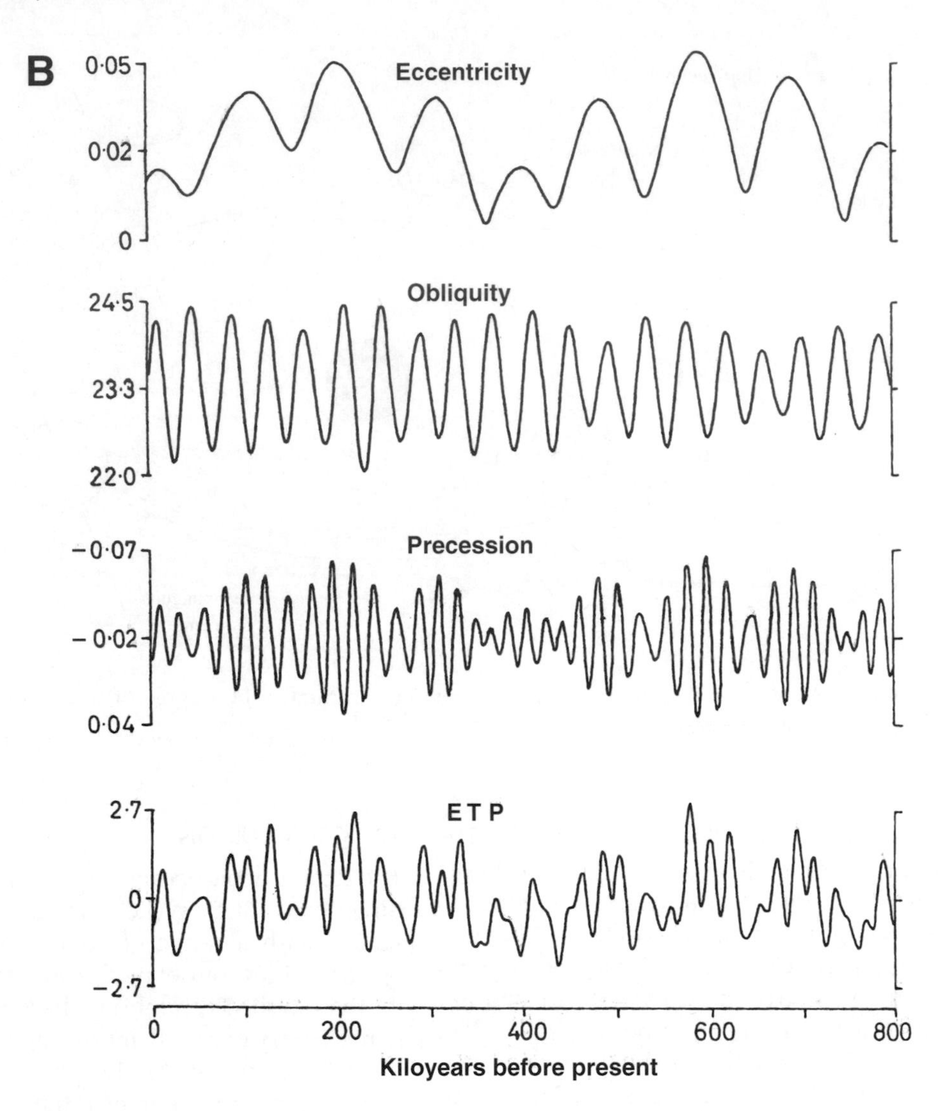

Figure 9.14 (*continued*) (B) Variations in solar energy arriving at the atmosphere as a result of cyclical change in orbital parameters shown in (A). Note that the three cyclical series interact, and produce a summed result of energy variation. (C) Detail of vertical variations in sediments and fossils in basin sediments of the Kimmeridge Clay in southern England (on opposite page). From House (1995), in House and Gale (1995). Reprinted with permission; © The Geological Society.

processes: unidirectional and cyclic, which are intermingled, but are dominated by cyclic features. The range of features demonstrates the fully interlinked state of the ocean–atmosphere–land system.

2. Unidirectional change includes the gradual drawdown of CO_2 from the ocean–atmosphere system into carbon stores, mainly carbonate (C_{CARB}) but also organic carbon (C_{ORG}). Burial of carbon is necessary for the maintenance of an oxygen-rich atmosphere, otherwise the oxygen would be converted back into CO_2. The result seems to have been a predominance of

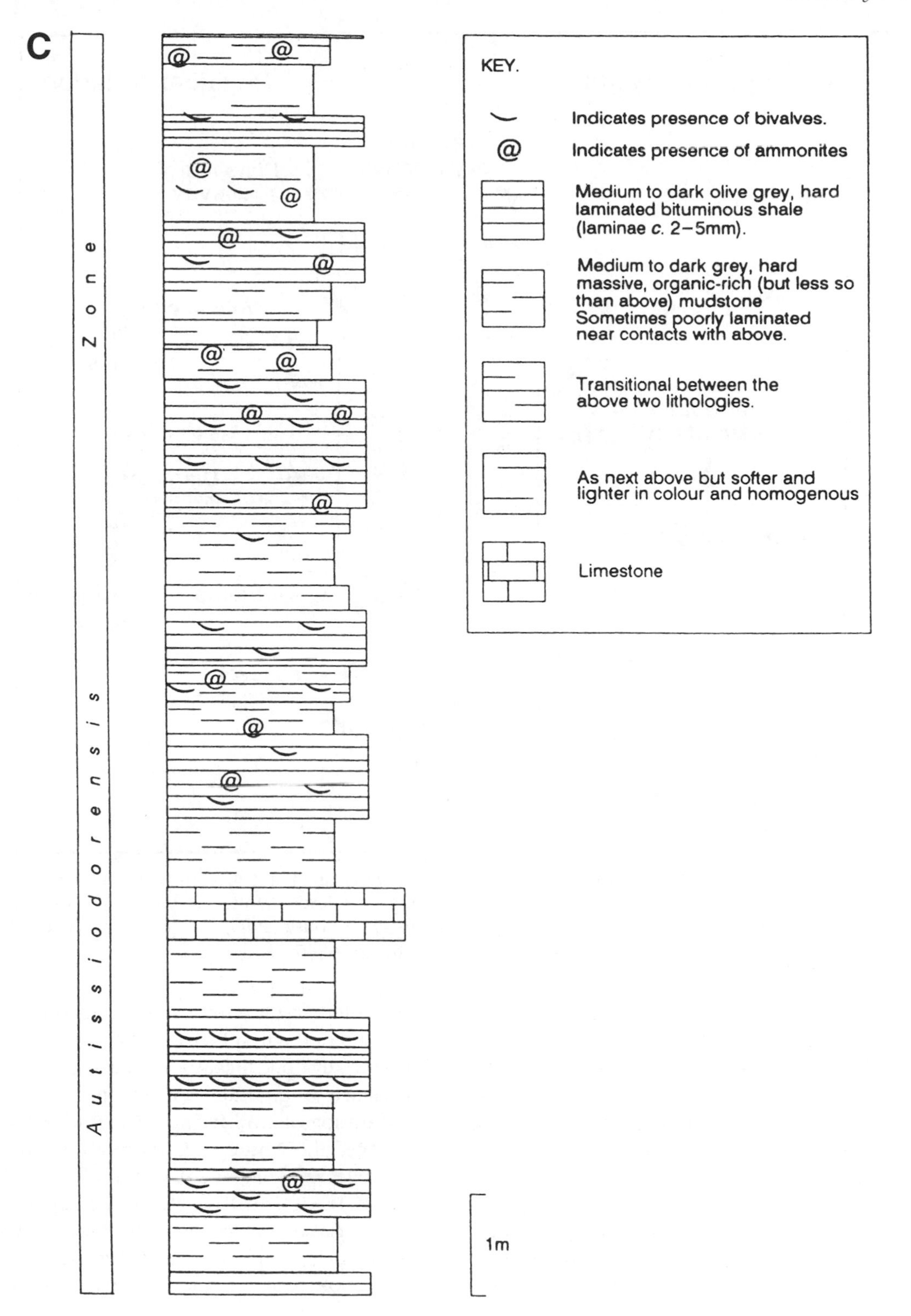

Figure 9.14 (*continued*) Figure continued overleaf.

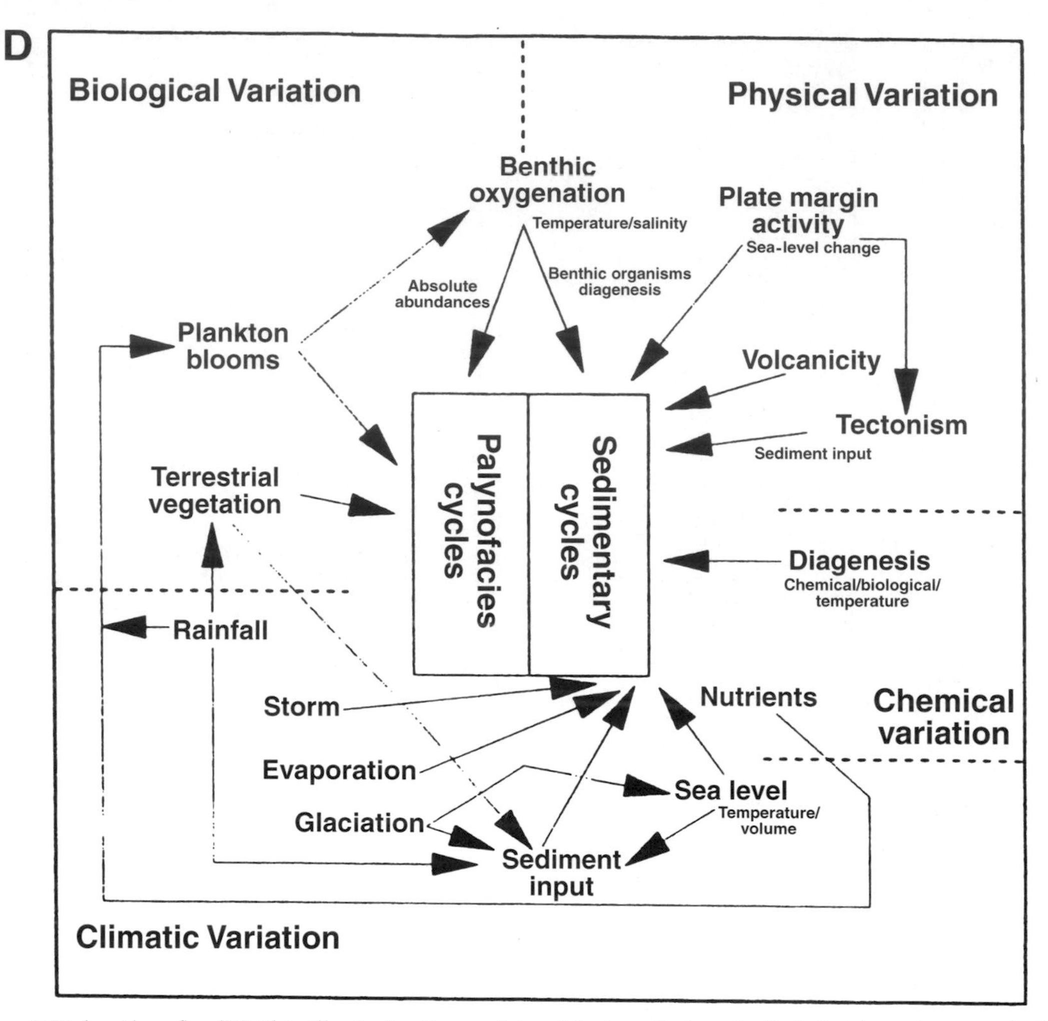

Figure 9.14 (*continued*) (D) This illustrates the variety of factors that controlled the deposits recorded in (C). In (E), the impact of orbital processes is emphasised, with reference to the obliquity and precession cycles, the two which have the closest fit to the data. From Waterhouse (1995), in House and Gale (1995). Reprinted with permission; © The Geological Society (continued overleaf).

salinity-controlled ocean stratification in the Palaeozoic warm climates, with poor deep-ocean circulation, helped by the fact that sea level was higher as a result of the generally active plate tectonic system in the Palaeozoic. Vertical mixing seems to have occurred on a large scale during the mid-Phanerozoic cool phase with the assembly of Pangaea, and growth of glaciation in the southern (Gondwana) continent. Rodinia-to-Pangaea represents a single cycle of continent fragmentation and reassembly – the end of the first Wilson Cycle. A return to warm stratified oceans in the Mesozoic as Pangaea fragmented lasted only until the separation of Antarctica in the mid-Tertiary allowed the Antarctic Circumpolar Current to develop. That event eventually isolated Antarctica from the poleward transport of equatorial heat, leading to the Late Tertiary cooling, and presumably contributed to glaciation and sea-level fall. Currently

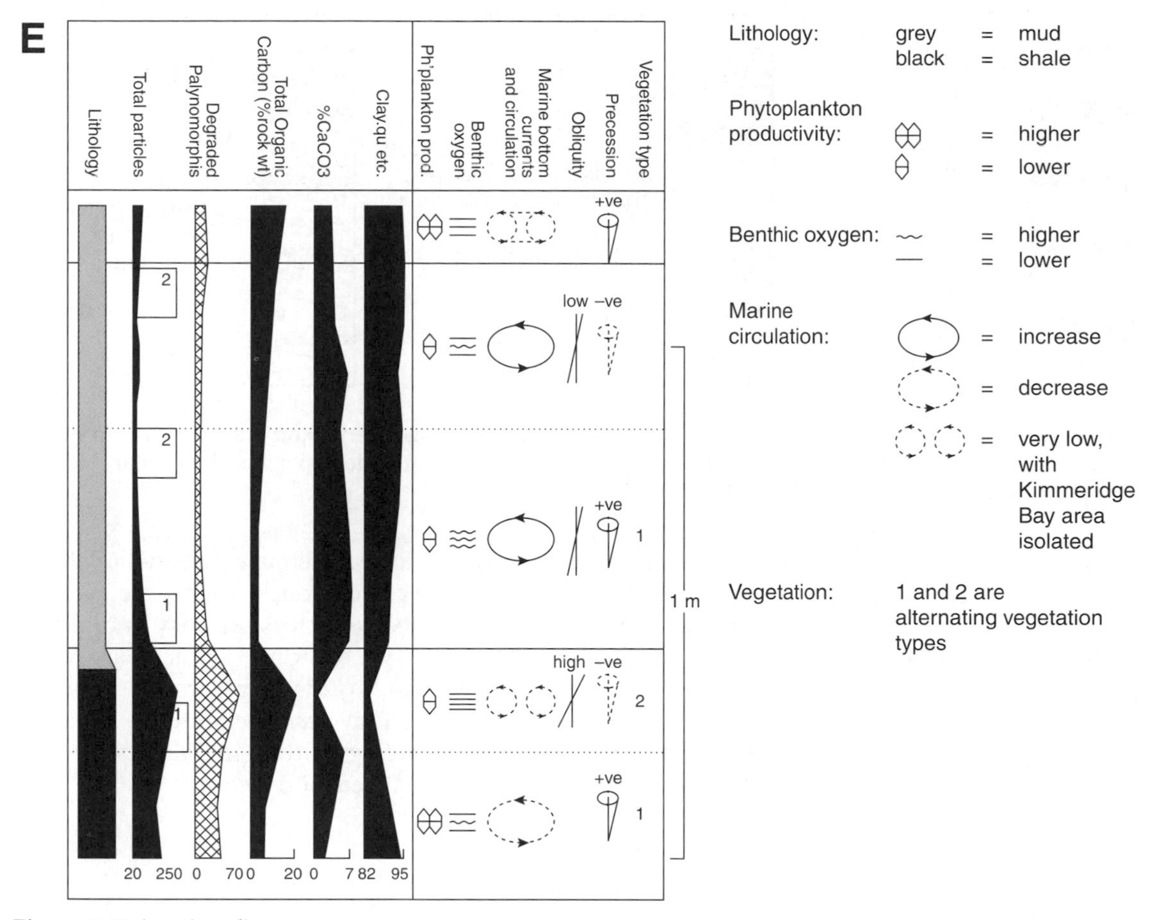

Figure 9.14 (*continued*)

Antarctica has 70% of the world's fresh water, and 90% of its ice.

3. Oxygen levels rose from the late Precambrian, but may have reached present levels in the lower half of the Palaeozoic Era. O_2 levels peaked (up to a possible 35% of atmosphere gases) by accelerated terrestrial photosynthetic productivity in the Late Carboniferous, and CO_2 % fell in response. These changes reversed in the Late Permian once the coal swamps disappeared as Pangaea assembled, and the rise in CO_2 may have contributed to the end-Permian marine and terrestrial extinction, assisted possibly by CO_2 emissions associated with the end-Permian gigantic Siberian flood basalts.
4. There seems to have been a unidirectional increase in marine diversity through time, but this depends on the validity of the database; figures taken only from the skeletal (fossil) record do not necessarily reflect the diversity of soft-bodied biota, which is preserved only in exceptional circumstances. Similarly, nutrient availability may have increased through time, according to the proxy provided by the increase in diversity of planktonic fossils, and trends of stable isotopes.
5. Continental reconfiguration is a continuous feature of Phanerozoic oceans. Not only did it influence sea level on a grand scale, but also variations in submarine weathering

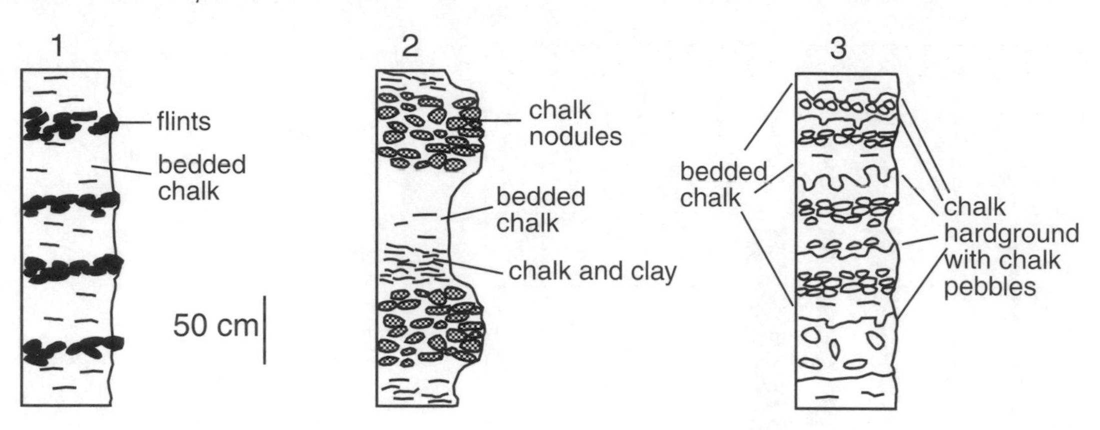

Figure 9.15 Examples of small sedimentary cycles of different sediments and trace fossils within the Cretaceous chalk in Northern Europe; their cyclicity suggests that the deposits may be orbitally forced. Adapted from Bottjer *et al.* (1986).

(which removes Mg from ocean water), and rate of chemical reactivity at subduction zones (where CO_2 is released from rocks into the ocean–atmosphere system). The result is the cyclic change between aragonite and calcite seas (Sandberg curve), with additional data from evaporites, implying a long timescale fluctuation of seawater chemistry.

6. Cyclicity in sedimentary deposits relates to ocean processes; much large-scale cyclicity is attributed to tectono-eustasy and glacio-eustasy. The concept of icehouse and greenhouse worlds is becoming accepted. Smaller-scale cycles are commonly linked to orbital forcing, although care is needed to filter out local and regional effects.

SHORT-TERM CHANGES IN PHANEROZOIC OCEANS

10

10.1 INTRODUCTION

Short-term environmental change, a subject of great current interest, has applications in ocean study, because the oceans have responded quickly to changes in the past, and almost certainly will do so in the future. Changes are revealed by the sedimentary and biotic record and had huge repercussions on surface environments. Sediments show sharp changes, and fossil assemblages show extinction. In this context, timescales of change may be regarded as *geologically* short (hundreds of thousands to millions of years), although modern studies of ocean systems suggest that the oceans were/are able to respond on the decade scale, considered more in Chapter 11. This chapter completes a survey of ocean processes in geological history by examining a selection of features at key points during the Phanerozoic. This chapter is thematic, rather than chronological, and examines processes causing ocean change.

10.2 CATASTROPHES, ORGANISMS AND OCEAN SYSTEMS

Catastrophes in the rock record are usually accompanied by mass extinctions. Biotic extinctions occur as two types: background and mass extinctions. Background extinctions take place on a continuous basis and may be regarded as the inability of individual species to survive continuous progressive changes in ecosystems. Mass extinctions relate to major environmental disruptions which are usually geologically rapid (100 000 years is considered rapid, but mass extinctions may occur over millions of years). Many mass extinctions are recorded, but the 'Big Five' (Late Ordovician, Late Devonian, end-Permian, end-Triassic and end-Cretaceous) are the best known. Causes of mass extinctions fall into a number of categories, which are extensively debated in a huge literature:

1. impact of extraterrestrial bolide (Greek for missile) – 70% chance of landing at sea, but must affect global climate irrespective of where it hits;
2. sea-level fall – decreases habitable areas of continental shelves, and shifts the heat balance of the planet's surface by exposing more land;
3. sea-level rise – may induce anoxia on shelf environments where most organisms live; also affects heat balance;
4. climate – especially cooling, which may be global, or polar region (Hallam and Wignall, 1997, p. 246); temperature is probably the most important limiting factor on organisms;
5. volcanic eruptions – dust, vapours and gases (especially increased CO_2) affect climate and possibly are poisonous.

Note that the last four are normally intimately associated with plate tectonics. The potency of ocean circulation and overturn, coupled with the heat capacity and solvent properties of water (Chapter 2), are major controls in the transfer of toxins and heat, which can either affect organisms directly, or indirectly by influencing climate change. Mass extinctions are involved in rapid ocean changes, and selected extinctions are included in the following survey of short-term ocean change, as well as oceanic events and processes which did not lead to mass extinction.

10.3 OCEANS AND GLACIATIONS I: THE ENIGMATIC LATE ORDOVICIAN EVENT

10.3.1 THE LATE ORDOVICIAN ROCK RECORD

Subject to recent intense scrutiny, study of the sedimentary, geochemical and biotic characters of the Late Ordovician extinctions has led to a consistent global picture of extinctions being related to, and probably caused by, the Hirnantian Stage glaciation, in which ocean systems were disrupted at the end of the Ordovician Period (Armstrong, 1995, 1996; Berry, 1996; Brenchley *et al.*, 1994, 1995; Harper and Rong, 1995; Kaljo, 1996; Owen and Robertson, 1995; Wang *et al.*, 1993). Dating indicates the glaciation lasted only 0.5–1 Ma (Figure 10.1), which is very short, and there were two extinctions. The first was associated with the onset of glaciation (the end-*pacificus* event), possibly by ice-induced regression across shelves, as global sea level fell by around 45–65 m, possibly as much as 100 m (Brenchley and Newall, 1984). The resulting widespread erosion left only a few sites globally which preserve a continuous record (Berry *et al.*, 1990; Brenchley and Newall, 1984). The second extinction seems to have coincided with a rapid return to warm conditions as the glaciation collapsed, at the boundary between the *extraordinarius* (= *bohemicus* in China) and *persculptus* zones (Armstrong, 1996), followed by rapid expansion of new faunas (Berry *et al.*, 1996) at *c.*440 Ma, a short distance below the Ordovician–Silurian boundary. The enigma is that the glaciation developed in a world of very high CO_2 levels (Figure 9.3), possibly as much as 14× present-day values (Berner, 1991; Crowley and Baum, 1991, 1995), which creates a considerable problem of explaining the cause of glaciation.

Following the exciting discovery of high levels of Pt-group metals (especially Ir) at the Cretaceous–Tertiary (K–T) boundary in 1980, with its likely asteroid-impact source, the search was on for evidence of impacts to explain earlier extinctions. There is indeed an Ir spike at the upper extinction horizon in the Late Ordovician, but the concentration of Ir is lower than expected for an asteroid impact, and is better interpreted as due to sediment starvation as the sea level rose rapidly when the glaciation collapsed (Wang *et al.*, 1992, 1993), although there is some claim for an impact (Wang Xiaofeng and Chai, 1990). The Ir level recorded is 0.64 ppb (Wang Xiaofeng and Chai, 1990), compared with 0.092 ppb by Wang *et al.*, (1992), but both of these are minor compared with the 9 ppb recorded for the K/T event (Alvarez and Asaro, 1990). Sediment starvation seems a more likely explanation for the increased Ir, and means that the sea level rose so quickly that sources of local sediment supply were covered by water, leaving condensed horizons in sedimentary basins. Although an impact effect is not completely ruled out, it could not have started the glaciation, because the Ir spike is at the *end* of the glacial phase, so there is still the problem of why the glaciation started. The most popular idea is that the Southern continent of Gondwana drifted across the South Pole, placing a landmass in the right position for the effects of a continental climate to be enhanced by its polar location. As glaciers formed, surface albedo increased, triggering global cooling and further build-up of ice. Thus the extinction began as a polar-cooling induced climate change. Consequent water circulation changes were then needed to carry cooled polar water to lower latitudes, so an ocean conveyor system was presumably needed, and may have also involved productivity and nutrient changes (Brenchley *et al.*, 1994). The continental configurations (Figure 9.2) are suitable for that scenario, and may have led to a climate not unlike the present (except for the high CO_2).

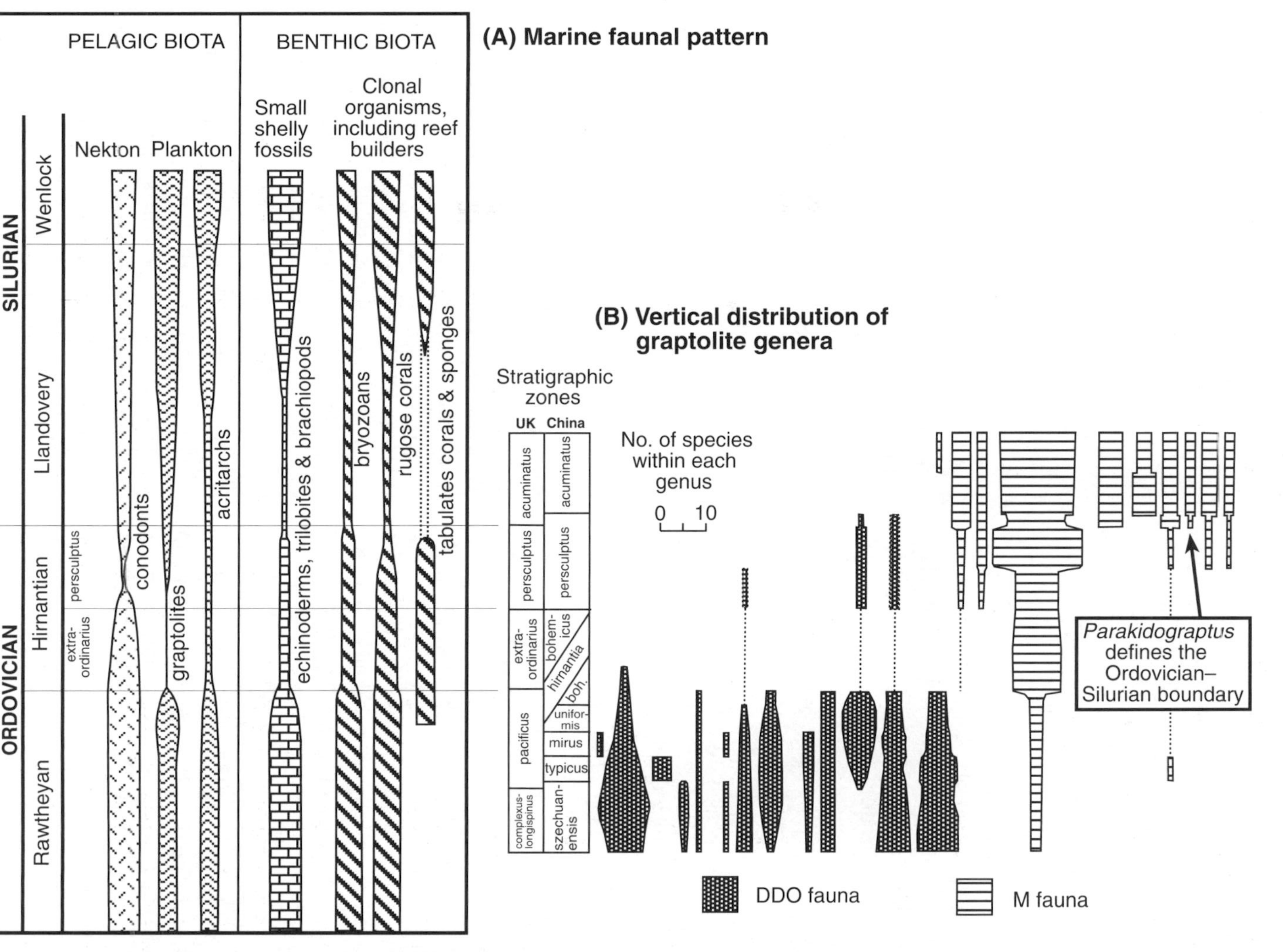

Figure 10.1 (A) Stratigraphy of the Late Ordovician to Early Silurian, showing the temporal changes in key fauna throughout the boundary interval. The *extraordinarius* zone corresponds to the late Ordovician glaciation. (B) Detailed changes in graptolites are shown by important faunal patterns; the DDO fauna is named after the three major genera (*Didymograptus*, *Dicranograptus*, *Orthograptus*), and the M fauna corresponds to monograptids. Distributions of different species are shown, but details of species names are not included. (A) Reproduced from Hallam & Wignall, *Mass Extinctions and Their Aftermath* (1997), by permission of Oxford University Press, Oxford; (B) adapted from Melchin and Mitchell (1991).

The glacial interval is represented in most places by a limestone, often less than 1 m thick. Graptolites and conodonts are the mainstay of Ordovician and Silurian stratigraphy, and from graptolites, Berry *et al.* (1990) suggested that the extinctions were diachronous from *pacificus* zone (high latitudes) to *extraordinarius* zone (low latitudes), such that the glaciation may therefore also have been diachronous. The pattern of extinctions, especially of conodonts (Armstrong, 1996), graptolites (Melchin, 1996), and brachiopods (Jin and Copper, 1996; Sheehan and Coorough, 1990), and associated facies changes, are currently interpreted as reflecting the effects of the glaciation, although not all fossil groups reflect the extinction events so well, e.g. acritarchs.

Rocks beneath the limestone are usually shales rich in organic matter, with a poorly developed benthic fauna of brachiopods in particular, and the seabed was not suitable for a rich diversity. The limestone, in contrast, is full of brachiopods, indicating the seafloor was easily habitable, and that the poisoning effect of poor oxidation, plus the potential for input of toxins (e.g. Leary and Rampino, 1990), was suppressed during the limestone phase. The top of the limestone changes sharply upwards into anoxic organic black shales, indicating that the glaciation collapsed rapidly. The environmental change was severe, from a well-oxygenated setting while the limestone was forming, to a poorly aerated seafloor shown by the black shale lacking any benthos in many sites; usually only graptolites are found, which fell onto the seabed from the aerated upper waters. There was rapid evolution of graptolites after this level. It is important to realise that deeper waters of the Palaeozoic ocean system were generally anoxic as revealed by ^{32}S data (Martin, 1995), described in Chapter 8, and highlights the unusual nature of this glaciation, as a rapid but major disruption to an otherwise relatively stable ocean system.

10.3.2 LATE ORDOVICIAN SEDIMENTARY GEOCHEMISTRY

Geochemical data, from a number of sites globally, aid understanding of the environmental setting, in conjunction with facies and fossils. Most important are carbon and oxygen isotopes, so often used in oceanographic studies. $\delta^{13}C$ values have been meaured from well-preserved brachiopod shells; remember that these are regarded by most workers as consisting of unrecrystallised calcite, and therefore still contain the isotope values they precipitated in their shells in equilibrium with seawater. The $\delta^{13}C$ data show a strong positive shift (Brenchley *et al.*, 1994) between the two extinctions (Figure 10.2), and this has been attributed to a substantial raising of ocean productivity resulting from ocean overturn and the consequent upwelling of nutrients from the oxygen-poor ocean floor to the surface ecosystems. Thus ^{12}C is selectively removed from the seawater in photosynthesis by plankton, leaving an ocean enriched in ^{13}C, incorporated into brachiopod shells as they grew on the seafloor. The $\delta^{13}C$ values fell rapidly to normal Palaeozoic levels after the extinctions, and faunas show rapid evolution. $\delta^{18}O$ also shows a positive shift during the glaciation, attributed to removal of the lighter ^{16}O in glacial ice, with a corresponding return to normal values afterwards. However, a recent review of the isotope data (Hallam and Wignall, 1997) suggested that the magnitude of the $\delta^{13}C$ shift (4–6‰) was too great to be accounted for by removal of ^{12}C by photosynthesis, because the amount of productivity required would be excessively large. Also, the organic matter would need to be buried as C_{ORG} long enough for the $\delta^{13}C$ shift to be stored in carbonate sediments; but there is no large increase of C_{ORG} in the Hirnantian sediments, and none would be expected if the ocean waters were mixed during the cool phase. Furthermore, the same review indicates that the positive $\delta^{18}O$ shift is so large (2–4‰, in comparison with the Quaternary 1.2‰) that

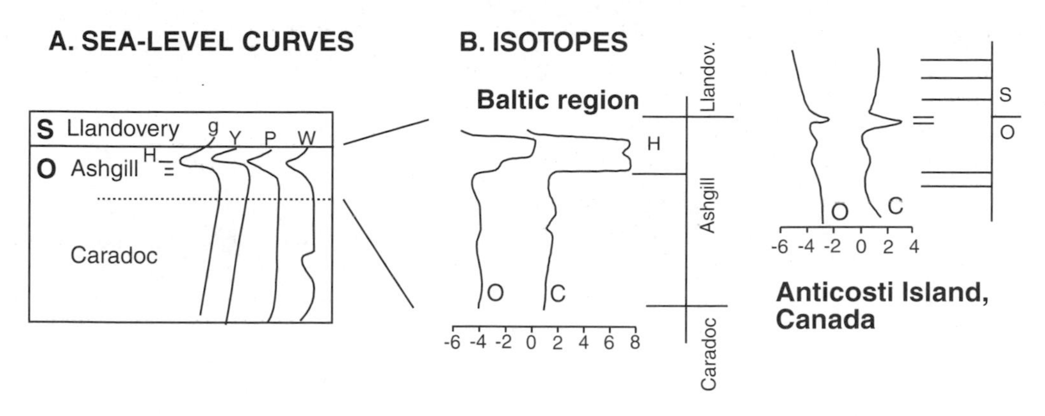

Figure 10.2 (A) Global (g) and regional sea-level changes during the Late Ordovician and Early Silurian; Y = Yukon area, P = Poland, W = Wales. (B) C and O isotope fluctuations in two areas during the Late Ordovician show the key features of shifts, with important implications for palaeoceanographic change; O = Ordovician; S = Silurian. Reproduced with permission from P. J. Brenchley *et al.*, *Bathymetric and Isotopic Evidence for a Short-Lived Late Ordovician glaciation in a greenhouse period* (1994), the Geological Society of America. Copyright © 1994 Geological Society of America.

the amount of ice storage required to remove sufficient ^{16}O from seawater would create ice sheets of unbelievable thickness and size. Thus the C and O isotope data are so extreme that their values are hard to interpret. Note, however, that the controls on carbon and oxygen isotopes in the Palaeozoic have been subject to uncertainty for many years.

If that is bad, there is another problem. If the glaciation was indeed triggered by Gondwana drifting over the South Pole, then within 0.5–1 Ma later, either it drifted away again for the glaciation to collapse rapidly and the oceans to return to stratified conditions, or some other climatic feature caused warming. For the former, it is asking a lot of continental drift to work so fast as to cause and then stop a glaciation in such a short time, and for the latter, there seems to be no mechanism. Even if the second extinction resulted from a bolide impact, the likely effect is increased dust and cloud cover, *increasing*, not decreasing, cooling. Thus, we are left with the impression that the Late Ordovician ocean–climate system was not playing strictly by the rules that scientists have invented; in other words, behaviour of the measured parameters must be different from other geological episodes.

In South China, the limestone formed in the glacial phase is accompanied by shifts in the concentration of certain chemical elements. Figure 10.3 shows how elements which are chemically inert (especially Th, Ta and Hf, which are called immobiles) decline during the glaciation and rise again afterwards, in a site known to be near land. This is curious because the expected result of a sea-level fall is increased erosion, rather than the decrease seen. The rare earth elements also respond in the same way, except that in the shales above the glacial interval, there is separation of the atomically lighter La from the heavier Tb and Lu; this suggests the elements have fractionated, and because they are controlled by the type of source rock, it indicates the source of sediment to this site was different from before the glaciation. This unexpected response of the immobile elements may indicate that the South China region was glaciated, and this has implications for both palaeogeographic and climatic models affecting that region. Element analysis can also be used to indicate oxidation states,

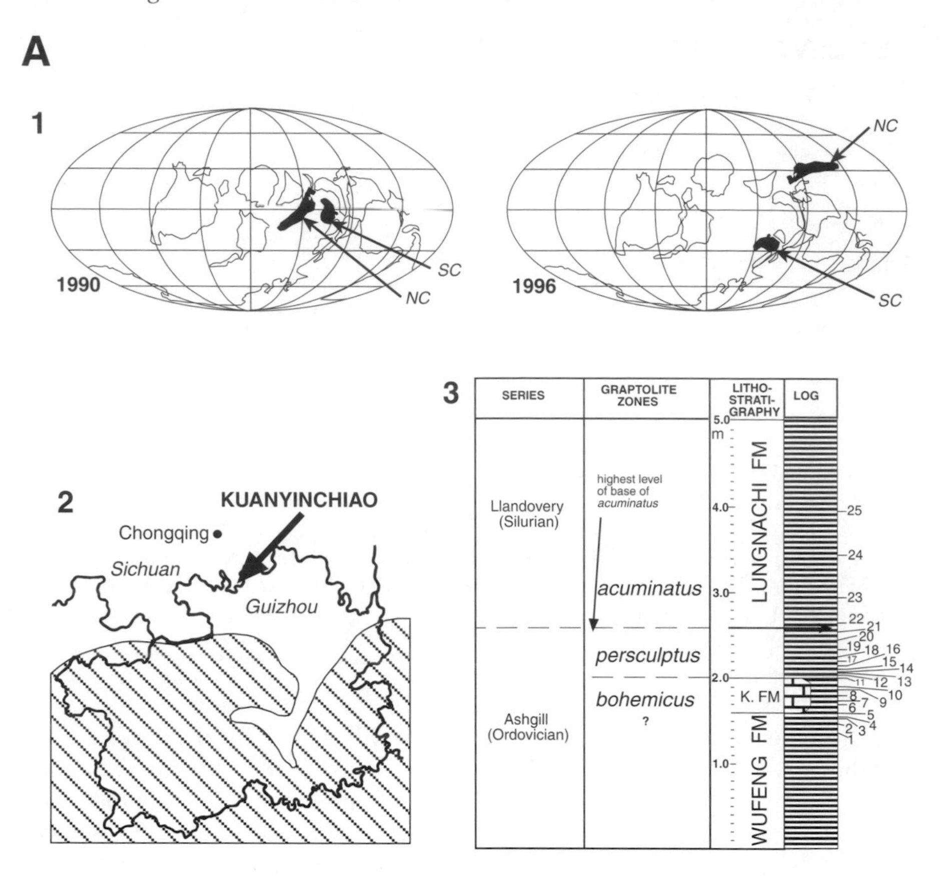

Figure 10.3 Case study of features of Late Ordovician environmental change in south China. (A) 1. The palaeogeography of late Ordovician times is uncertain, as are all continental reconstructions before the Devonian; the most popular reconstruction places North and South China close together; another shows them widely separated. Critically, the latitude of South China is unclear, important to determination of its climate. 2. Palaeogeography of one area near a landmass, with one site (Kuanyinchiao) highlighted. 3. Stratigraphy and lithofacies of Kuanyinchiao, with sample locations shown in (B).

and this example shows how difficult it can be to interpret the results. U and Eu, for example, are well known to be affected by changes in oxidation state of the environment, by declining in abundance in sediments if the conditions are oxidising. That is because they can exist in different states such that in reducing conditions they are precipitated and in oxidising conditions they are dissolved. Their trends in Figure 10.3 broadly follow their expected response, but not cleanly, and it is known that other processes control these elements. Other elements in the suite vary, and the reasons are unclear; some elements are subject to diagenetic modification, and so do not follow the trends of the sedimentary facies. This example of the use of chemical elements in analysis of sedimentary suites shows how they must be examined along with the sediment types and fossils, in order to fully investigate the controls.

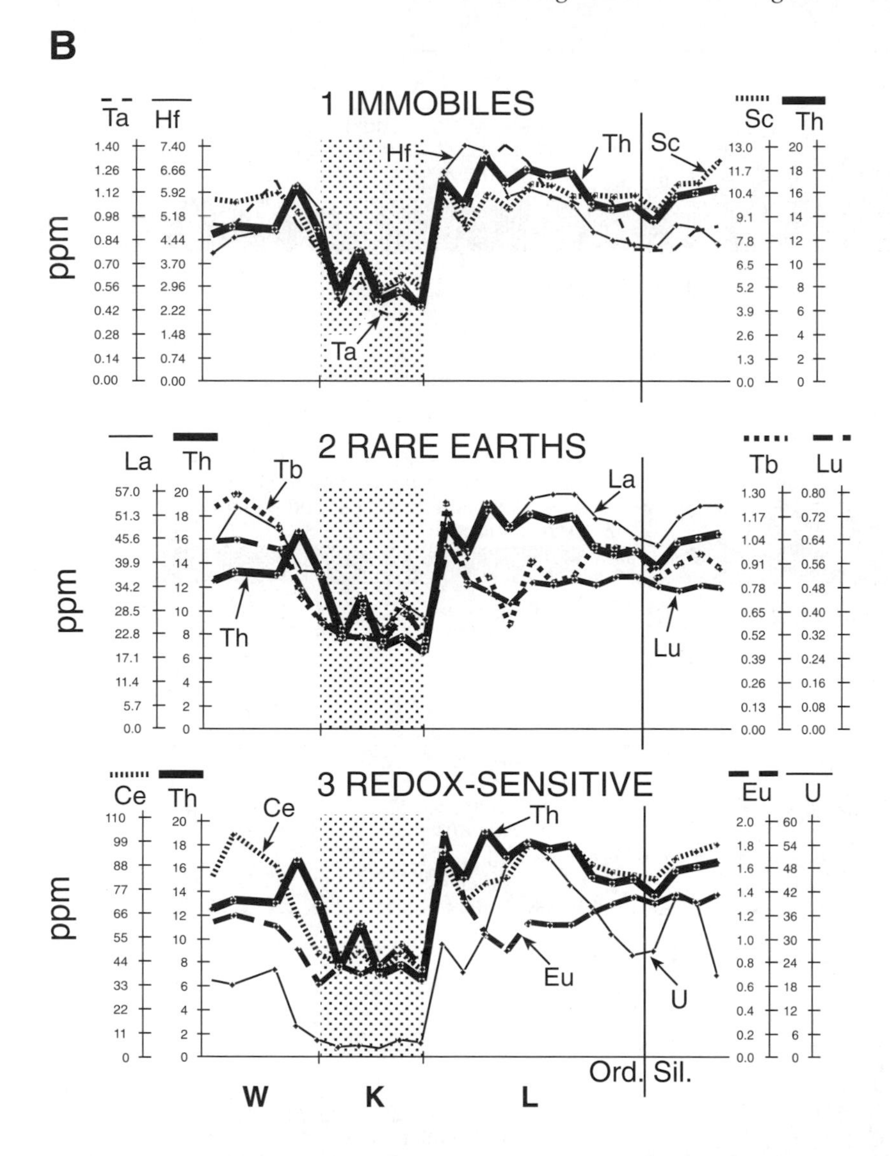

Figure 10.3 (*continued*) (B) Suite of elements across the Hirnantian glacial interval and Ordovician–Silurian boundary at Kuanyinchiao. 1. Immobile elements show good covariance, and indicate that the rate of supply from the land fell during the glacial phase; this corresponds with a large sea-level fall (not illustrated). 2. The light rare earth element La shows reasonably good covariance with the immobile element Th, but the heavy rare earths Lu and Tb are partially fractionated after the glaciation, and suggest that there is a change in the source rock from which these elements were derived. 3. Elements which are potentially redox-sensitive (Ce, Eu and U) show partial covariance with Th, but there is sufficient departure to suggest composite controls on these elements. These data can be interpreted differently, depending on the location of South China; if it was on the equator, then the fall in sedimentation rate associated with the Kuanyinchiao Formation limestone could relate to shallowing and lessening of clay deposition on the shelf, so that limestone formed instead of shale; if it was further south, it may have been within reach of the glaciation, and it is possible that sediment starvation may be due to an ice cap, preventing erosion. W,K,L = Wufeng, Kuanyinchiao and Longmaxi formations. Adapted from Zhang *et al.* (2000).

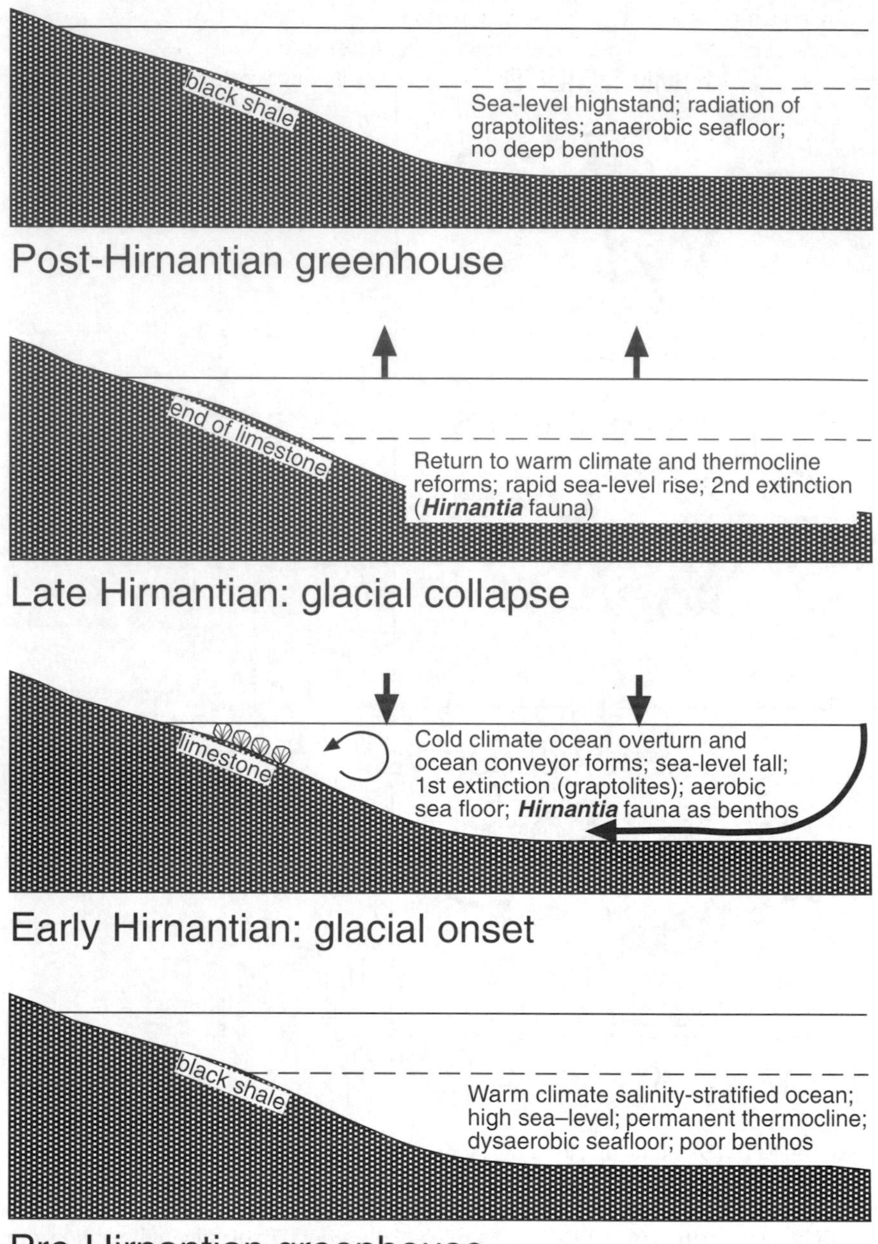

Figure 10.4 Model of Hirnantian glaciation history. The four stages suggest that the salinity-stratified oceans of late Ordovician time were interrupted by ocean conveyor-type circulation, leading to the well-oxygenated sediments of the glacial interval. End of glaciation led to re-establishment of the stratified ocean. This popular application of global conveyor deep ocean circulation depends on the accuracy of interpretation that these processes did operate in the Palaeozoic; given that the precise mode of operation of the modern conveyor is unclear, its application to ancient ocean must be considered with care.

The late Ordovician glaciation took place when the world was a warmer place. The rapid decline of the glaciation is consistent with the instability of a glacial phase in a world of high temperatures of the order of 20–25°C at the relevant mid-latitudes (Tobin and Walker, 1997, Figure 13) and enhanced CO_2 levels (Crowley and Baum, 1991, 1995), although what triggered its demise remains a mystery. Figure 10.4 gives a summary of the changes associated with the Hirnantian glaciation.

10.4 OCEANS AND GLACIATIONS II: THE LATE QUATERNARY YOUNGER DRYAS COOLING AND RELATED EVENTS

The last glacial maximum (around 18 000 years BP in NW Europe) was followed by a general climatic amelioration. However, an event of possible global oceanographic significance, which has captured great interest amongst Quaternary Earth scientists, is the northern European Younger Dryas cooling of the period between 13 000–11 700 calendar years ago (i.e. real-time as opposed to radiocarbon years). Ice core, ocean sediment and terrestrial records indicate a rapid re-cooling in NW Europe of around 5°C over a decade or so, which lasted for *c.*1300 a, temporarily reversing the trend of warming. The possible global nature of this climatic reversal is supported by data from eastern North America, Barbados, the Andes and Africa, and it is even (weakly) in the Vostok ice cores from Antarctica (Roberts, 1998). This is the last event of the latest glacial cold stage, for which an agreed model remains elusive, yet is thought to contain evidence of changes in thermohaline circulation (Broecker, 1994; Broecker and Denton, 1990a; Broecker *et al.*, 1985), which has implications for current thought on climate change. The Younger Dryas has become important for models of modern ocean system change, not least because the temperature drop (5–7°C) is little different from the 8°C changes noted in the last 100 years on the coast of Greenland (Lehman, 1993), close to the site of NADW formation. The modern ocean conveyor system transports warm, saline water northwards to high latitude areas of the North Atlantic, and releases enormous amounts of heat to the overlying colder atmosphere before sinking as NADW (Chapter 3). Disruption to the circulation may consequently lead to climate change in high latitude regions of Northern Europe, Greenland and beyond.

The history of climate change throughout the whole of the last glacial period reveals a range of climatic features, some of extraordinarily short durations of 5–20 a, shown by variations in the electrical conductivity of Greenland ice cores due to fluctuations in airborne dust content settled as ice accumulated (Taylor *et al.*, 1993). Of current particular interest are Heinrich Events (HEs) and Dansgaard–Oeschger Events (D–OEs) (see Figure 10.5). HEs are represented by sheets of ice-rafted debris deposited on the North Atlantic seafloor released from melting icebergs (Broecker, 1994), suspected as originating from the episodic collapse of the Laurentide ice sheet. Further work on the latest (called H1) of the 6 or 7 HEs indicates that other ice sheets around the north Atlantic underwent some melting (McCabe and Clark, 1998). D–OEs are fluctuations of temperature, ice-accumulation rate and gas (CO_2 and CH_4) content recorded in ice cores; the most recent D–OE is the Younger Dryas cooling (Broecker, 1994). The relationship between HEs and D–OEs is not clear; while atmospheric concentration of CO_2 varied by < 10 ppm by volume with D–OEs, it varied significantly (by *c.*20 ppm by volume) with HEs, particularly for HEs which began with a long-lasting D–OE (Stauffer *et al.*, 1998). Nevertheless, the effects of both seem to be widespread; for example, results from sediment cores in the Arabian Sea show fluctuations in monsoons which correlate with both D–OEs and HEs (Schultz *et al.*, 1998). Guo *et al.*, (1998) linked changes in the Loess Plateau of China with NADW. Such work not only shows the broad climatic nature of these events, but also illustrates the importance of shifting climate in areas subject to extremes, such as the

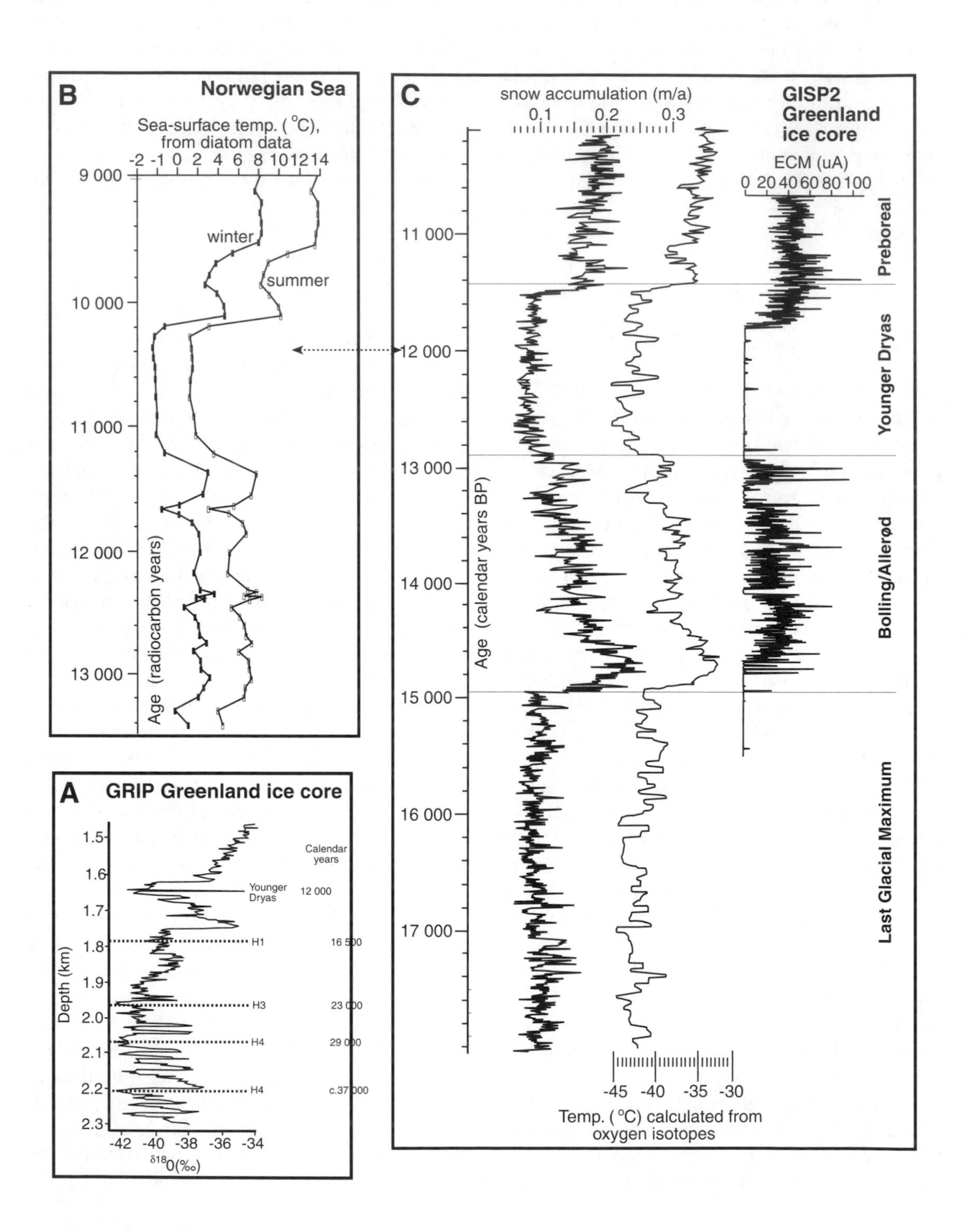
B
Norwegian Sea
Sea-surface temp. (°C), from diatom data
winter
summer
Age (radiocarbon years)
C
snow accumulation (m/a)
GISP2 Greenland ice core
ECM (uA)
Age (calendar years BP)
Preboreal
Younger Dryas
Bolling/Allerød
Last Glacial Maximum
Temp. (°C) calculated from oxygen isotopes
A
GRIP Greenland ice core
Calendar years
Younger Dryas
Depth (km)
H1
H3
H4
$\delta^{18}O$(‰)

effect of monsoons on surface water circulation in the Indian Ocean (Chapter 3). Cooling in the northern Mediterranean Sea has also been correlated to HEs (Rohling *et al.*, 1998). Furthermore, ODP data from the Amazon fan in the South Atlantic also record Heinrich events (Maslin, 1998), and it has also been argued that there was an increase in deep-sea ventilation in the Southern Ocean during the beginning of the Younger Dryas, because of a rapid rise of $\delta^{18}O$ in Antarctic ice (Broecker, 1998).

The possibility that these two categories of climate shift affect ocean circulation is important because of their rapidity of change. In the case of the HEs, the discharge of cold fresh water and icebergs into the North Atlantic surface waters is interpreted to cool the surface water and, importantly, to dilute it so that the normal situation of sinking of cold dense saline water stops or is disrupted. If that happens, the release of vast amounts of heat to the atmosphere is suppressed, cooling the North Atlantic region; transport of water around the globe declines, and also oxygenated water is not carried to the deep ocean.

The cause of the Younger Dryas cooling may have been a similar shut-down of NADW production due to catastrophic drainage of the meltwater from the North American Lake Agassiz into the North Atlantic (Figure 10.6). Because NADW is considered to be the main driving force of the ocean conveyor (Chapter 3), the concept of switching off the conveyor is currently popular. An alternative idea is that the vertical circulation in the North Atlantic 'flipped' to another circulation mode, where deep-water circulation operates at shallower depths. Deep-water corals from a seamount at 1600 m depth are bathed in North Atlantic intermediate-depth waters, and oxygen isotope data from dated corals show changes during the Younger Dryas (Smith *et al.*, 1997), interpreted as disruption of intermediate water circulation during that time, and if the dating of the corals is accurate, the results also suggest that the change occurred in as little as 5 years. Furthermore, the same coral data are used by those authors to support an earlier suggestion (Fairbanks, 1989) that the Younger Dryas event took place when there was little ice sheet melting, a view which seems to be counter to the Lake Agassiz theory. If that was the case, then the Younger Dryas cooling could have been caused by, not the cause of, disruption of the thermohaline circulation. The link between atmosphere and ocean during the Younger Dryas cooling is revealed in the rapid response of alpine vegetation in the Colorado Rocky Mountains (Reasoner and Jodry, 2000).

Complexity of control of the Younger Dryas seems to grow with each paper published. One of the joys involved in dating oceanic records is ensuring that the ^{14}C radiodating system is properly calibrated, so that real dates can be obtained, and note that radiocarbon dating is not linear. The normal ocean conveyor carries ^{14}C away from surface waters at a (presumed) steady rate, so if the conveyor stops, then there is an accumulation of ^{14}C in shallow waters. This ^{14}C is taken up into datable materials (shells) and deposited, and therefore would give radiocarbon dates which would depart from the expected trend. Observed departures in the Younger Dryas interval in Venezuela suggest that the ocean conveyor ceased for a few hundred years at the start, then it resumed for the remainder of the interval. Because the

Figure 10.5 Diagrams of recorded changes in the Younger Dryas event. (A) Oxygen isotope curve from Greenland, showing the significant excursion during the Younger Dryas. (B) Diatom data show the cooling in both winter and summer in the Norwegian sea during the Younger Dryas phase. (C) Combined data from GISP2 Greenland ice core. During the Younger Dryas, cooling reduced air humidity and therefore snowfall rates were also reduced; temperatures fell; electrical conductivity measurements (ECM) of the ice showed a fall because the air was drier and more dusty, inhibiting current flow. (A) Reprinted by permission from *Nature*, **372**, pp. 421–424 (1994), Macmillan Magazines Ltd. (B) and (C) adapted from Lowe and Walker (1997, p.352 and 353).

North Atlantic region (and even as far away as New Zealand) remained cold throughout the 1300 a of Younger Dryas time (therefore no conveyor?), one interpretation is that the conveyor was operating in the Southern Ocean, so that Antarctica was being warmed by heat release as deep water was created adjacent to it (Broecker, 1998). The suggestion is extended to suggest that the conveyor may operate in a seesaw fashion between the North and South Atlantic. During the period immediately before the Younger Dryas (called the Bolling–Allerød phase) heat release in the Antarctic area was reduced while in the North Atlantic it was strong; in the Younger Dryas it was the reverse. The final point here is that the scale of these changes is in the order of millennia; at a larger timescale of 100 000 to a million years, Antarctica shows climate changes in phase with the Earth as a whole (Broecker, 1998).

Finally, other work on the glacial effects on oceans raises another spectre. It has been indicated recently that seafloor conditions during glaciations was *highly reducing* (Broecker, 1997), in comparison with interglacial times. If you consider this in relation to the discussion in Chapter 9, and in Sections 10.3 (the end-Ordovician event) and 10.5 below, for the rest of the Phanerozoic, that greenhouse worlds lead to stratified oceans and anoxic bottom conditions and icehouse worlds lead to ventilation of the seafloor, then such a revelation shows how sensitive the ocean is to change. It emphasises the difficulty of applying models in too broad a fashion to what is really a highly complex and dynamic ocean system. A telling commentary on the current state of understanding of the Younger Dryas event (Broecker, 1999) draws attention to the inadequacy of models applied to it.

10.5 OCEAN CHANGE AND ANOXIA I: PRINCIPLES

10.5.1 BACKGROUND: GENERAL PATTERNS OF MODERN OCEAN OXYGEN DEFICIENCY

Oxygen deficiency is commonly recorded throughout the Phanerozoic oceans, and illustrates the crucial controls of plate tectonics and degree of ocean stratification on oxygen (and aerobic life) in bottom waters. In modern oceans, oxygen levels may be measured directly from bottom waters and sediments, but in the sedimentary record it is necessary to use proxy indicators, such as sediment type and structure, bioturbation in the sediments (disturbance by animals needs oxygen), benthic shelly fossils, and stable isotope data; each is important in different settings. Oxygen deficiency frequently results in preservation of laminated sediments because of the lack of burrowers, one of the best examples of which is the Permian Basin of West Texas; uniformity of sediments, and therefore process, in the deep floor of this isolated basin have been revealed by matching sets of laminae up to 24 km apart (Anderson, 1996). Anoxia occurs at a range of scales, from isolated basins to global events. The abundance of oxygen deficiency in ancient sediments is a direct reflection of the higher sea levels, and therefore wide shelves and epeiric basins, in which oxygen-poor sediments could accumulate. Thus the modern models of the Black Sea and Baltic Sea represent a tiny fraction of basin anoxia through time.

The Black Sea (2200 m deep) is a silled basin (i.e. a basin with a shallow threshold (sill) to connect to the open ocean), morphologically suited to development of anoxic deeper water, and is unusual in being relatively simply

Figure 10.6 Schematic diagrams of a popular model of the Younger Dryas event. Before and after the Younger Dryas phase, North Atlantic Deep Water (NADW) formed by cooling and sinking of warm surface waters, as a principal component of the ocean conveyor system. The warming resulted in glacial meltwater flow down the St Lawrence River to form a surface pool of cold fresh water in the North Atlantic which inhibited NADW formation during the Younger Dryas. The cooling which followed re-established the glaciation, shutting off the meltwater supply, and eventually led to renewal of NADW formation. The accuracy of this model is discussed in the text.

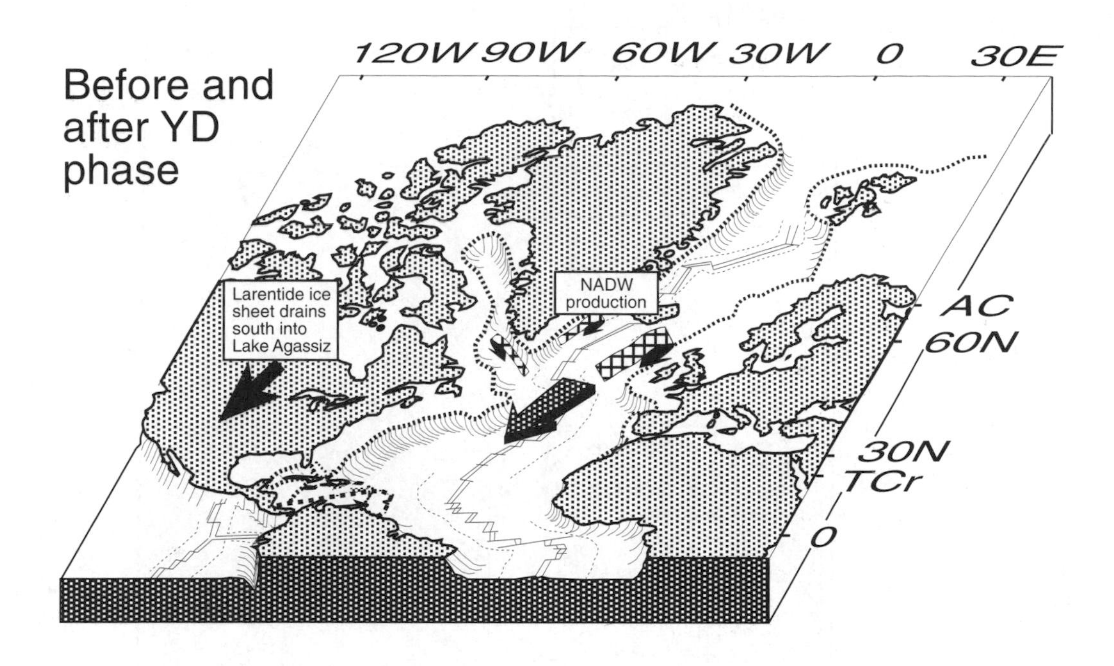
Before and after YD phase
120W 90W 60W 30W 0 30E
Larentide ice sheet drains south into Lake Agassiz
NADW production
AC
60N
30N
TCr
0

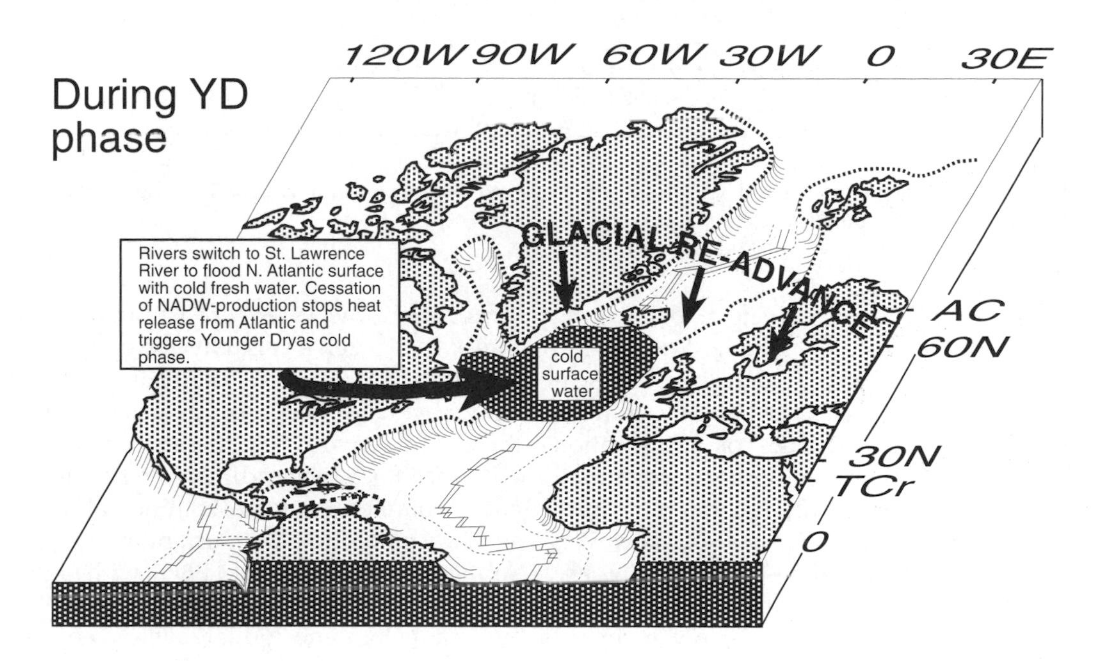
During YD phase
120W 90W 60W 30W 0 30E
Rivers switch to St. Lawrence River to flood N. Atlantic surface with cold fresh water. Cessation of NADW-production stops heat release from Atlantic and triggers Younger Dryas cold phase.
GLACIAL RE-ADVANCE
cold surface water
AC
60N
30N
TCr
0

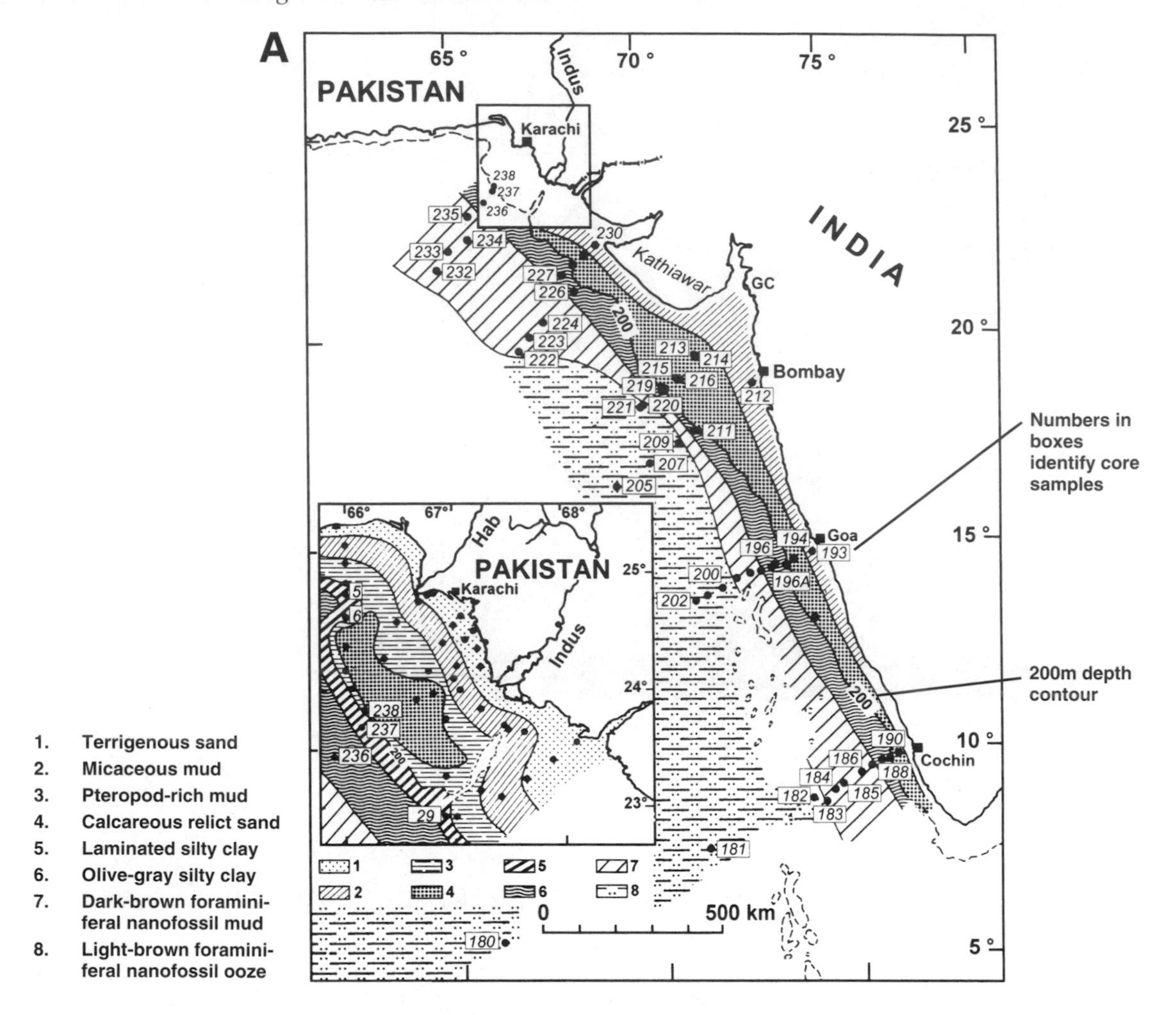

Figure 10.7 Example of vertical variations of dissolved oxygen in water, based on data from India-Pakistan. (A) Location and facies distribution in Northern Indian Ocean.

stratified, with deeper waters below the redox boundary at *c.*200 m being low oxygen (Leeder, 1982, p.253), although there are seasonal variations in water circulation within the basin, leading to subtleties of stratification which have only recently been identified (Özsoy and Ünlüata, 1997). A common feature in unrestricted oceans and seas, however, is 'oxygen-minimum zones' (OMZs) (Figure 10.7). Oxygen profiles from the modern Pacific and Atlantic Oceans show narrow to broad OMZ between 400–1000 m (Bearman, 1989b, p. 96); rarely, the water becomes anoxic. Below the OMZ, oxygen levels may rise up to those of the surface, because of underpassage of the OMZ by cold deep, oxygen-rich water of the ocean conveyor. If the seabed intersects with the OMZ, then sediments reflect its presence, commonly along continental borders such as in the Indian–Pakistani margin of the Arabian Sea (Schultz *et al.*, 1996), with high levels of organic carbon and laminated (probably seasonal) Holocene and late Pleistocene hemipelagic muds, between 250–800 m depth.

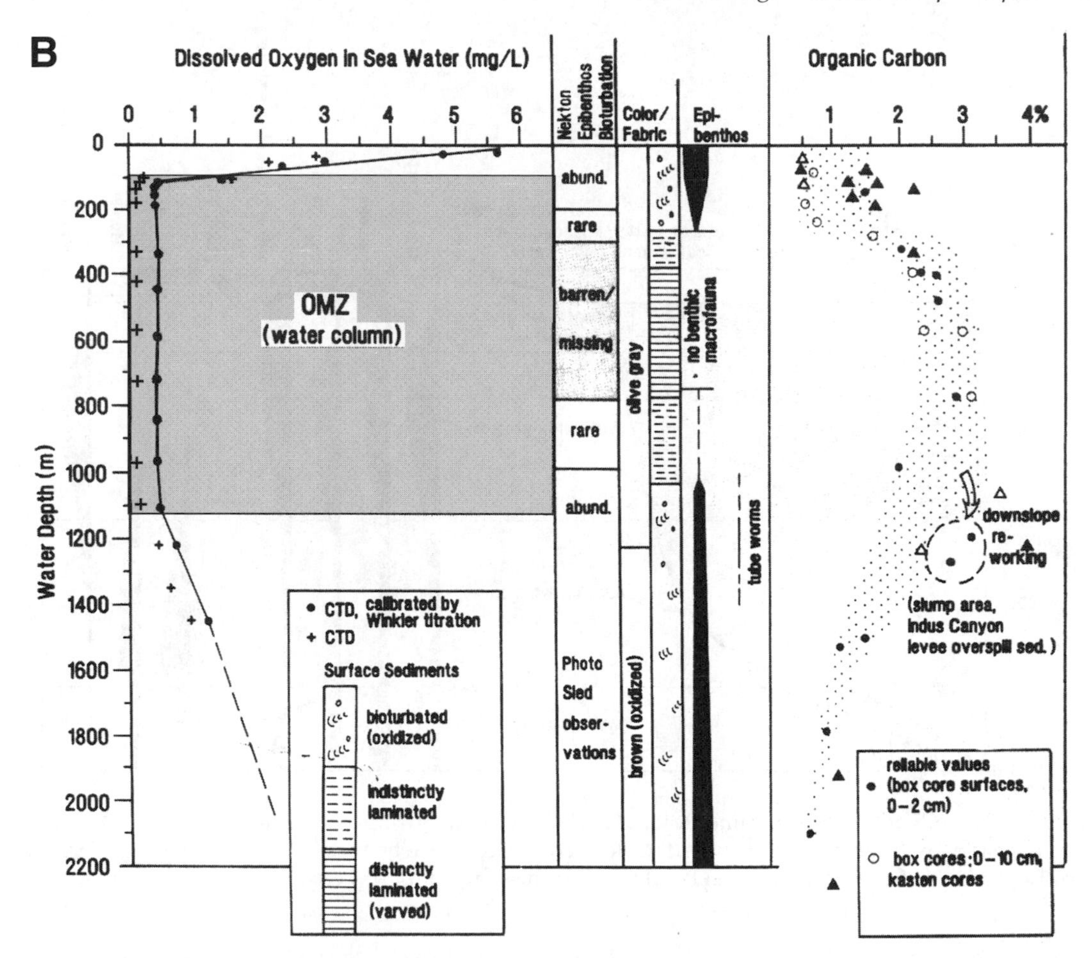

Figure 10.7 (*continued*) (B) Position of oxygen minimum zone (OMZ) in vertical transect, showing variations in dissolved oxygen, biotic habits and organic carbon content (continued overleaf).

OMZs seem to develop as a result of oxygen depletion due to degradation of plankton in the water column, beneath the photic zone. In the case of the Arabian Sea, spatial and temporal fluctuations over the last 30 ka in the OMZ were inferred from the changing patterns of dark and light coloured sedimentary layers (Schultz *et al*., 1996), which has obvious implications for the rock record: if the OMZ is in contact with shallow shelf sediments, then those will record the reduced oxygen; deeper sediments may be more oxygenated depending on efficiency of deep circulation. Therefore, in the rock record, although anoxic sediments may be recognised at one site in a basin, that does not automatically mean that such anoxia will be found all over the basin. One has to look to find out, and not to assume.

10.5.2 RECENT AND MODERN OCEAN OXYGEN DEFICIENCY: RECORDS FROM THE CALIFORNIA BORDERLAND SILLED BASINS

Because of low sea levels and active thermohaline circulation in modern times, the widespread occurrence of oxygen deficiency in epeiric seas is prevented, so models of ancient anoxia are based on evidence derived

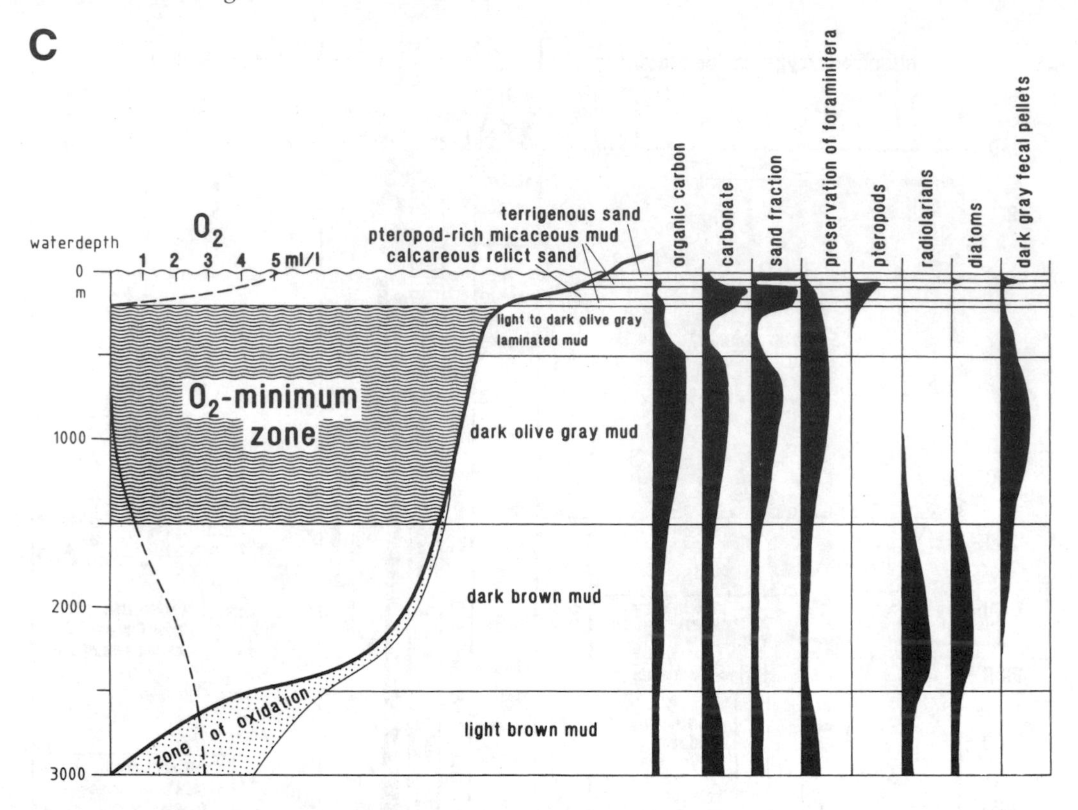

Figure 10.7 (*continued*) (C) Sedimentology and microfossil groups in relation to the OMZ of the India–Pakistan continental margin; note the close correspondence between oxygen levels, and facies and biota. Reproduced from Schultz *et al*: (1996) with permission from the Geological Society of London.

from silled basins, morphologies of which constrain circulation in deeper water. While the Black Sea is an atypical example because of its isolation within a continent (Section 4.3.5), and its lower-than-normal salinity, the silled basins of the Californian margin provide better natural laboratories to examine variability, determined by a variety of input factors. Importantly, the presence of sedimentary laminae does not automatically demonstrate anoxia; rapid sedimentation rate could be responsible instead. In both cases, benthic organisms are excluded by the unfavourable conditions, a feature of considerable potential importance in the ancient record. A review of differences between modern neighbouring basins off the coast of California (Gorsline *et al*., 1996) shows how in some basins or parts of basins (San Pedro, Santa Monica and Santa Barbara basins), lamination is due to cyclic seasonal fluctuation of suspended sediment supply into anoxic bottom environments, where anoxia also prevents biotic reworking of the sediments. In others, turbidite input via submarine canyons inputs fine-grained sediment at a rate exceeding the ability of bioturbation to disturb it (e.g. San Pedro, Santa Monica, San Diego, Santa Cruz and San Nicholas basins) (Figure 10.8)

Holocene sediments are subject to scrutiny on a millimetre scale, for example in cores from the Santa Monica Basin (Hagadorn, 1996); the data indicate organic carbon input (from primary productivity and some terrestrial input) is the main control on lamination, rather than seasonal fluctuations in the oxygen

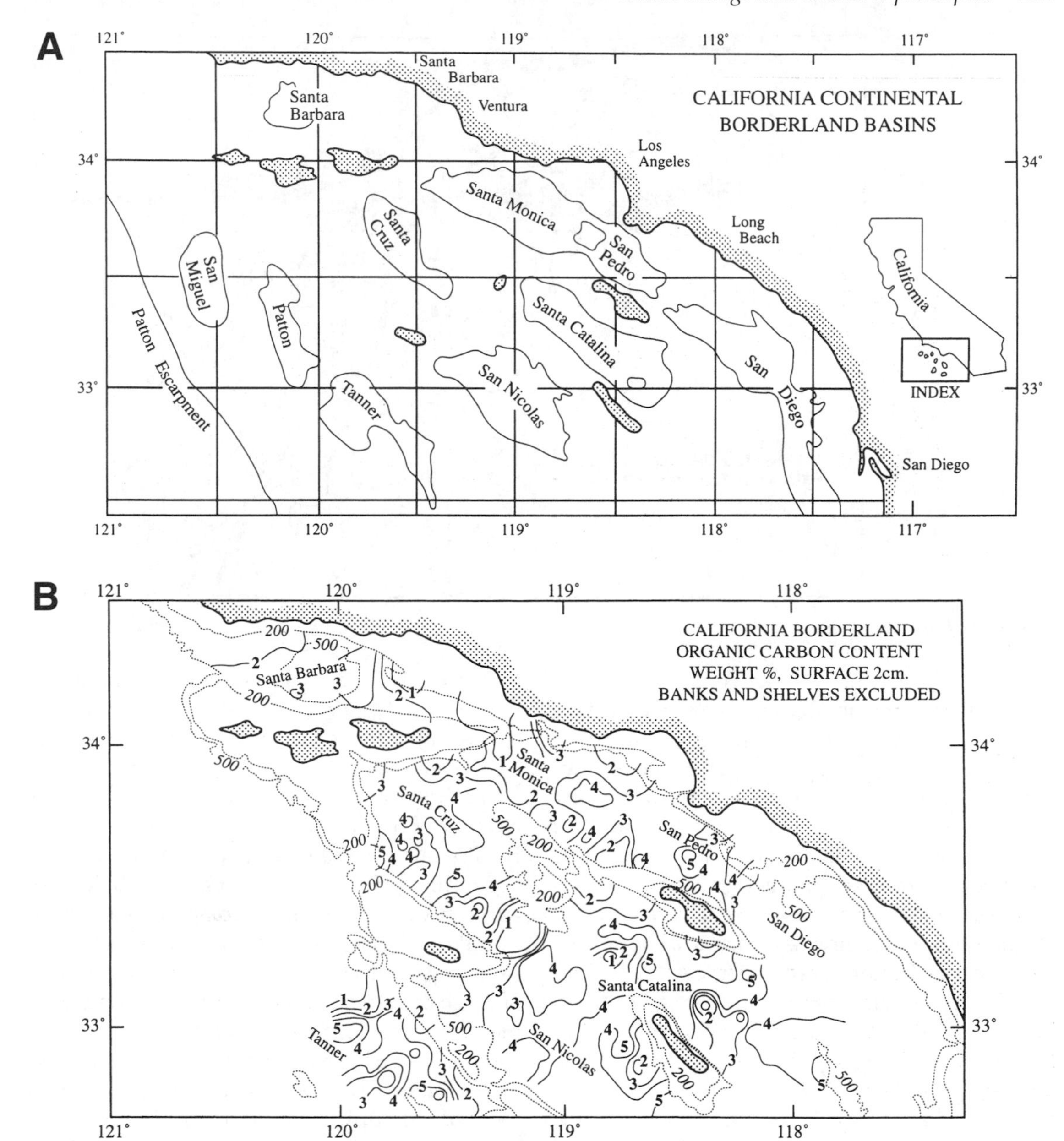

Figure 10.8 Features of the California continental borderland basins, showing a suite of topographic lows, and variations in amount of deposited organic carbon, related to reduced circulation within the basins. Reproduced from Gorsline *et al*: (1996) with permission from the Geological Society of London (continued overleaf).

levels. Seabed anoxia appears to have been a long-lived feature of this basin, so that organic carbon arrived in an already low-oxygen setting at the seafloor. Palaeontological data are part of the information used to demonstrate that late Quaternary and Holocene organic fluxes to the seabed of the Santa Barbara Basin were related to upwelling (Bull

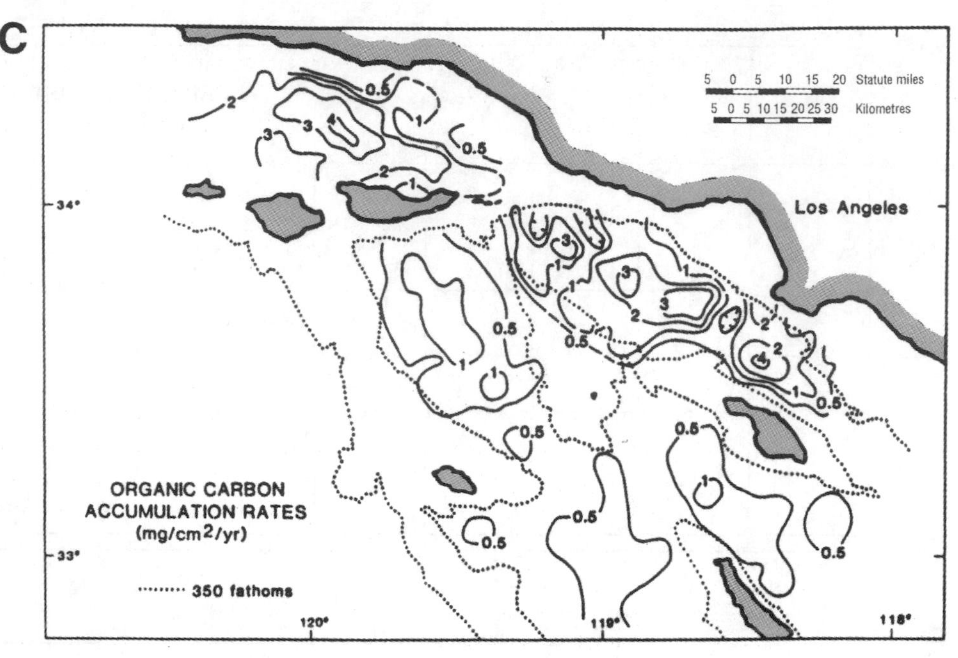

Figure 10.8 (*continued*)

and Kemp, 1996): during cool phases, the laminae are rich in diatom ooze, suggesting upwelling of nutrient-rich waters, coinciding with lack of benthic foraminifera, consistent with oxygen depletion of bottom waters by decomposition of the descending mass of organic matter. Glacial sediments in these deposits are not laminated, because of increased basin-floor oxidation, presumably due to sinking of northern and southern cold oxygen-rich waters in the Pacific.

10.5.3 A WARNING FOR PALAEOCEANOGRAPHY: MODERN DIATOM OOZES

Certain aspects of deep-sea oceanography rely on simple principles, and the development of laminated sediments is interpreted as representing deep anoxia, whereby sea-floor organic activity is inhibited. Of considerable interest, therefore, is the discovery by Kemp *et al.* (1996) of laminated diatom oozes on the Neogene Pacific Ocean floor in areas of well-oxygenated deep water. These oozes are interpreted to be caused by frontal zones within the upper ocean, due to changes in ocean state during La Niña events (where cold waters enter the warm waters generated in El Niño times, and represent an overshoot of the return to normal conditions after the end of an El Niño event); their evidence is based on 1992 observations in the Pacific, when floating near-surface diatom mats developed. The interpretation is that these may sink to the seafloor, retaining their integrity, leading to a strengthened seabed mat which was not susceptible to biotic reworking. Preservation of such structures in the rock record may lead to incorrect interpretations of ocean-floor anoxia. Another example of application of these ideas is in the eastern Mediterranean Sea, where Late Quaternary sapropels (fine-grained organic-rich sediment) have formed. These are usually considered to be due to deep anoxia (e.g. Béthoux and Pierre, 1999), but Kemp *et al.* (1999) suggested that diatom production was responsible for the high levels of organic matter in diatom-rich sapropels, a potential explanation of why the eastern Mediterranean seafloor is well oxygenated today; thus there is no need to invoke a change in circulation. However, note that this applies only to diatom-rich sapropels. This is current research and its implications are yet to be fully assessed.

10.6 OCEANS AND ANOXIA II: OCEANIC ANOXIC EVENTS (OAES)

OAEs were recognised more than twenty years ago (Tucker and Wright, 1990) in Jurassic and Cretaceous sediments when short-lived episodes of ocean productivity were so rapid that the OMZ became oversaturated with degradable organic matter. Undigested material collected on the seafloor and buried in sediments, and therefore betrays the event.

Work on DSDP and ODP data globally show that the late part of the Early Cretaceous suffered several OAEs, and despite integrated biostratigraphy, lithostratigraphy and geochemistry, the cause of the events remains unclear (Bralower *et al.*, 1994). Nevertheless, this was a time of increased ocean crust development as continental separation accelerated and sea level rose, and may therefore relate to a complex crust–ocean interaction (Chapter 4). The OAE around the Cenomanian–Turonian boundary time (middle Cretaceous), coincides with a minor mass extinction. The significant increase in ocean crust formation at that time might be responsible for volcanic outgassing of CO_2 and other gases such as SO_2, H_2S and halogens (Kerr, 1998). CO_2 could assist the development of the anoxic seafloor, and together with the others lead to acidification of oceans, and release of more CO_2 by carbonate dissolution. The known sharp increase in $\delta^{13}C$ at the Cenomanian–Turonian boundary has been interpreted as indicating burial of $\delta^{13}C$-depleted organic matter, and therefore a *fall* in atmospheric CO_2 (Kuypers *et al.*, 1999). Altogether, five OAEs have been identified in the Cretaceous (Johnson *et al.*, 1996) (Figure 10.9), and OAEs are a feature of transgressive settings, important in the Jurassic and Cretaceous; OAEs probably link to the fact that higher sea levels tend to correlate with warmer climates (see Chapter 9), which themselves relate to the concept of calcite seas during phases of active tectonics (when the sea level would be higher). In turn, the deep-ocean circulation systems tend to be suppressed, in contrast to icehouse conditions of cold deep-water sinking and ocean conveyors. In greenhouse settings, ocean productivity is lowered because of suppressed nutrient cycling (especially upwelling), so that it is perhaps not surprising that estimates of Cretaceous productivity suggest rates of only 10% of modern values (Bralower and Thierstein, 1984). Results from the Caribbean Cretaceous show a correlation between three episodes of reduction in reef systems and OAEs, and there may be a link between OAEs and bottom water formation. The possibility of episodes of formation of warm, saline bottom water, which then acted as a heat transport mechanism to remove heat from the tropics during the Cretaceous, may have resulted in increased surface water productivity and consequent burial of organic matter, proxy for the OAE (Johnson *et al.*, 1996).

OAEs are important but complex markers for rapid ocean change, but it remains to be seen to what extent they correlate with phases of plate activity, and why they are prominent features of the Cretaceous period, but less well known in other parts of calcite sea history.

10.7 OCEANS AND ANOXIA III: THE END-PERMIAN MASS EXTINCTION

This complex event killed more than 90% of ocean biota and severely affected terrestrial ecosystems (Erwin, 1994; Hallam and Wignall, 1997). Traditionally regarded as having occurred over a long time, due to recorded gradual organic diversity reduction during the later Permian, later evidence suggests a two-stage extinction. The first may well have been related to severe regression in the Late Permian, but that regression was replaced by

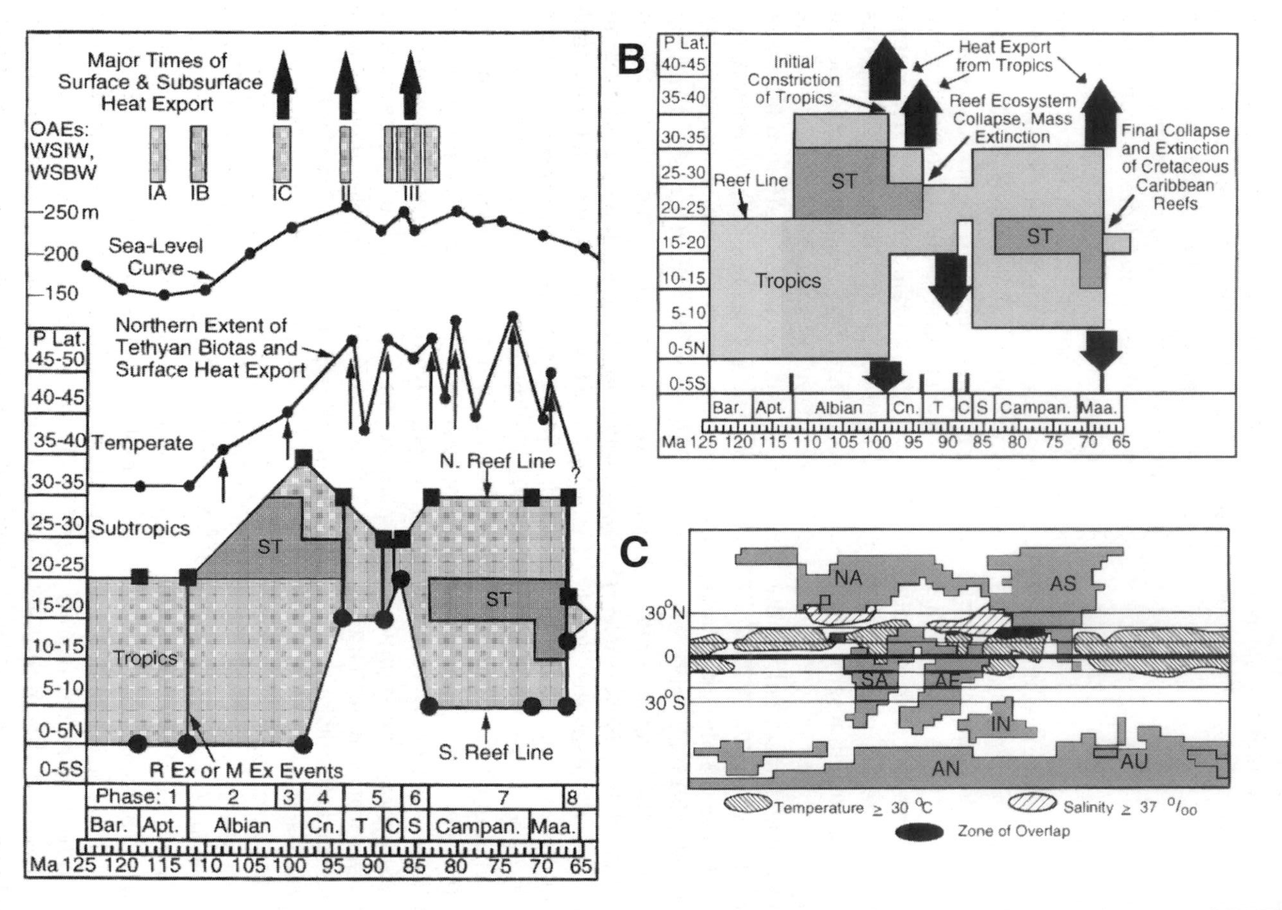

Figure 10.9 Cretaceous Oceanic Anoxic Events (OAE's). (A) Data of sea-level change, distribution of climate, and timing of OAE's for the Caribbean area. R Ex = regional extinction; M Ex = mass extinction; WSIW = warm saline intermediate water; WSBW = warm saline bottom water. (B) The palaeolatitudinal extent of Caribbean Cretaceous tropical environments using corals and rudist bivalves (reef-building bivalves). The darker area labelled ST (super Tethyan) contains unusually high diversity levels of rudistids. Note the sudden collapse of the reef systems towards the end of the Cretaceous, linked to cooling and sea-level fall prior to the Cretaceous/Tertiary extinction event. (C) A General Circulation Model (GCM) of salinity and temperature at 100 Ma, based on four times the modern CO_2 level. The model shows the very warm tropics and increased salinity neighbouring continents in the tropical belt. The opportunity for anoxia to develop in deep water is increased during such settings. Reproduced with permission from C. C. Johnson, *Middle Cretaceous Reef Collapse Linked to Ocean Heat* (1996), the Geological Society of America.

transgressive facies in the last stage (the Changxingian Stage in China) of the Permian System which was accompanied by increases in diversity right up to the boundary, followed by a rapid extinction event (Bowring *et al.*, 1998; Wignall, 1992a,b) (Figure 10.10). Several sources detail the organisms involved (Briggs and Crowther, 1990; Erwin, 1994; Hallam and Wignall, 1997); curiously, nektonic organisms, especially fish, largely escaped the effects. The end-Permian extinction's causes have been extensively debated, but brackish oceans, cosmic radiation and glaciation are discounted. Although those times were represented by enormous amounts of evaporite accumulation, not enough salts were removed as evaporite deposits to influence global salinity sufficiently to upset the salt balance of marine organisms. There is no evidence for cosmic radiation, and the late Palaeozoic glaciation finished well before the extinction. CO_2 was relatively low in the Late Permian (Berner, 1991), largely because the Late Carboniferous forests had drawn down most atmospheric CO_2. Nevertheless, CO_2 levels were rising throughout the Late Permian, and oxygen levels falling, as discussed in Chapter 9. Approximately coincident with the extinction is large-scale basalt eruption in Siberia (the largest of the Phanerozoic), the bulk of which seems to have been lava, and the amount of dust generation is presumably limited, because basalt eruptions tend not to be explosive and therefore produce little dust compared with acidic volcanoes, but a large eruption is likely to have produced massive quantities of CO_2. The most compelling data comes from the widespread evidence for ocean anoxia present in sedimentary rocks, coupled with low diversity biota of types that seem suited to stressful environments (disaster biota). There is weak evidence of a bolide impact at the boundary, so that idea remains, but lacks the strength of evidence of the Cretaceous–Tertiary event.

Prominent in Permian–Triassic boundary sediments is the change from diverse communities in shelf settings to poor communities in largely non-bioturbated sediments rich in pyrite. Alone, these features are clearly indicative of anoxic water, but there is supporting geochemical evidence of anoxia (Wignall, 1992b). A positive Ce anomaly in early Triassic sediments indicates anoxic seafloors, because Ce is incorporated into sediments in anoxic conditions. An accompanying rapid positive change in sulphur isotopes is interpreted as due to anoxic seafloor. Finally the carbon-to-sulphur (C/S) ratio was lower in the earliest Triassic sediments relative to beds above and below; like the sulphur isotopes this ratio is taken to indicate widespread pyrite (which is indeed common in the sediments following the extinction), and carbon levels were low presumably because of lower productivity in these poor conditions. Taken overall, these geochemical data indicate the earliest Triassic oceans had low oxygen, but note that its effects are largely restricted to shelf environments, because the deep ocean does not seem to show effects of the extinction. Moreover, the extinction correlates in low latitude settings with rapid transgression across shelves, although why that should automatically lead to widespread anoxia is more complex (see below).

The concurrent demise of many terrestrial faunas and floras emphasises the crucial interaction of the ocean and atmosphere as linked systems; an appreciation of the prior geological history is relevant. One feature is the very high peak in oxygen estimated as a result of the Late Carboniferous forests (Berner and Canfield, 1989). The photosynthesising mass appears to have resulted in a huge injection of oxygen into the atmosphere, which would naturally have found its way into the oceans. Part of Carboniferous time was an icehouse phase, so that ocean conveyor systems most likely operated, with the seafloor oxygenated. This situation changed as Pangaea finally assembled, sea level fell, and climates warmed as the glaciation waned. Some rise of sea level was inevitable as the glaciers melted, and appears to have contributed to the flooding of the

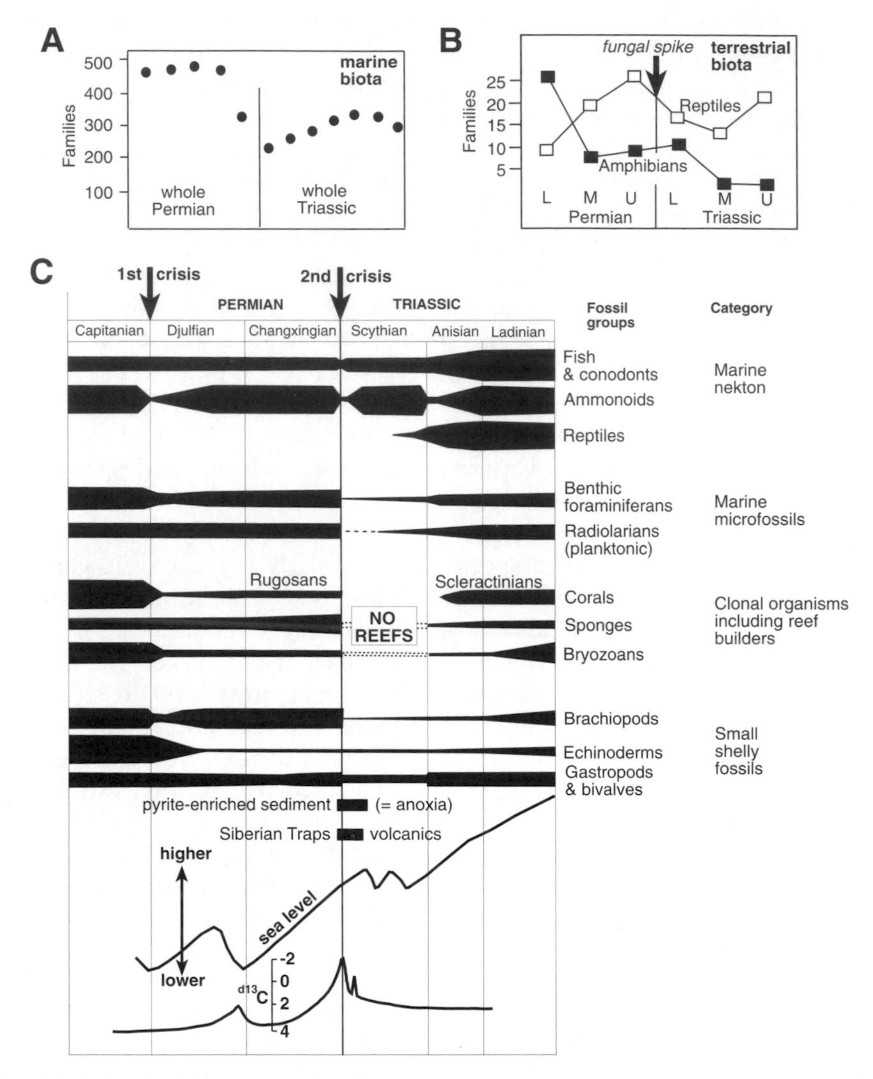

Figure 10.10 Late Permian extinction data. (A) During the late Permian there was a notable decline in marine biota diversity, and the recovery phase in the Triassic was prolonged. (B) Terrestrial vertebrates do not show such dramatic change as marine fossils, but there was a curious large increase of fungae in the latest Permian to earliest Triassic (the fungal spike), taken to indicate drastic alteration of terrestrial ecosystems. (C) Details of marine taxa show that the Permian extinction event was a two-stage process, with expansion and diversification in between the two; the second crisis is the more devastating. Global sea level was (probably) rising throughout the P/T boundary, although note that sea level was generally low. The approximate coincidence of anoxia (shown by pyrite in sediments), increased CO_2 (from volcanics) and productivity collapse (^{12}C spike is a proxy), in marine sediments, testifies to rapid ocean deterioration, although the precise timing of the volcanics is not sufficiently known to prove that volcanic eruptions caused the extinction.

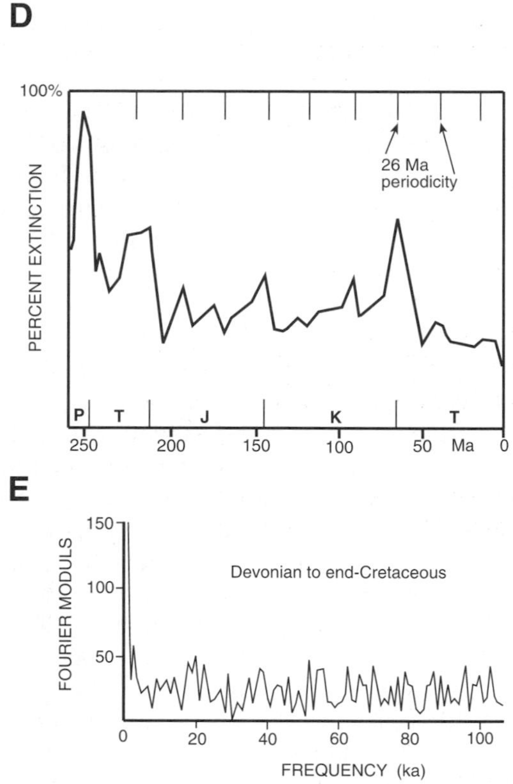

Figure 10.10 (*continued*) (D) The idea that mass extinctions in the Phanerozoic follow a periodicity of 26 Ma has been challenged by reprocessing of the data using Fourier analysis (E), which indicates that there is no pattern. Oceans are involved in mass extinctions, but the influence of processes external to the Earth is therefore not confirmed by this new analysis. (A), (B) and (D) adapted from Briggs and Crowther (1990), *Palaeobiology: a synthesis*, Blackwell Scientific Publications; (C) adapted from Hallam and Wignall (1997) *Mass Extinctions and their Aftermath*, Oxford University Press, Oxford; (E) adapted from House and Gale (1995) *Orbital Forcing Timescales and Cyclostratigraphy*, Geological Society of London Special Publication 85.

North Sea basins in the middle Permian (Chapter 9), but overall the collapse of plate activity caused seafloor subsidence on such a large scale that net sea-level fall was several hundred metres. Coupled with this is the suggestion of uplift along the southern margin of Gondwana, to expose the Carboniferous coal (Faure *et al.*, 1995). The result of these changes depleted terrestrial vegetation, and most likely resulted in reoxidation of any exposed dead plant matter, together with forest fires, leading to rapid fall in oxygen levels, if Berner's graph is correct. Thus the coincidence of a sea-level rise across shelf environments and the decline of ocean–atmosphere oxygen levels in a warm climate and supercontinent setting, may have contributed to shelf anoxia and mass extinction. Note, however, that there is a timescale problem involved in explaining the extinction; the latest Permian shows very rapid demise of biota, which would require a catastrophic event, which is why the idea of a bolide impact remains a possibility, as do the Siberian flood basalts, which erupted quickly over a short period of time. The more gradual changes of CO_2 and O_2, characteristic of most of the Permian, are not consistent with such rapid changes.

Ideas of anoxia causing ocean extinctions were extended in an ambitious theory linking the oceanic setting of the Permian with that of the Precambrian. Grotzinger and Knoll (1995) drew comparison between the inorganic $CaCO_3$ precipitates, which seem to characterise parts of the Precambrian, and the large amount of apparently inorganic cement which occupies a major part of the Late Permian reef complexes, such as the famous Capitan reef in southwest USA. Grotzinger and Knoll (1995) interpreted this cement as signalling a large increase in the stored CO_2 in anoxic deep-ocean waters of the Late Permian. This idea was further developed (Knoll *et al*, 1996) whereby ocean overturn in the boundary interval resulted in upwelling of this poisonous water onto shelf environments, causing the extinctions. Such theories produced mixed reactions in the scientific press (see *Science*, 1996, **274**, 1549–52), and the arguments get complicated. The concentration of CO_2 in deep waters (having been drawn down from the atmosphere) would presumably produce cooling (Hallam and Wignall, 1997), yet the global temperature trend was increasing then. However, there certainly needs to be an explanation of the large amounts of inorganic $CaCO_3$ cement in the Late Permian reefs, and its presence suggests something unusual about Late Permian seawater chemistry, such that there was an excess of carbonate in the ocean system. In the Carboniferous Period large amounts of $CaCO_3$ were deposited in shelf settings, and exposure of these in the Permian could have led to increased $CaCO_3$ saturation of the sea. One suggestion is that the carbonate-secreting organisms of the Permian did not have the capacity to remove sufficient $CaCO_3$ from the water, leading to inorganic precipitation of cements on the seafloor as the $CaCO_3$ content of the waters became very high (Edwards, 1990). A remarkable crust, capping reef complexes, reported from the boundary in South China, comprises a digitate structure of recrystallised $CaCO_3$ (Kershaw *et al.*, 1999); whether this represents an organic microbialite structure which grew in the aftermath of the extinction, or an inorganic deposit, possibly related to upwelling, awaits confirmation.

If the deep ocean could store large quantities of CO_2, then this is reminiscent of the highly speculative soda ocean scenario discussed in Chapter 8. Mixing the anoxic CO_2-rich water with aerated surface waters would be expected to precipitate large amounts of $CaCO_3$ inorganic precipitates at the boundary. However, an *atmospheric* source of the CO_2, to cause shallow marine anoxia, is possible instead from the Siberian eruptions; however, its expected effect would be acidification of the oceans, dissolving, not precipitating, $CaCO_3$. Thus, much is yet to be explained about the late Permian oceans, and the work of Grotzinger, Knoll and co-workers has highlighted the complexities involved.

A remaining problem is the precision of dating the boundary, and an enlightening discussion (Bowring *et al.*, 1998) shows that the causes of the end-Permian event therefore remain speculative; it may have involved oceans, but certainly took place very quickly. Figure 10.11 summarises the various models which have been applied.

10.8 OTHER EXTINCTIONS

The O/S and P/T extinctions were selected to demonstrate the importance of ocean processes, and these can be seen in other extinctions. For example, anoxia is invoked to explain an Early Cambrian extinction (Zhuravlev and Wood, 1996), whereby black shale deposition in unbioturbated laminae represents anoxic seafloor conditions. Zhuravlev and Wood (1996) suggest that this was caused by eutrophication due to upwelled currents, leading to phytoplankton blooms, and then deoxygenation of bottom waters. The late Devonian (Kellwasser event) and Late Triassic extinctions, for example, have a range of explanations, with anoxia being high on the list (Hallam and Wignall, 1997), although climate change and plate tectonic relationships play their role. Recent

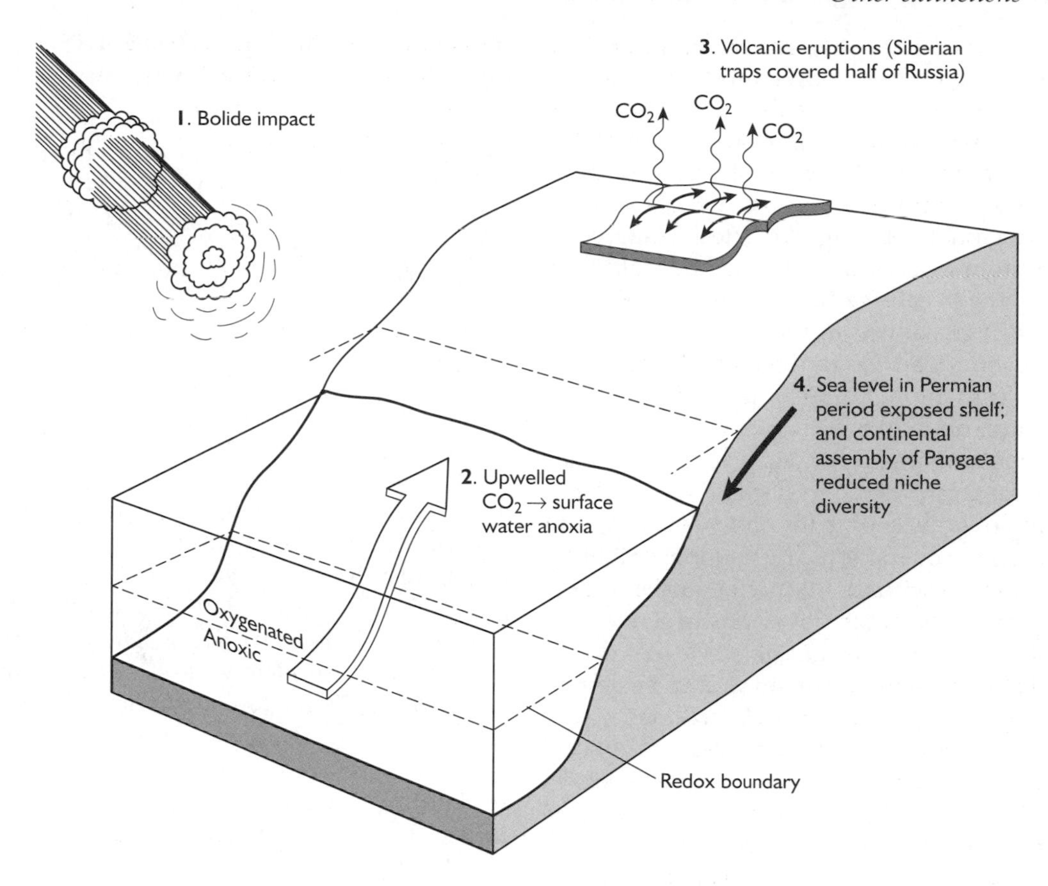

Figure 10.11 Models of alternative possible causes of the late Permian extinction event. This composite figure contains geologically instantaneous processes (1–3) but process number 4 must have developed over a long time. The three main contenders as instantaneous extinction vectors are: a volcanic eruption (Siberian flood basalts); a meteorite impact; and ocean overturn leading to marine shelf anoxia. All three causes have some evidence, although clearly this event was complex, and one of the biggest problems is precise dating and therefore correlation of events in different parts of the world. For example, the age of the Siberian flood basalt has not been sufficiently constrained for it to be unequivocally linked to the extinction event.

results for the Late Triassic suggest that a multiple impact was possible (Spray *et al.*, 1998), although whether that caused the extinction requires more data.

Details are not given here for the famous Cretaceous–Tertiary (K/T) event because it is widely covered elsewhere (Hallam and Wignall, 1997), but its popularisation in the media would lead you to think of it in only simple terms. Thus the current contenders described in popular media are bolide impact (the Chicxulub Crater of Yucatan) and volcanic eruption (the Deccan Traps in India), which have focused minds on instantaneous destruction that the fossil record (where much of the evidence lies) does not so conveniently reflect. The loss of half the planktonic foraminifera of the latest Cretaceous times at the boundary (Hallam and Wignall, 1997, p. 191) indicates there was indeed a geologically instantaneous event; that is, it happened over a period of time which cannot be resolved by stratigraphic methods to less than a few thousand years, so it could indeed have been instantaneous (over weeks or months). However, there is plenty of evidence that there was a deterioration in global climate during the latest Cretaceous, with cooling and sea-level fall, that could have contributed to the extinction; and note that not all species involved in the extinction died out at the boundary. But the data change with new information. For example, evidence from dinosaur finds over the years suggested that the dinosaurs died out just before the impact, but recent intensive collecting of the latest Cretaceous sediments in North America indicates dinosaur remains are abundant right up to the boundary, and shows the importance of data collection in the development of theories. However, the fact of falling sea level leads to problems of data collection in any geological setting, because terrestrial sediments can become eroded. Even though dinosaurs are found up to the boundary in parts of North America, the deteriorating climate of the latest Cretaceous are likely to have restricted dinosaurs to only the favourable areas; the youngest dinosaur fossils are from North America, so their extinction before the boundary in most sites remains an observed fact. Such information is important because it is all too easy to simplify these events in your mind, forgetting that the real world does not work like that. Palaeontology is full of examples of biotic refuges, where the last remaining populations survive in small areas while the rest of the species have become extinct. The last word has not been said about the K/T event; recent studies of sediments of the K/T interval in Mexico (Ekdale and Stinnesbeck, 1988) revealed calcitic spherules which pre-date the extinction event by at least months or years, because of the nature of the trace fossils present. This work suggests the possibility that, if the spherules were the result of impact, then impact occurred a significant time interval before the marine extinctions. Where does that leave us?

10.9 SALINITY FLUCTUATIONS: THE MESSINIAN SALINITY CRISIS

Throughout the Tertiary Period, progressive closure of the Tethys Ocean led to the formation of the Mediterranean Sea, and involved collision of Arabian and Eurasian plates in the early Miocene Epoch (20 Ma), cutting off the Pacific connection, and the climate of the Mediterranean region became drier. Later, during the Messinian stage of the Miocene *c.*6 Ma, the Mediterranean Sea basin also became isolated from the Atlantic Ocean, by a combination of the closure of the Gibraltar Straits due to collision of Africa and Spain and a fall of sea level. Then, as now, the rate of evaporation exceeded input of water to the basin, resulting in severe desiccation (Hsü *et al.*, 1973, 1977). Calculations show that if the modern Mediterranean was isolated and dried up, it would completely desiccate in *c.*1000 a, and deposit *c.*70 m of salt (Crowley and North, 1991, p. 201). Evidence from drilling the Mediterranean Sea floor, and seismic work, revealed at least 1400 m of evaporites accumulated, perhaps as much as 2–3 km (Crowley and North, 1991, p. 201), overlain by normal marine clays, and presumably created a dramatic landscape because the Mediterranean Sea has average water depth of 1.5 km, maximum nearly 5 km. Drilling of the Mediterranean crust, however, did not completely penetrate the evaporite sequence, so the total thickness is yet to be determined (Riding *et al.*, 1998), but the deposit comprises, broadly, a lower halite and an upper gypsum unit (Rouchy and Saint-Martin, 1992). Because the desiccation of the entire volume of that sea would yield only 70 m of evaporites, the preserved thickness is best explained by repeated evaporation and replenishment of seawater. Because the evaporites are widespread on the deep Mediterranean floor, this is called deep desiccation, and it is possible that the entire Mediterranean Basin did dry up several times, constituting the Messinian salinity crisis (Figure 10.12). However, it is yet

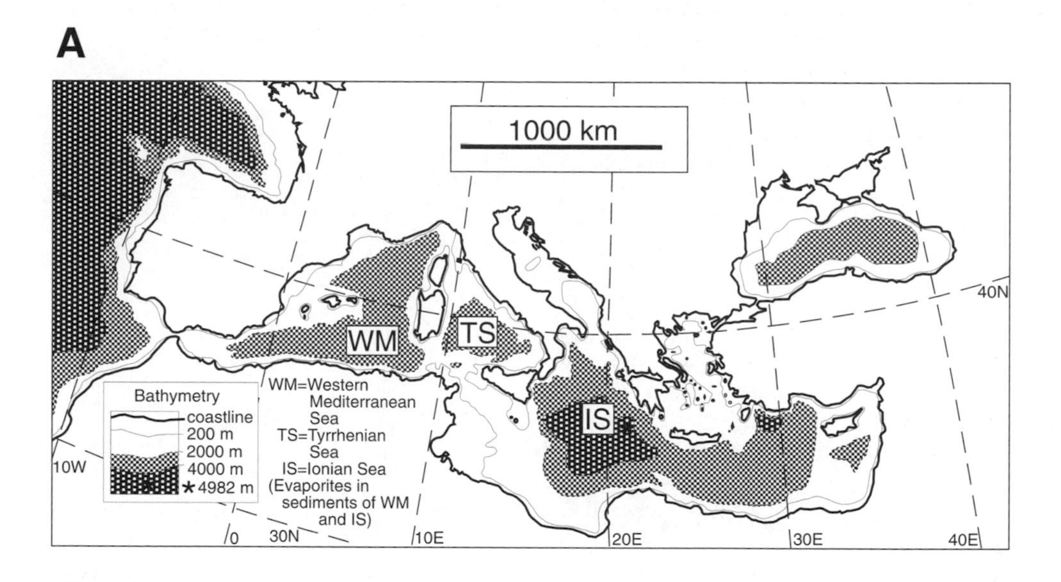

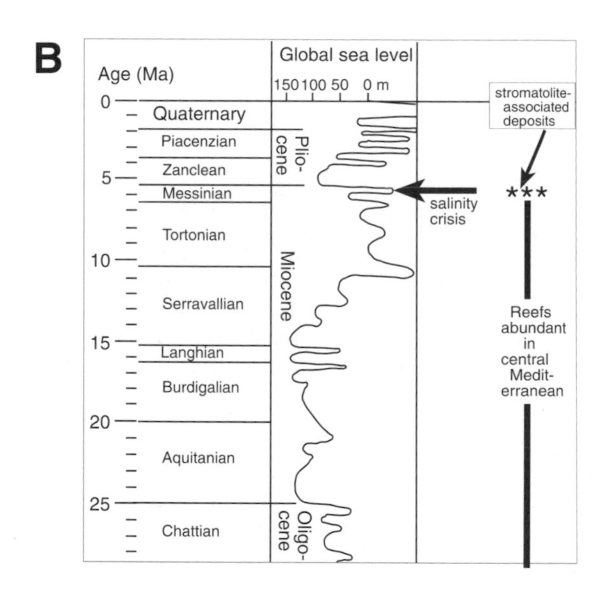

Figure 10.12 The Messinian salinity crisis in the Mediterranean. (A) Map of the Mediterranean area, showing the location of evaporite deposits across the Mediterranean basins. (B) Stratigraphy of the Miocene Epoch, during which the salinity crisis occurred (adapted from Pedley, 1996).

(continued overleaf).

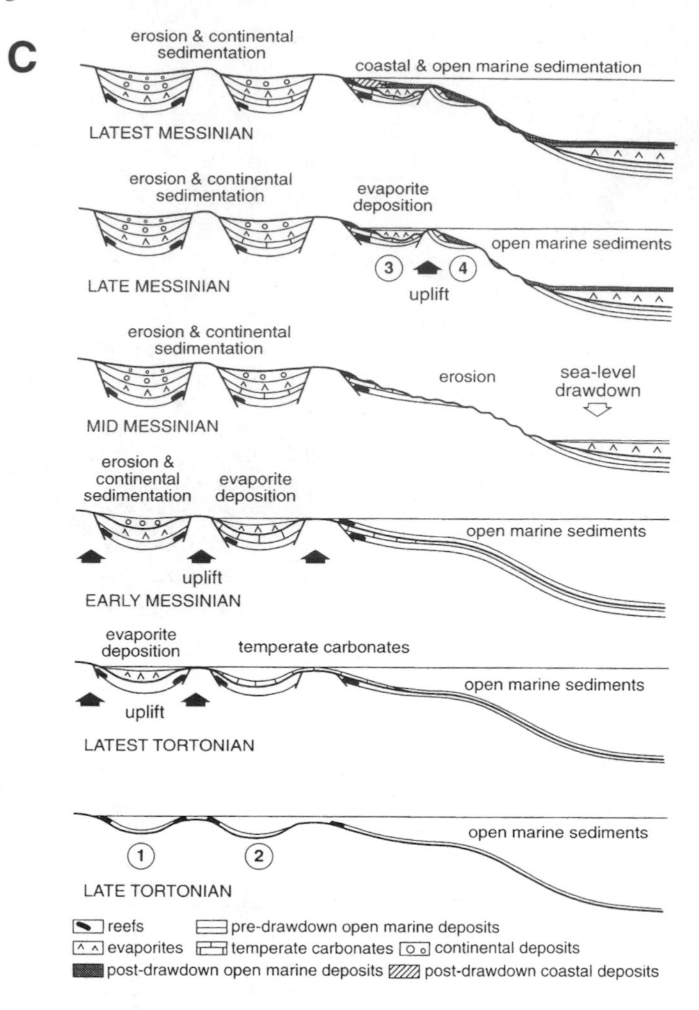

Figure 10.12 (*continued*) (C) General model of the process of development of the crisis, showing isolation of the Mediterranean basin, deep evaporation and sea-level fall, and replenishment once the crisis was over. Note that marginal basins, which were isolated before the main onset of the event, contain evaporite deposits unrelated to the main event, and demonstrate the importance of isolating processes in basin development. Reprinted with permission from *Marine Geology*, **146**, Riding *et al.*, Mediterranean Messinian Salinity Crisis: constraints from coeval marginal basin, Sorbas, southeastern Spain. Copyright (1998), with permission from Elsevier Science.

to be confirmed that complete desiccation did occur.

Much useful information has been derived from small marginal basins dotted around the Mediterranean (Riding *et al.*, 1998), particularly in Spain and Sicily, which reveal normal marine sequences before and after the crisis, indicating that it started and stopped quickly. The evaporation led to drawdown of the dissolved salts, creating a possible drop in Mediterranean water level by as much as 1500 m, although more conservative estimates only allow 150 m (Clauzon *et al.*, 1996). Deep erosion then followed, before refilling with normal marine water occurred.

The crisis appears to be due to a combination of two processes: a global sea-level fall of as much as 50 m (Crowley and North, 1991,

p. 199), caused by cooling and growth of the Antarctic ice sheet, and tectonic collision between Africa and Spain to cause the isolation of the basin, with final closure of the Atlantic gateway via southern Spain (Soria *et al.*, 1999). The timescale is very short, but precision of dating is always a problem when geological events are fast, and the evolution of the event may be more complex, and has two phases because there are two evaporite units (Bearman, 1989a; Clauzon *et al.*, 1996). Firstly, between 5.75–5.60 Ma, a global sea-level fall led to isolation of only marginal basins, not the Mediterranean itself. Secondly, between 5.60–5.32 Ma, during a *warming* phase, the global sea level may have actually risen (Clauzon *et al.*, 1996), also indicated by the eustatic sea-level curve (see Figure 10.12B). Isolation of the Mediterranean was achieved by closure of Africa and Spain, and the Mediterranean underwent deep desiccation. Recent reassessment of dates (Riding *et al.*, 1998) suggests that the entire episode took place *c.*5.9–5.5 Ma, over a period of less than 0.5 Ma. Furthermore, data from marginal basins around the Mediterranean show that desiccation occurred both before and after the Messinian crisis in the Mediterranean itself, and demonstrate the effects of isolating small sub-basins, which then undergo their own evolution. There is some evidence that the 400 Ka eccentricity term influenced biotic and geochemical change during the onset of the crisis (Kouwenhoven *et al.*, 1999).

In detail, therefore, the exact nature of the salinity crisis is not fully resolved amongst authors, but of wider interest is the effect of the crisis on the global oceans. Within the Mediterranean area, the salinity in the Miocene seems to have been normal, apart from the salinity crisis, because of well-developed reefs (Pedley, 1996). Calculations suggest that *c.*8% of the global ocean total of salt was removed during this time, and global salinity fell by *c.*2‰ (Crowley and North, 1991), but this was spread over *c.*0.5 Ma (Bearman, 1989a), so that the overall effect on the oceans seems minimal. It might be expected that the global salinity was raised by the growth of the ice sheets, removing some water from oceans, during cooling, but there is no adverse effect on communities. Depths at which Pacific and Atlantic pelagic foraminifera lived appear to have shifted, but this is attributed to an increase in AABW as the Miocene build-up of the Antarctic ice sheet progressed (Murray, 1995), and there were fluctuations in the generation of AABW throughout the history of the Miocene (Sykes *et al.*, 1998).

10.10 OCEANS AND EARTHQUAKES

Geologically instantaneous events, occurring in minutes, have superficial effects on the oceans, but the interactions of ocean water and coastlines are dramatic. This survey is concluded by drawing attention to their results.

Ocean sediments not only record deposition of material which has oceanographic and climatic significance, but also record tectonic activity. Earthquake energy contains a significant component of surface waves, which disturb sediments, resulting in contorted laminations as a result of liquefaction. Recognition of this in the rock record requires distinguishing it from the results of slumping, of sediments deposited on slopes, but the field of palaeoseismology is growing. For example, laminated sediments laid down in tectonically active regions are likely to record earthquake shock and therefore not only may aid in the interpretation of tectonic events, but also, in a recent context, may be used in earthquake prediction studies via the relatively new field of neotectonics. Where contorted laminations are widespread and well constrained stratigraphically, such as in Jurassic–Tertiary marine sediments in the Polish Carpathians (Haczewski, 1996), they are most reasonably attributed to earthquakes, and in this case, occur between major nappe units, thereby placing them in the appropriate tectonic context.

Tsunamis are large waves generated at sea by seafloor earthquakes or marine slumping and transport suspended material onto continental fringes. Such sediment becomes interbedded with coastal, intertidal and supratidal deposits and their study therefore may be used to record the record of past earthquakes. The best examples, on the western North American seaboard, associated with the active margins there (Long and Shennan, 1994; Nelson *et al.*, 1996), are providing models for recognising the effects elsewhere, in the UK and Mediterranean regions, for example.

During earthquake activity, the crust is severely shocked, causing local relative sea-level changes of up to several metres per event, by moving the land rather than the sea. Accumulated events can lead to dramatic features on rocky coasts; the best known are coastal notches, formed during sea-level still-stands, which then become stranded several metres above sea level. Recent work on these in the Mediterranean helps to reveal that the northeastern coastline of Sicily has been uplifted over the last 9000 years, while the nearby Calabrian coast of mainland Italy has hardly been affected (Pirazzoli *et al.*, 1997; Rust and Kershaw, 2000; Stewart *et al.*, 1997). Although such features are marginal to the study of oceanography, they have both an academic and an applied value in the understanding of processes at ocean borders.

10.11 CONCLUSIONS

The short-term nature of many features of Phanerozoic oceans shows the capacity of ocean systems to change quite quickly, and draws attention to the fact that, despite the enormous volumes of water involved, seawater properties are dynamic. What remains unclear, however, is the degree to which the oceans themselves cause short-term events, in contrast to simply responding to external forces. The Younger Dryas phase is an example of this conundrum, and unravelling the causative factors is what oceanography is striving to achieve, to aid the field of prediction of climate change.

10.12 SUMMARY

1. Oceans are shown to be able to change quickly, and understanding the reasons is of practical value in predictive oceanography. Climate, plate tectonics, sea-level change and bolide impacts all can have short-term effects, with the first three being related, although the fourth influences the others. Several examples of major events in the Phanerozoic record demonstrate the range of ocean responses.
2. A two-stage mass extinction in the Late Ordovician is attributed to onset, then collapse, of glaciation in the Southern Hemisphere, as revealed by a combination of facies, faunas and geochemical datasets. A short phase (0.5–1.0 Ma) of interpreted thermohaline circulation in an otherwise poorly circulated ocean system may have been sufficient to trigger extinction, and provide empty ecological niches for renewed evolution afterwards. Anomalous characters include: a high level of atmospheric CO_2 within which the glaciation developed, and oxygen and carbon isotope shifts seem to be larger than predicted for the scale of the glaciation.
3. The Younger Dryas phase of the Late Quaternary shows short-term change in the Atlantic Ocean system, seemingly related to fluctuations of the ocean conveyor system by disruption of NADW production, and possibly part of a seesaw switch of deep-water formation; the Southern Ocean may be the alternative source of deep water while the NADW 'factory' was closed down.
4. Anoxic processes in deep oceans, related to fluctuations in productivity and/or crust–ocean chemical and tectonic interactions, can develop rapidly and generate oceanic anoxic events (OAEs). These are especially well recorded in the Jurassic and Cretaceous, associated with transgressions.
5. The complex end-Permian mass extinction may be related to widespread anoxia,

although the reasons are unclear. Massive volcanic eruptions in Siberia may play a part in generating large volumes of CO_2 to initiate an anoxic event, although a bolide impact has not been ruled out.

6. A global salinity decrease of *c*.2‰, spread over *c*.0.5 Ma, is calculated from salt deposits in the Mediterranean, laid down during the Messinian salinity crisis, although this seems to have had minimal effects globally. The entire Mediterranean dried up at *c*.6 Ma, having been isolated from the global ocean system by tectonic closure of Tethys Sea, coupled with a sea-level fall attributed to growth of the Antarctic ice sheet.
7. Geologically instant features of the oceans are reflected by the results of earthquakes, which generate tsunamis, and local shifts in crustal elevation, causing local relative sea-level change. The relationship between such events and the records at coastlines is of practical value in determining rates of vertical movement of crustal masses.

PART C

HUMAN IMPACTS ON THE OCEAN SYSTEMS

Parts A and B of this book examine how the ocean system functions now and in the geological past. The range of information available is vast, and provides an important backdrop to the final section of our study of ocean systems: assessment of current change in marine systems wittingly and unwittingly instigated by mankind. The aim through the book is to inform the reader of the nature and history of ocean systems to aid the formulation of a perspective on human-induced change. A great deal written in the scientific and popular media does not permit such a perspective, and so, if you have not already done so, you are advised to read the introductions, conclusions and summaries of all the preceding chapters before reading Part C.

Despite our short geological history, the impact of humans on the ocean system has in some cases been severe, and far outweighs some recent natural changes. The oceans were largely protected in the past from human impacts due to their remoteness and vast size. However, technological developments, particularly since the 1940s, have allowed an increase in human use of the oceans and an increase in the potential damage we can cause to the ocean system. Most current problems relate to:

1. marine pollution impacts;
2. over-exploitation of biological resources, such as overfishing and destruction of coral reefs by the tourist industry;
3. the wider issue of anthropogenically enhanced global warming.

Chapter 11 provides an overview of items 1 and 2, while Chapter 12 addresses item 3.

DEGRADATION OF THE MARINE ENVIRONMENT 11

11.1 INTRODUCTION

Marine pollution and human overuse of the oceans' store of food threaten the stability of ocean ecosystems, with a potential impact on the economy of many developing and developed countries. This chapter examines the susceptibility of different marine environments to degradation, and also addresses the only way in which damage can be managed, i.e. the legal framework which may control human use of the oceans. The balance to be struck is between the conflicting needs of nature conservation and human resource requirements.

11.2 MARINE POLLUTION: THE NATURE OF THE PROBLEM

The oceans have been used as a site to dump waste for thousands of years. The overriding philosophy of waste disposal in the oceans seems to be that the oceans are large enough to store our waste without significant change, and implies that pollutant wastes are rapidly diluted to innocuous levels with no lasting effect on the ocean system as a whole. While this may be true for certain pollutants and certain marine areas, it is by no means universally applicable. The impact of any pollutant depends largely on its behaviour in the marine environment, the scale of its input, and the type of environment to which it is discharged.

Human activity has caused significant alteration of some fluxes in the crustal-ocean factory, and has introduced many completely artificial substances into the sea. An exhaustive list of all types of pollutant, and their marine behaviour, is beyond the scope of this book, but information is available elsewhere (Hollister and Nadis, 1998; Jenkins, 1992). Instead, specific pollutants are selected to illustrate how ocean processes and chemical behaviour determine their impact on marine ecosystems. A pollutant can be thought of as any substance (natural or artificial) which has a deleterious or harmful effect on the environment. Pollutants have a detrimental impact by virtue of overloading the capacity of the marine environment to break them down or remove them. A little oil in the sea is not a problem and may even stimulate the growth of organisms; however, when that oil is released in vast quantities by something like a tanker accident, it tends to have major harmful effects on the marine environment (Figure 11.1). The next section illustrates this idea of 'system overload' by examining three types of major pollutants, and finishes by examining wholly artificial substances.

11.3 MARINE POLLUTION I: NUTRIENTS

11.3.1 NUTRIENTS AND ENVIRONMENT CHANGE

Nutrients essential for ecosystem functioning occur unevenly in the oceans, being largely concentrated in coastal runoff zones, and areas of upwelling (Chapter 6); both regions are controlled by continental configuration and ocean movement. Lack of sufficient nutrient supply therefore limits phytoplankton growth in the open ocean. In some coastal areas, however,

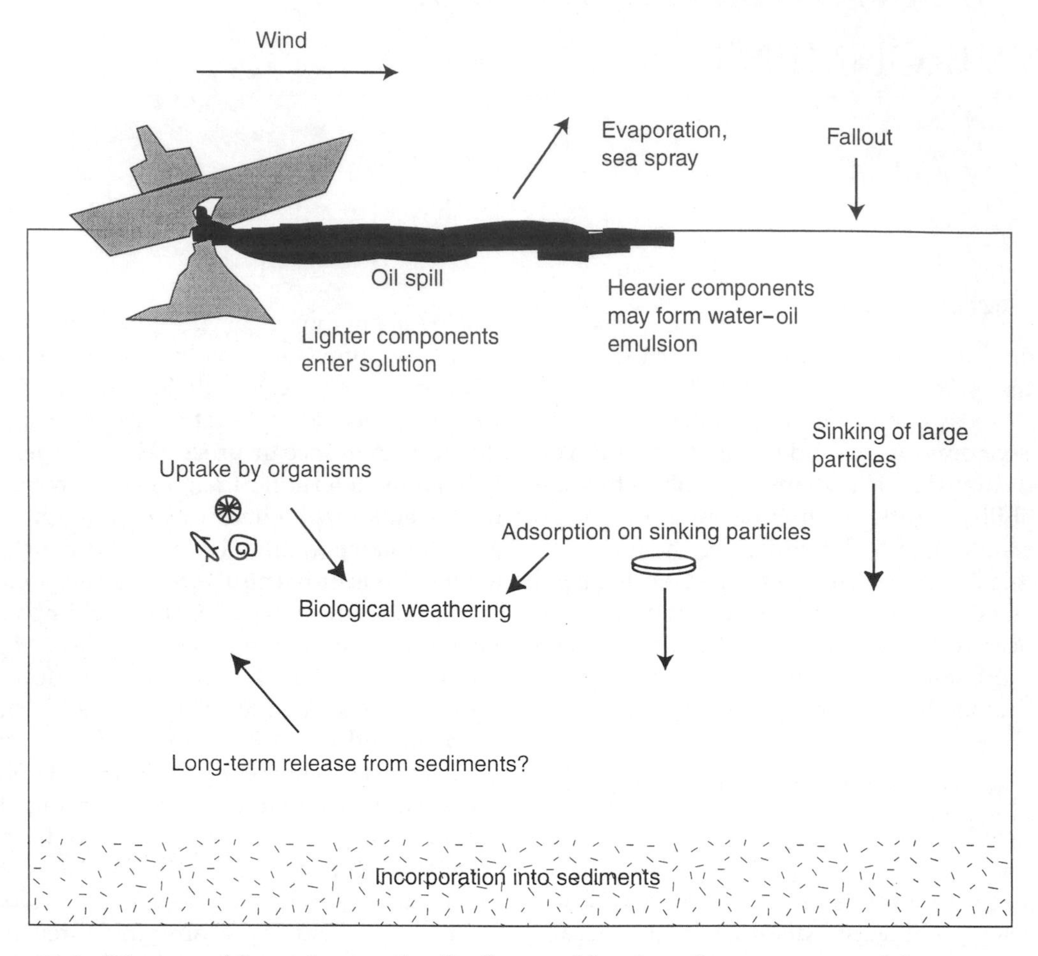

Figure 11.1 Diagram of the pathways of spilt oil at sea. Note how the open nature of the ocean ecosystems allows transport of oil products between ocean, atmosphere and sediment. © Andy Cundy.

enhanced nutrient input from land runoff, and dumping of sewage and fertilizers, may change the species composition in phytoplankton communities, and in extreme cases may cause eutrophication and bottom-water oxygen depletion. It is an interesting perspective that humans have influenced global N cycling processes to such a large extent that more atmospheric N_2 is being converted into biologically reactive forms by anthropogenic activities than by the sum of all natural processes Nadelhoffer *et al.*, 1999).

A severe flood of the Mississippi and other rivers within its catchment area of the United States Midwest in 1993 is believed to be the cause of the expansion of a hypoxic zone in the part of the Gulf of Mexico near the mouth of the Mississippi (Rabalais *et al.*, 1996, 1998) (Figure 11.2). The hypoxia (where bottom-water oxygen concentrations are 2 ppm or less) covered *c.*18 000 km^2. Hypoxia is a fairly common phenomenon in water bodies which have restricted mixing, and seems to be due to development of the summer thermocline in

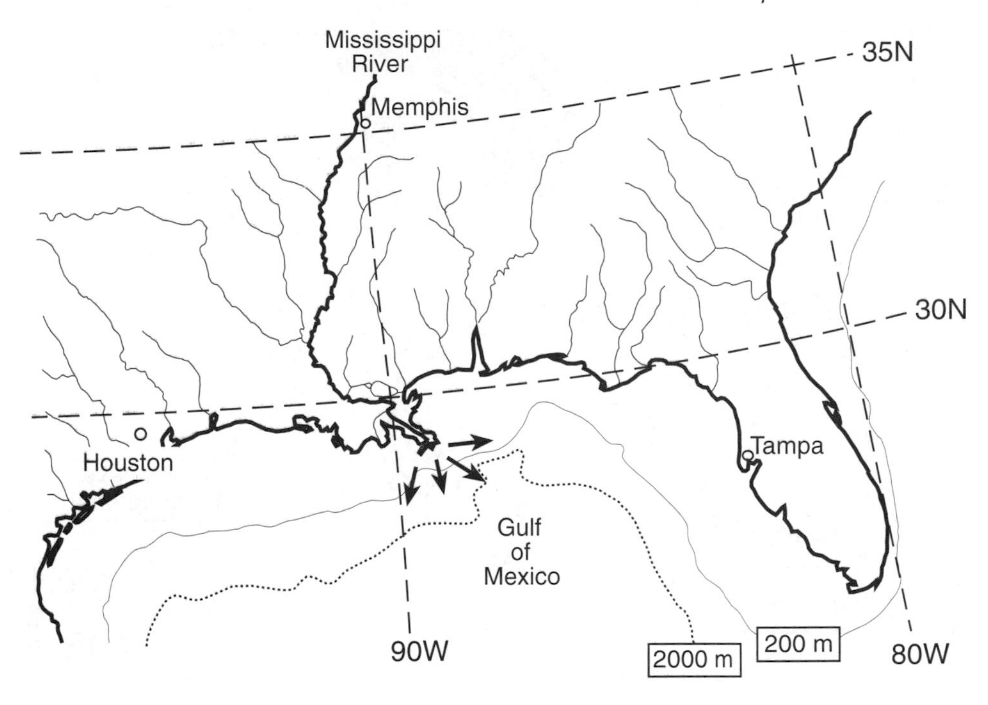

Figure 11.2 Distribution of hypoxia in the Gulf of Mexico, resulting from the 1993 Mississippi flood, which flushed huge quantities of agricultural nutrients into the sea.

water rich in nutrients derived from the Mississippi. Nutrients promote algal blooms via this eutrophication, resulting in a large increase in the rate of supply of organic detritus to deeper waters. Bacteria feeding on the resultant organic detritus then deplete bottom waters of oxygen. Although the hypoxic zone forms each summer, the 1993 flood flushed massive quantities of nutrients from agricultural, urban and industrial areas in the river system into the Gulf of Mexico, thereby doubling the normal area of the hypoxic zone, and degraded the local marine ecosystem, which supports a major US fishery. In response, researchers are calling on local bodies to undertake low cost measures to reduce nutrient input into the Gulf. Research on the Mississippi catchment shows that close proximity of nitrogen sources to large streams and rivers increases the flux of nutrients to the Gulf of Mexico (Alexander *et al.*, 2000).

11.3.2 NUTRIENTS IN THE NORTH SEA AND BALTIC SEA

Eutrophication is a potential problem wherever water mixing is restricted, as seen by comparing the North Sea and the Baltic Sea, which are both continental shelf seas in NW Europe and are subject to broadly similar human pressures (Figure 11.3).

The North Sea (40–100 m deep) receives a large input of nutrients from surrounding land areas; for example, the River Rhine alone currently contributes 1 million tonnes of nitrate per year, and so far the amount has doubled every 15 years. Bottom-water oxygen depletion has been observed around The Dogger Bank and areas close to the coast of The Netherlands, although this is relatively short-lived (Royal Society, 1993), because strong tidal action stirs up the water and mixes oxygen throughout the water column. In contrast, the Baltic Sea (7–459 m deep,

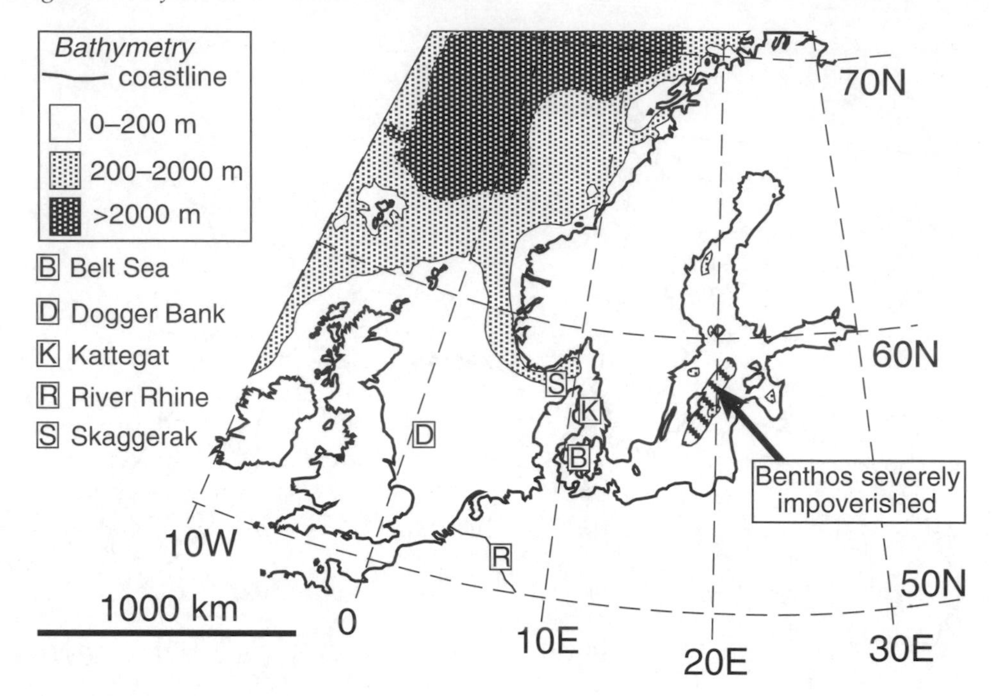

Figure 11.3 The restricted circulation of the Baltic Sea is in strong contrast to the open access of the North Sea to the Atlantic Ocean. Baltic Sea surface salinity of 6–7‰ rises to 12–13 on the seafloor, and contrasts with the normal *c*.35‰ of the North Sea. The Baltic Sea has a permanent halocline, with little mixing between the upper and lower waters; the deeper water is therefore poorly oxygenated. Fluctuations of salinity cause the halocline to move vertically, and decreased salinity in recent years may lead to improved oxygenation in parts of the Baltic.

consisting of a series of sills and basins), which receives at least 1 million tonnes of N and 50 000 tonnes P each year, began to show signs of eutrophication and decreased bottom-water oxygen in the 1950s and large areas of seafloor are now permanently anoxic (Gerlach, 1994). Since 1900, the exported production in the Baltic (that part of photosynthesised material that gets incorporated into the sediment) has shown a 5- to 10-fold increase (Rahm, 1987; Stigebrandt, 1991). The differences between these two seas are due to differences in water circulation and mixing; these are restricted in the Baltic Sea by three main controls:

1. strong water column stratification (caused by a layer of less saline surface water derived from river runoff);
2. weaker tidal mixing (due to a restricted connection with the North Sea and the Atlantic);
3. the presence of sills and basins (which inhibit physical mixing with North Sea waters).

Hence the Baltic Sea system is naturally prone to oxygen depletion. However, part of the reason for anoxia in deeper areas of the Baltic is the presence of a permanent halocline, which limits mixing between surface and deeper waters. An observation of potential importance by Gerlach (1994) is that in recent years the Baltic Sea salinity has lessened, leading to re-oxygenation in places. Gerlach suggested that projects such as the proposed bridge between Denmark and Sweden across the Belt Sea may restrict input of salt water from the North Sea, and freshen the Baltic Sea further, with beneficial results. The contrast between the Baltic and North Seas illustrates the importance of vigorous mixing in maintaining bottom-water oxygen; the current rarity of deep ocean anoxic areas is largely attributable to the vigorous thermohaline circulation which aerates the open ocean, described in Chapter 3.

11.3.3 RED TIDES

A final issue regarding nutrient input to coastal waters is the 'red tide' phenomenon. This is caused by phytoplankton blooms of principally red-pigmented dinoflagellates, which kill fish by clogging their gills and other structures, and which may synthesise substances that are toxic to higher organisms. Such toxins are then accumulated up the food chain by bivalves and some fish, causing paralytic shellfish poisoning in sea birds and fish, and in humans unfortunate enough to eat the contaminated organisms. Red tides occur naturally, but may be enhanced by high nutrient inputs (Anderson, 1994).

11.4 MARINE POLLUTION II: ANTHROPOGENIC RADIONUCLIDES

Radioactive nuclei of atoms (radionuclides) occur naturally throughout the ocean system, and many have been used as tracers of ocean processes, and as dating tools. Artificially produced radionuclides have also been added to the marine environment, mainly from nuclear weapons testing and from nuclear reactors (in the form of effluent discharges or released during accidents). Keeping them in perspective, such radionuclides represent only a tiny fraction of the global radioactivity to which we are exposed, and have little general effect; life evolved in the presence of appreciable background radioactivity from the air, water and rocks around us. However, where artificial radionuclides are concentrated, human health implications can arise. The largest source of anthropogenic or artificial radionuclides is from nuclear weapons testing and reactor accidents, with peak inputs occurring in 1963 (due to an increase in the above-ground testing of nuclear weapons immediately prior to an international test ban treaty) and, around Europe, in 1986 (due to the Chernobyl accident). These radionuclides were widely dispersed in the atmosphere, and have subsequently been similarly widely dispersed around the marine environment by current action and biological processes. In fact radionuclides released by nuclear weapons testing (particularly tritium, ^{3}H) have been used to assess ocean mixing, e.g. mixing between surface and deep waters in the North Atlantic (Jenkins, 1992), and fallout from the Chernobyl accident can be used as a stratigraphic marker in studies of modern salt marshes.

Nuclear power station discharges are more localised in nature, but can cause considerable regional contamination of the marine environment. A wide variety of radioactive waste is produced during the operation of nuclear power stations and nuclear reprocessing plants, much of which is stored or disposed of on land. Some waste, however, has been discharged to sea, and its behaviour and accumulation in the marine environment depends on the chemistry of individual components of the waste, and how well the waste is dispersed following its release. As an example, the Sellafield site in NW England (Figure 11.4) has discharged effluents with low levels of radioactive contamination into the Irish Sea since the early 1950s. Those components of the waste which are soluble (e.g. Cs and Sr isotopes) are dispersed by ocean currents, and have been detected in the North Sea, the Baltic Sea and the Arctic Ocean (Jefferies *et al.*, 1973; Livingston and Bowen, 1977). While these radionuclides are present at extremely low concentrations, problems have arisen around the immediate outfall area where marine organisms absorb the radionuclides and concentrate them. For example, seaweeds may concentrate radionuclides by 10 000 times, and shellfish by 100 000 times. Eating large quantities of shellfish (or seaweed) may result in an appreciable radioactive dose. Other components of the waste are much less stable in dissolved form and are rapidly removed by

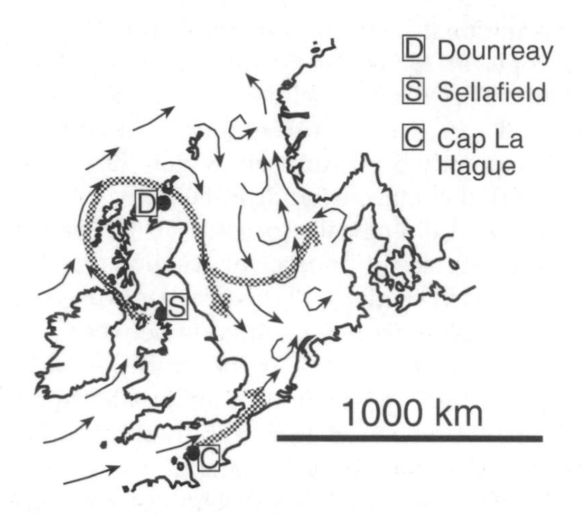

Figure 11.4 Distribution of ^{137}Cs resulting from leakage out of the Sellafield nuclear reprocessing facility and Dounreay nuclear reactor. Its distribution is influenced by surface currents, and its passage into the North Sea is clearly demonstrated (adapted from Libes, 1992).

adsorption or precipitation onto sinking sediment particles (e.g. isotopes of Pu, Am and Ru). These components have contaminated sediments around the Sellafield outfall pipe, and this contaminated sediment may be transported by marine currents into estuaries and coastal depositional environments such as salt marshes. When contaminated sediment is deposited near the high tide mark and dries out, humans may be exposed to the contamination by breathing in wind-blown material.

While these problems with nuclear wastes are relatively localised, a potential global issue regarding radioactive pollution is the ocean dumping of high level nuclear waste, the most problematic form of nuclear waste, being both physically hot and highly radioactive. The amount of high level waste produced by the nuclear industry is fairly small, and mostly has been kept isolated from the ocean environment. Plans by major nuclear powers to dispose of high level wastes in deep ocean trenches or in abyssal plain sediments have not been implemented, due to public opposition. Unexpectedly vigorous deep ocean currents could redistribute any released waste rapidly to surface layers, and chemical reactions in sediments may corrode containers and remobilise waste components. In the current political climate, large scale dumping of radioactive waste into the deep sea is unlikely to take place. Land-based disposal, however, has proven problematic due to geological and political difficulties (Haszeldine and Smyth, 1997; Strong *et al*. 1994), so deep sea disposal remains a possible future option (Hollister and Nadis, 1998).

11.5 MARINE POLLUTION III: HEAVY METALS

11.5.1 HEAVY METAL TOXICITY

Metals whose atomic weights exceed 20 are termed heavy metals. In trace amounts, many are essential to life, but most become toxic at slightly higher concentrations, so have a low threshold of toxicity. Their toxicity depends mainly on three factors:

1. the chemical form of the element, because not all chemical forms are toxic: some metals (e.g. mercury, arsenic, lead) become more toxic through methylation, an organic process leading to an organo-metal complex;
2. environmental factors: oxidation state and pH, for example, can influence whether a contaminant is mainly found absorbed or adsorbed onto sediments, or is in solution; salinity can influence the uptake of heavy metals by sediments;
3. condition of organism (age and size, stage in life history, existing stress): examples are (i) juvenile organisms may be more susceptible to contaminants than adult forms, (ii) uptake of contaminants may be correlated with size of organism, (iii) existing environmental stress may increase the toxic effects of heavy metals and other contaminants.

11.5.2 HEAVY METAL DISPERSAL

Some metal pollutants are widespread, particularly those which are highly volatile, such as Hg, or those that are released from diverse sources worldwide, such as alkylPb from car exhaust emissions. It is thought that elevated heavy metal concentrations tend to add to the environmental stress imposed on marine organisms, and indirectly lead to mortality. Nevertheless, cases exist where gross contamination of the marine environment, such as the (in)famous example of Minimata Bay in Japan has resulted in direct mortality of organisms, including humans. In the 1950s mercury was introduced into Minimata Bay as an industrial effluent, and was bioaccumulated by marine algae. At the top of the marine food chain were humans, who ingested contaminated fish and shellfish, so local inhabitants developed neurological problems, with birth defects and high infant mortality. In the open ocean, where heavy metal concentrations have been diluted by current action, gross contamination is far less common. Thus the scale of degradation associated with pollution depends on what part of the marine environment is involved; estuaries and enclosed shelf seas are usually far more contaminated than more remote areas which are distant from major pollutant sources. However, some pollutants may have major impacts even in remote areas, which include artificially produced organic compounds, discussed in Section 11.6.

11.5.3 HEAVY METALS AND THE ESTUARINE FILTER

The areas most affected by heavy metal pollutants are often those nearest to urban and industrial sources; estuaries in particular are susceptible to heavy metal contamination. Estuaries are popular sites for both industrial installations and urban areas because of ready availability of water for effluent disposal, and because they provide ready access to marine shipping routes. Furthermore, estuaries are susceptible to contamination due to the sedimentary processes that occur where seawater and fresh water mix. In estuaries, river-borne, clay-rich sediment derived from terrestrial erosion encounters the higher salinity of seawater and often flocculates into larger particles which sink and are incorporated into mudflats and saltmarshes (Figure 11.5). The clay-rich particles involved in flocculation are electrically charged, have a large surface area/mass ratio and are plate-like, which means that they have lots of charged sites for a comparatively small mass of sediment. Clays become coated with such substances as reactive Mn, Fe and organic compounds, and therefore act as a sponge for charged pollutants (principally heavy metals but also radionuclides, some hydrocarbons and pesticides).

Estuaries are thus prime sites for sediment accumulation, and also for the trapping of pollutants. Estuaries therefore may operate as a filter, preventing adsorbed pollutants from entering coastal waters. An example is Southampton Water (Figure 11.6), a major estuarine area on the south coast of England, which contains sediments labelled with local industrial effluent. Copper was released over a number of years from a local oil refinery (Cundy and Croudace, 1995), which adsorbed onto sediments around the outfall pipe and was subsequently redistributed around the estuary by currents. Maximum sediment concentrations were over 1000 ppm, over 50 times the natural background concentration (Croudace and Cundy, 1995). The concentration declines significantly even a short distance away from the discharge pipe, and so the Cu contamination is local rather than regional, and as it is derived from a single major source (a point-source) the contamination is fairly easy to control. A series of effluent-improvement measures implemented by the refinery in the early 1970s led to a significant reduction in sediment-bound Cu (Cundy *et al.*, 1997).

Non-point-source pollutants are far more difficult to control. Pb around the Southampton Water area is derived from a range of

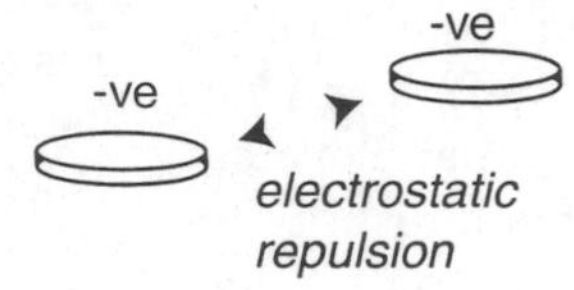

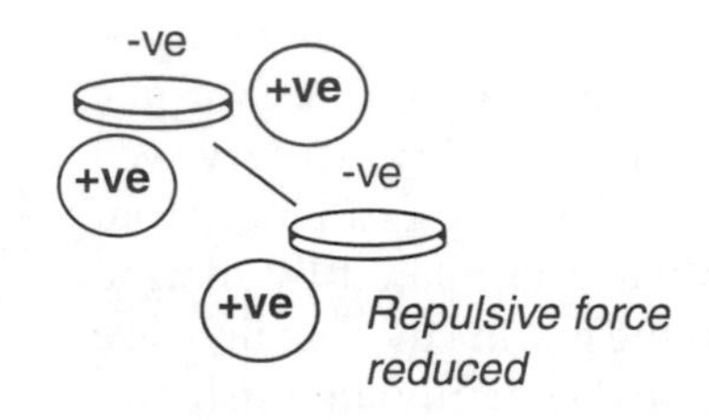

Figure 11.5 Estuaries draw heavy metal pollutants from river water. Saline water is rich in dissolved ions, whereas river water is not. Clay particles in suspended river load have charged surfaces capable of attracting ions, but because such ions are in short supply in river water, the charges on suspended clays repel. On mixing with saline water in estuaries, the dissolved ions in water attach to clays, which then flocculate because the repulsing forces between clays are reduced. Sedimentation of clays onto the estuary floor removes (filters) much heavy metal from the water. © Andy Cundy.

sources such as urban runoff, car exhausts and industrial processes, and it is not clear which source the bulk of the Pb comes from. In both point and non-point sources of this example, sediment further away from urban and industrial areas is less contaminated due to sediment mixing and dilution of contaminants, and so gross contamination with these heavy metals is fairly limited in areal extent.

11.6 MARINE POLLUTION IV: BIOACCUMULATION OF PESTICIDES AND PCBs

As technology has developed, many thousands of artificial compounds have been created as a consequence of human activity. Many are now present in the marine environment due to atmospheric input and river runoff,

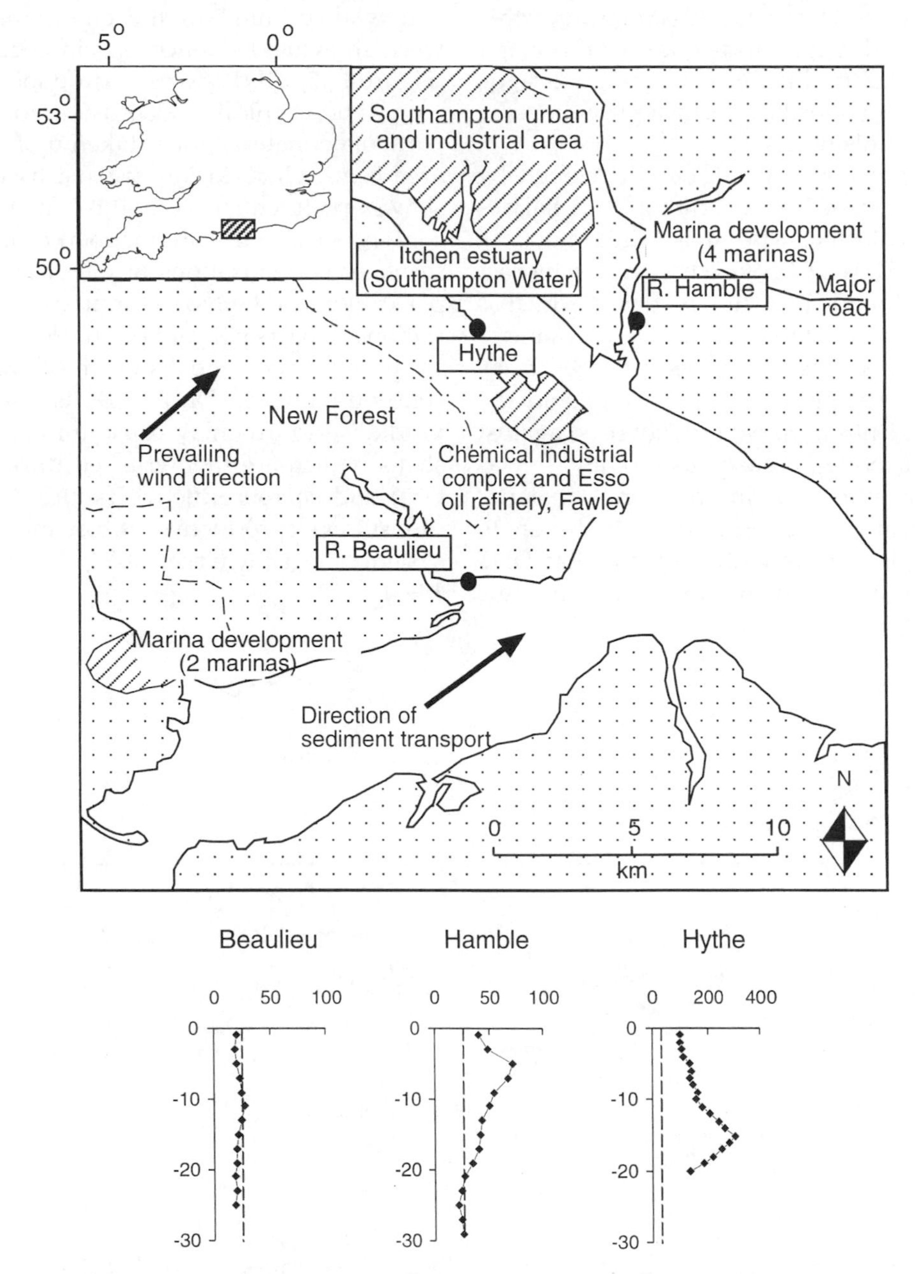

Figure 11.6 Southampton Water: an example of estuarine pollution. Copper released from the Fawley oil refinery (centre of map) was distributed around the estuary but concentrations fall not far from the site, for example in places upcurrent, as at Beaulieu. Sample profiles from cores show Cu peaks between 5 and 15 cm below the surface; the peaks relate to periods of heaviest outflow, but the decline in the upper part of the cores shows the effectiveness of measures to reduce the pollution since the 1970s. This example indicates the local nature of estuarine currents, and also how quickly the system recovers when the source is removed, because of the effectiveness of the flocculation process in removing sediment from suspended load. Adapted from Cundy *et al.* (1997).

usually at very low concentration. Nevertheless, artificial compounds may have a major impact on marine life due to their high toxicity, particularly where they are bioaccumulated by marine organisms.

A wide range of pesticides developed by organic chemists contain carbon and halides such as chlorine. The most (in)famous of these is DDT (dichlorodiphenyltrichloroethane), a broad-spectrum pesticide which is also relatively stable in the environment, reducing cost because it does not need to be frequently re-applied. First exploited in the 1940s, DDT has been extensively used against malaria-carrying mosquitoes, and was extremely effective in reducing malaria-related deaths. Concern, however, began to mount in the 1960s with evidence that DDT and its breakdown products had become widely distributed in the environment and were showing evidence of bioaccumulation (Figure 11.7). DDT shows hydrophobic behaviour, and rapidly adsorbs onto sinking particulate material, or is taken up by marine organisms. Once inside organic tissue, DDT shows preferential solubility in fats, and concentrates up the marine food chain. Initial seawater concentrations of the order of parts per trillion can be bioaccumulated to concentrations of parts per million in predators such as sea mammals and fish-eating birds. High concentrations of DDT and its breakdown products in birds may cause thinning of egg shells (which are then crushed during gestation) and so reproductive failure. Mammals can suffer reproductive stress and immune system effects (Clarke, 1993; Lahvis *et al.*, 1995).

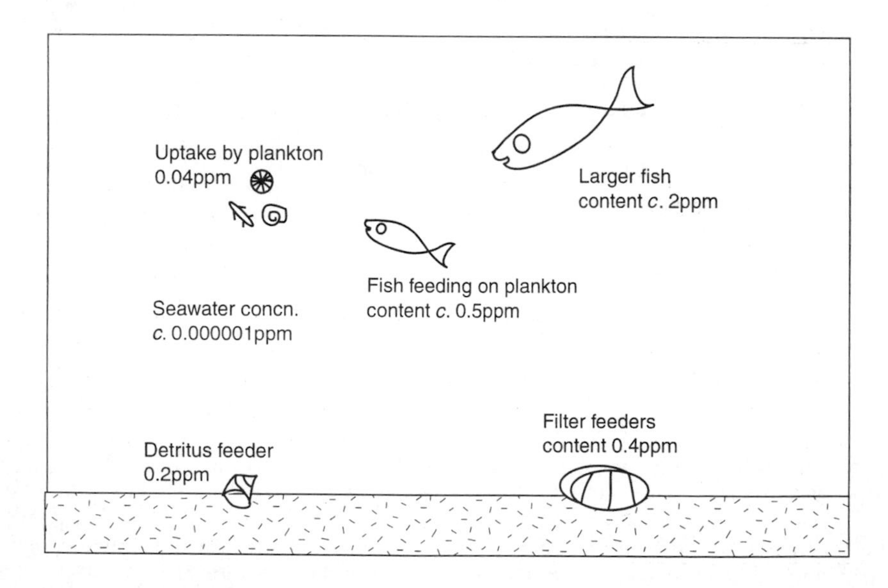

Figure 11.7 Pathways and bioaccumulation of DDT. This diagram illustrates the concentrating effect of ecological processes on accumulation of such organic pollutants up the food chains, which therefore also have potential dangers for humans. ©Andy Cundy.

Concurrent bioaccumulation of other artificial compounds, such as PCBs (polychlorinated biphenyls), may worsen these impacts. PCBs are synthetic compounds which are very stable at high temperatures, and so are used mainly as dielectrics in capacitors and transformers. Typically introduced into the marine environment by industrial discharges and from dumping of electrical components, elevated concentrations of PCBs are of great concern because the compounds are carcinogenic. PCBs, like DDT, rapidly adsorb onto sediments or are taken up by organisms, and tend to bioaccumulate in fats. Sea mammals are in particular danger from contamination by these compounds due to their huge fat stores (up to 40% of body weight). Recent work examining beluga whales in the St Lawrence estuary indicates that these mammals contain enough PCBs (500 ppm) to be classed as toxic waste (Motluk, 1995). Due to their solubility in fats, high concentrations of these pollutants occur in mothers' milk, and consequently are passed on to young animals.

High levels of PCBs and DDT are thought to depress the immune system so that the mammals become more susceptible to disease, and also disrupt reproductive systems. As individual animals become more stressed they mobilise more fat reserves, causing further release of pollutants into the metabolic system. Ultimately, the mammals may be killed by disease or have less reproductive success. Despite the depressing nature of such a narrative, it is very difficult to link deaths directly back to pollution. For example, during the North and Baltic Sea harbour seal epidemic of 1988, 12 000 seals (two-thirds of the regional population) were killed by a strain of viral distemper (Clarke, 1993; Motluk, 1995). The seals were heavily contaminated with PCBs, which *may* have weakened their immune system sufficiently to allow the disease to spread, but there was no proof that PCBs were really responsible. Bioaccumulation of persistent artificial compounds is of major concern; high pollutant concentrations may be found in organisms great distances from the source of the pollutant, even in remote polar regions.

11.7 OVEREXPLOITATION OF BIOLOGICAL RESOURCES – 'OVER-FISHING'

11.7.1 THE SCALE OF THE PROBLEM

The oceans' biological resources have been exploited for thousands of years; e.g. shell middens show that coastal Palaeolithic communities lived largely off shellfish, and offshore fishing had developed by at least 10 000 years ago in the Aegean (Robert, 1998). The scale of exploitation was largely limited by boat size, navigation and fish-finding technology, although over-fishing (fish are caught at a faster rate than stocks can regenerate) still occurred in some areas, so that a conference on overfishing in 1902 led to the establishment of the International Council for the Exploration of the Sea. A dramatic increase in fishing technology has, however, taken place since the 1950s, when military technologies (such as radar, open-ocean navigation aids and sonar) began to be utilised by the fishing industry, and industrial fishing using large factory ships expanded. The estimated global total catch of fish increased from around 20 million metric tons in 1950 to a peak of 82 million metric tons in 1989. Since then, this trend in growth has been replaced by general stagnation or decline (Safina, 1995). In the Atlantic Ocean, fishing grounds from north to south have all reported reduced catches in recent years; the same is true, but less profoundly, for the Mediterranean Sea and Pacific Oceans. Only in the Indian Ocean are catch figures still rising (Safina, 1995).

The United Nations Food and Agriculture Organisation (FAO) estimates that 25% of the world's wild fish stocks are overexploited or have crashed, while another 44% have reached the maximum rate of exploitation at which fish stocks can regenerate (Holmes, 1996). So, the global fishing industry is removing fish too effectively from the ocean; it is estimated that the world's industrial fishing fleet has twice the capacity needed to extract what the oceans can sustainably produce (Safina, 1995). Fishery collapse has often been dramatic: the Grand Banks, off Newfoundland, were formerly one of the world's major fisheries. In 1968, 810 000 tonnes of cod were caught around the Banks, and the fishing industry was Newfoundland's major employer. By 1992, the fishery had closed, with the loss of thousands of jobs. The cod, formerly the major fish species, was commercially extinct. The factors behind this collapse are complex (Mackenzie, 1995): fish stock models were (arguably) applied too simplistically, over-capacity in the fishing fleet, and there has been a lack of political will to severely cut fish catches while the fishery was in decline. Such rapid collapse of a major fishery brought the issue of over-fishing into the public arena: the failure of one of the world's major fishing powers to manage adequately a major fishing resource within its own territorial waters suggests that preserving fisheries in areas where ownership of fish resources is disputed is an extremely complex task. This is shown by the example of the North Sea. Fishing pressures from surrounding countries mean that as much as 60% of the fishable cod stock is removed annually: only 4% of fish aged one will survive to the age of 4, which is a typical maturity age (Cook *et al.*, 1997). Without a reduction in the rate of exploitation through effective catch controls or a reduction in fishing effort, the North Sea cod stock may well collapse. Controls of this nature, however, are extremely difficult to implement, even where a framework for international agreement exists (in the European Union and the Common Fisheries Policy). Short-term national and economic interests often take priority over the long-term viability of the fishery – recent efforts to control fishing effort in Europe by limiting the time individual boats spend at sea have been unsuccessful. Where strong management has been implemented, as in the case of striped bass along the eastern USA, fish stock regeneration has occurred (Safina, 1995), but it remains to be seen whether regulation and international cooperation will prevent the world's major fisheries suffering the same fate as the Grand Banks fishery.

11.7.2 THE WAY FORWARD, AND THE ROLE OF AQUACULTURE

A potential solution to the problem of overfishing is the development of large-scale aquaculture. Approximately 20% of fish consumed are now produced by aquaculture (Holmes, 1996), a system where fish are farmed under (semi-)controlled conditions. Common species farmed are those which can tolerate crowding, e.g. salmon, shrimp and tilapia. Expansion of aquaculture is largely inevitable as 'wild fish' such as cod and haddock become more difficult to catch and economically less viable, despite the large subsidies which currently support the global fishing industry. Aquaculture, however, does have numerous associated problems, as the following list shows.

1. Pollution of surrounding waters. Uneaten food and wastes from fish farms increase organic detritus and may promote anoxia, while pesticides used may affect natural ecosystem structure.
2. Loss of coastal habitat. Mangrove forests occur along low-energy tropical and subtropical coasts and are prime aquaculture sites; mangroves are breeding grounds and nurseries for many marine species, and their destruction further reduces the ability of wild fish stocks to regenerate.
3. Biomass fishing, a procedure where drift nets with fine mesh are used to catch feed for shrimp farms, can result in removal of juvenile forms of wild fish, further limiting stock regeneration.

An alternative is investment in sea ranching, where fish are raised in hatcheries, and released into the open ocean once they pass their vulnerable juvenile stage. The fish are then caught as adults. While this system has been successful in Japan due to their largely closed fisheries, large-scale investment in sea ranching is unlikely in areas with open fisheries, such as in the European shelf seas.

Recent developments of genetic modification (GM) have produced rapidly growing fish. There is an uncertain future for this work on health grounds(with the inevitable political implications).

11.8 CORAL REEFS

There has been a huge increase in interest in coral reefs following the highly publicised 1997 International Year of the Reef, and the Global Coral Reef Monitoring Network. Coral reefs are enigmatic structures; misleadingly, the high diversity of biota which form the complex interrelationships of a reef give the impression of a need for substantial nutrient supply to support it. Research in the last twenty years has revealed, however, that the concentrations of important nutrients (P and N) are actually in critical short supply in reef systems, and that the complex interactions between biota are developed to retain nutrients in the system, minimising loss to the environment. Reef systems are therefore developed in 'blue-water' conditions, clear blue-coloured water with minimal suspended matter and low nutrients (Hallock and Schlager, 1986). Input of nutrients to reef systems therefore disrupts a delicate balance of interflow between reef components, and causes reef degradation.

Reefs are shallow water systems because of their dependance on photosymbiotic interactions between green algae (zooxanthellae) growing in coral tissues, aiding coral growth by providing O_2 *for,* and receiving CO_2 *from,* coral respiration, as well as other nutrient interchanges. Because reefs are constructed by secretion of $CaCO_3$ to form the hard skeletons of corals, algae and shelly organisms, the energy budget of reefs can therefore be estimated by the distribution of $CaCO_3$ amongst the various components (Stearn *et al.*, 1977). Reefs grow in three settings:

1. near-shore fringing reefs, which form minor barriers adjacent to coastal sites such as Tahiti;
2. offshore barrier reefs, growing on submerged topographic features, close enough to sea level to develop;
3. atolls, growing on submerged volcanoes in oceanic settings, not necessarily located near land.

By far the greatest proportion grow near land. Nutrient input into any of these settings may be achieved naturally by upwelling, or unnaturally by oil pollution, but the biggest problems for reefs are created by enhanced nutrient and fine sediment input to shallow waters. These inputs are created by such processes as increased industrialisation in coastal areas, and increased terrestrial runoff related to deforestation, as well as disturbance of fine sediment by dredging activities. The results of increased nutrient input, in particular, are:

1. the growth of algae as coats on the corals, smothering them;
2. silting of corals, blocking the light from the critical algae;
3. response of corals by their production of mucus, which therefore draws energy from other important energy requirements of corals, reproduction in particular.

The result of the problems is that approximately 60% of global reefs are threatened or declining. Apart from the danger to the corals themselves, reefs not only have an accessory biota supported by the reef, but the young of a large number of economically important fish species develop in reefs. Death of reefs therefore has economic implications along two lines: tourism and fishing industries associated with reefs. As with other aspects of oceans, the only way of survival of reefs is management.

11.9 MARINE PROTECTION: THE LEGAL FRAMEWORK

There is a publicly recognised need to protect the marine environment, but laws are difficult to implement. In general, coastal nations have sovereign rights over a 200 mile (= *c*.300 km) Exclusive Economic Zone (EEZ), subject to conservation constraints. The remainder of the oceans are beyond the limit of national jurisdiction, although any exploitation by nations must give due consideration to the rights of other nations. Various global conventions have been ratified, creating rules of international law which bind the states party to them. Examples of legislation put in place include the following.

1. Various regional conventions: the North Sea (Bonn Agreement 1969); the Baltic Sea (Helsinki Convention 1974); the Mediterranean Sea (Barcelona Convention 1976); all these oblige member states to cooperate to control and minimise pollution.
2. United Nations Conventions: United Nations Convention on the Law of the Sea (1982) provides a comprehensive framework covering all sources of marine pollution. States have the obligation to protect and preserve the marine environment, from pollution sources both on the mainland and from vessels or structures operating under their authority. The recent UN Convention on High Seas Fisheries aims to protect fisheries in international waters from over-exploitation.

While these conventions have had some success, as the previous sections illustrate there is a need for much stronger international control, particularly in global fisheries exploitation. With further technological development and an expanding world population the stresses on the ocean system are likely to increase rather than decrease.

11.10 CONCLUSION

The depressing story of degradation of the marine ecosystem is a reflection of two global processes:

1. The interconnectedness of terrestrial, aquatic and atmospheric physical systems, so that flow of mobile components is unstoppable; thus the open nature of ecosystems applies thoroughly to the oceans, and the nature of ocean currents as transport agents means that ecosystems far from the source may be affected.
2. The nature of food chains causes concentrations of pollutants into organisms commonly eaten by humans.

It is perhaps ironic that estuarine trapping of sediment, plus the geological accident of the location and configuration of shallow marine epicontinental seas (e.g. Baltic and North Seas), has led to the accumulation of pollutants in the places where they do the most harm.

11.11 SUMMARY

1. The nature of ocean current systems and the interconnectedness of various parts of the ocean–atmosphere–land linked system means that pollutants inevitably find their way to the oceans.
2. There are four major groups of pollutants: (a) nutrients (cause eutrophication, oxygen-depletion and toxicity); (b) natural and artificial radionuclides (cause a range of diseases); (c) heavy metals; and (d) pesticides/PCBs (may be bioaccumulated and have increased toxicity at upper ends of food chains, affecting humans who eat such food).
3. Coastal regions and shallow marine basins become traps for pollutants. Estuaries operate a natural filtering system whereby clays encountering seawater flocculate and drawdown pollutant metals and organic complexes adsorbed onto charged clay particles; although this partly protects the ocean, the result is concentration of pollutants in populated regions.
4. Geological controls operating millions of years ago resulted in configuration of epicontinental basins, which may have good circulation and only occasional pollution crises (e.g. North Sea), or poor circulation

and well-established, long-lasting crises (e.g. Baltic Sea).

5. Fishing decline is due to over-fishing and pollution.
6. Danger of loss of coral reefs due to overuse and pollution and increased sedimentation due to runoff will only be controlled by management.
7. Although national measures are well developed in some countries for marine management, the open oceans belong to nobody, and so global management is virtually non-existent; it is hard to sec how things will improve.

OCEAN SYSTEMS AND THE ANTHROPOGENIC GREENHOUSE

12

12.1 INTRODUCTION: THE GREENHOUSE IN PERSPECTIVE

In recent years, concern has risen about the effect of human activity on global climate, via an enhanced 'greenhouse effect'. The so-called 'greenhouse effect' is a natural phenomenon which acts to keep the Earth around 30°C warmer than its position in the Solar System suggests it should be. The Earth receives radiation from the Sun, and re-radiates this back to outer space. The Sun radiates energy at shorter wavelengths than the Earth does, due to its higher surface temperature (Figure 12.1). This difference in wavelength has important implications for the radiation budget of the Earth – the atmosphere is largely transparent to the incoming, short-wave radiation but absorbs some of the outgoing, longer-wavelength radiation. This absorption is caused by a range of greenhouse gases (especially CO_2, CH_4, N_2O and water vapour) which show strong absorption at longer (infrared) wavelengths. This absorption of radiation causes the Earth to warm, maintaining the Earth's surface temperature at levels equable for life. The atmospheric concentrations of the major greenhouse gases (CO_2, CH_4 and N_2O) have dramatically increased over the last 200 years or so, and this increase is generally thought to be a result of increased emissions due to human activity (particularly industrial and agricultural development). Large increases in the atmospheric concentrations of greenhouse gases are likely to result in a net warming of the Earth's surface (see Section 12.2), i.e. global warming.

It is difficult to predict the impact of global warming, due to the complexity of the ocean–atmosphere system, and the lack of detailed understanding of biogeochemical cycles and ocean circulation. The response of the oceans to global warming will be of major climatic, biological and economic importance; consequently, predicting how the oceans will respond is critical in assessing the potential effects of enhanced global warming. This has stimulated a vast amount of research, using computer modelling of future ocean change, studies of the present ocean system, and studies of past ocean processes as analogues of future change. It is important to realise that the oceans are capable of both reducing and amplifying the effects of climate change, and this chapter examines the potential impacts of, and role of the oceans in, global warming.

12.2 THE ANTHROPOGENIC GREENHOUSE EFFECT

Since the industrial revolution, human activity has dramatically increased the atmospheric concentrations of a number of gases which are known to absorb infrared radiation in the range 7–13 μm, through which >70% of the radiation emitted from the Earth's surface escapes into space. These increases have coincided with an upward trend in temperature over the last 120 years (e.g. Jones *et al.*, 1986), with several of the warmest years on record occurring in the 1980s and 1990s. While these temperature increases may be a function of natural variability within the Earth's climate

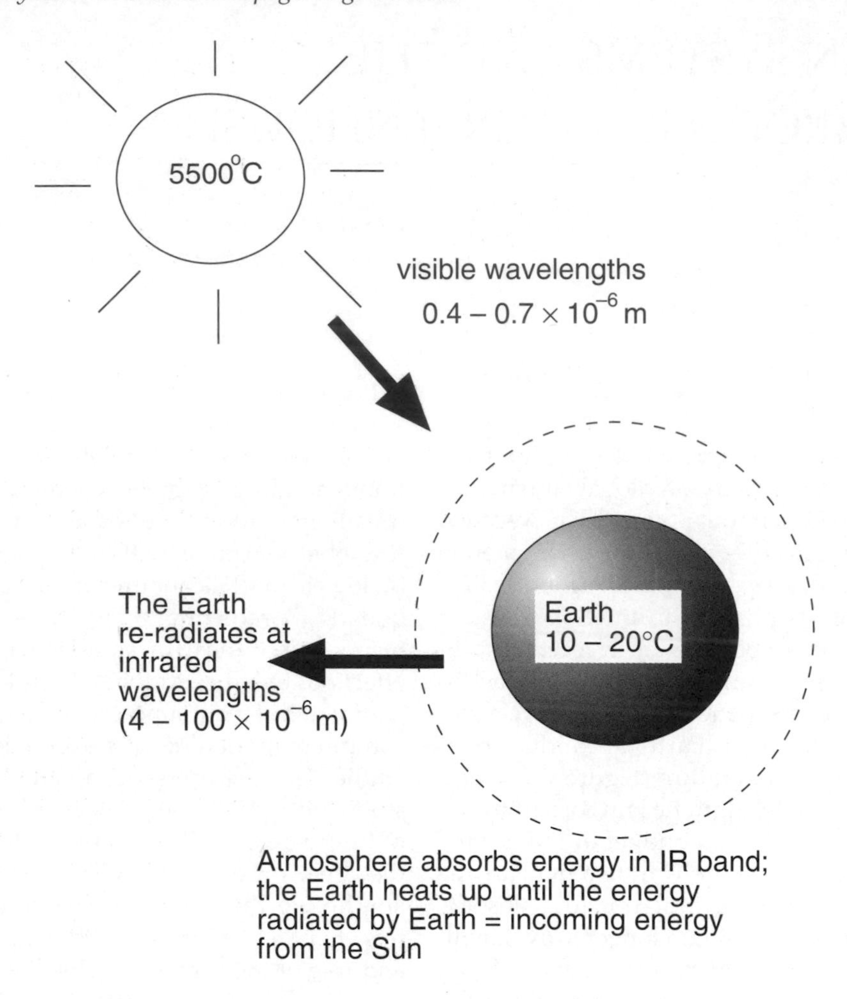

Figure 12.1 Principal features of greenhouse process. © Andy Cundy

system, it is clear that past rises in CO_2 concentrations have driven rises in sea surface temperature (SST), so there does seem to be a link between greenhouse gas concentrations and global temperature change. A quotation from the Intergovernmental Panel on Climate Change (IPCC) summarises the position: 'We are certain of the following: there is a natural greenhouse effect which already keeps the Earth warmer than it would otherwise be; emissions resulting from human activities are substantially increasing the atmospheric concentrations of the greenhouse gases: carbon dioxide, methane, chlorofluorocarbons (CFCs) and nitrous oxide. These increases will enhance the greenhouse effect, resulting on average in an additional warming of the Earth's surface . . .' (IPCC, 1990). The most recent consensus of climate models, again published by the IPCC, projects an increase in global mean surface temperature of 1–3.5°C by 2100 (IPCC, 1990). This represents a very rapid rate of temperature change compared with other fluctuations over the last 10 000 years.

12.3 RESPONSE OF THE OCEAN SYSTEM

12.3.1 IMPACTS OF GLOBAL WARMING ON THE OCEANS

Global warming is likely to increase both sea surface temperature (SST) and global sea level through thermal expansion of seawater and land-ice melting. Predicted rates of sea-level rise are constantly being modified, but the latest IPCC estimate is a rise of 15–95 cm by 2100, allowing for average ice melt (IPCC, 1996), although the actual sea-level rise observed at a particular location will depend on regional (e.g. tectonic) and local (e.g. changes in tidal regime and coastal configuration) effects. Note that, because of the high heat capacity of water, a small change in seawater temperature gives a large change in atmospheric temperature. A change in SST and sea level is likely to affect ice cover and ocean circulation, with likely knock-on effects on global biogeochemical cycles. While our understanding of the potential impacts of global warming on the oceans is still relatively rudimentary, a number of changes can be predicted:

1. An increase in SST and reduction in sea ice around areas of deep-water formation may cause a weakening of the thermohaline circulation (e.g. Manabė and Stouffer, 1993) – it might even shut down.
2. Changes in SST may alter ecosystem structure and cause migration of biological zones.
3. Changes in precipitation and runoff may affect nutrient availability and productivity. The significant warming of polar latitudes predicted by climate models (IPCC, 1996) may affect local and regional wind systems, and ultimately trade wind intensity due to a reduced latitudinal temperature gradient. Such a scenario would cause changes in upwelling and mixing, with implications for ocean productivity.
4. An increase in SST is likely to lead to an increase in coral bleaching and reduction of coral productivity (as observed during warming associated with ENSO events) (e.g. Glynn and Croz, 1990).
5. An increase in SST may cause changes in surface circulation patterns, and the frequency of events such as El Niño may change (although this is by no means certain and is the subject of ongoing research) (e.g. Knutson and Manabe, 1994).

12.3.2 THE OCEANS AS MEDIATORS OF CLIMATE CHANGE

Despite the potential changes outlined above, the ocean system may act to reduce the magnitude of global warming, through the physical and biogeochemical processes at work in the oceans. The oceans have a much longer response time to climatic changes than the atmosphere, due to the high heat capacity of water (see Chapter 2), and the long timescales of deep circulation in the ocean (centuries) mean that heat stored within the oceans takes a long time to return to the atmosphere. The result is a tendency to smooth out atmospheric variability. In addition, the oceans are an important sink for anthropogenically produced CO_2, and may have absorbed around 30% of the extra CO_2 produced by human activity. In the short term (tens to hundreds of years) CO_2 is dissolved in cold waters at polar latitudes, which sink and carry CO_2 into deep-ocean storage. In the medium to long term (thousands to millions of years), CO_2 removed by organic production as C_{ORG} solids reaches the sediment, possibly to be stored in petroleum reservoirs in the future.

Hence, the oceans play a major role in regulating the concentration of CO_2 in the atmosphere, and any changes in global ocean productivity and the rate of deep-ocean circulation has a major impact on atmospheric CO_2. This has led to an interesting concept of fertilising the oceans and providing a 'quick fix' for global warming. In the eastern Equatorial Pacific, and parts of the Southern Ocean and North Pacific, the nutrient nitrate is present in relatively high concentrations, but plankton productivity is low (Frost, 1996). In these so-called 'high-nitrate, low-chlorophyll' (HNLC) regions, the limiting factor for phytoplankton growth has been proposed as the availability of iron, which is an essential micro-nutrient (Martin, 1992). Adding iron to ocean areas may thus cause a bloom of phytoplankton, which take up CO_2. Studies in the equatorial Pacific (Coale *et al.*, 1996; Frost, 1996) reveal that the addition of iron to surface waters causes phytoplankton blooms. It has been argued also that only small changes in dissolved iron concentration drove important fluctuations in productivity, and that productivity may be influenced by proximity to an iron source, as in the tectonically active area of Papua New Guinea (Wells *et al.*, 1999). If iron was downwelled instead (e.g. around the Antarctic), where the rate of sinking of organic material is higher, then organic material (containing C_{ORG}) may be expected to be carried swiftly to the seabed and locked in ocean floor sediments. This mechanism possibly amplified cold conditions during the last glacial maximum; cold, arid and windy conditions around the continents might have carried more iron to the oceans on dust, stimulating higher productivity, and leading to CO_2 drawdown (Ittekkot, 1993). Estimates of a 60 ppm reduction in CO_2 in the atmosphere could be achieved by putting iron continually into the Southern Ocean. The potential destabilising effects of large-scale, deliberate tampering with the ocean and global climate system are not known, however, and so this idea will probably not be put into practice.

12.3.3 NEGATIVE FEEDBACK PROCESSES

A large number of ocean processes exert a negative feedback (reduce or reverse) on climate change, and help to maintain the stability of the ocean systems. A simple example of negative feedback processes shows how these mechanisms may work. Increased heating of seawater is disadvantageous to many plankton. A product of organic activity is dimethyl sulphide (DMS), for example by the coccolithophore *Emiliana huxleyi* (Simó and Pedrós-Alló, 1999), and it is known that DMS aerosols aid cloud seeding. Cooling results from the blocking of sunlight, by reflecting much back to space.

The nature of these (often complex) feedback processes is only just beginning to be understood, but their importance over geological times cannot be underestimated. They provide support for the Gaia concept that the Earth operates as a kind of superorganism which regulates planetary conditions through negative feedback processes.

12.3.4 POSITIVE FEEDBACK IN THE OCEANS: AMPLIFICATION OF CLIMATE CHANGE AND THE 'CLATHRATE GUN' HYPOTHESIS

We have already seen in Section 10.4 how rapid changes in deep-ocean circulation may cause major shifts in climate, both regionally and (possibly) globally. While one of the major concerns related to global warming is the future behaviour of the thermohaline circulation under increased global temperatures, the oceans may also be responsible for rapid shifts in climate due to the sudden release of greenhouse gases. One way to amplify climate change through a positive feedback process is the release of methane from the seafloor.

Clathrates (also called gas hydrates) are methane compounds, which, when buried in sediments, are kept in solid form by cool temperatures and high pressures. However, they decompose with pressure decrease or temperature increase, releasing methane, a potent greenhouse gas. Clathrates are found in large quantities in deep-ocean sediments and buried on the continental shelves, and destabilisation (by, for example, seafloor slumps) and decomposition of clathrates should release methane, adding to global warming. There is an estimated 10^{19} g of carbon stored as methane beneath the seafloor worldwide (Kvenvolden, 1998), which is more than the estimated total carbon in the atmosphere, ocean and land-surface reservoir (excluding lithospheric store) (see Figure 6.6A). Carbon isotope evidence in sediments (methane is very strongly ^{13}C-deficient giving distinctive highly negative $\delta^{13}C$ values) suggest that methane gas bursts occurred at the beginning of abrupt SST shifts during the last glacial period (Carlowicz, 1996). If changes in ocean circulation led to deep-water warming and release of methane from clathrates, the consequent rapid global warming could release more methane from clathrates, thereby forming a positive feedback process – the 'Clathrate Gun' idea proposed by James Kennett and co-workers. The clathrate gun idea has been interpreted as one cause of a huge-scale turbidite (22 000 calendar years old) in the west Mediterranean Sea, flowing from the Rhone fan down to the Balearic abyssal plain (Rothwell *et al.*, 1998) during the low sea-level stand at that time, with its consequent release of pressure off the seabed. The possible role of methane release from clathrates in acting as a positive-feedback mechanism for future global warming remains under debate (e.g. Loehle, 1993; Harvey and Huang, 1995). Gas hydrates may be a time-bomb for world climate if they are destabilised too rapidly, and current ideas of drilling into them are considered dangerous by some authorities.

12.4 CURRENT EVIDENCE FOR CHANGES IN THE OCEAN SYSTEM

The principal question here is whether global warming-induced changes are apparent in the modern ocean system. The best answer is that it is difficult to say, because long-term measurements of ocean processes are few and many ocean processes are naturally very variable. A recent example of a possible global warming-induced shift in ocean circulation comes from the Mediterranean. Measurements begun in 1908 reveal that deep water in the eastern Mediterranean is derived from sinking of Adriatic water; but from 1987 deep water has originated from sinking in the Aegean instead (Roether *et al.*, 1996). The reason seems to lie in surface salinity shifts; salinity has increased in the Aegean, possibly due to long-term climate change (increased evaporation) or increase in intensive agriculture (decreased freshwater runoff), leading to a change in the physical character of the seawater, in relation to S–T parameters. The impact of this on the Mediterranean region is likely to be quite minor, but this example is significant as the first direct observation of a rapid deep-water circulation 'flip'.

Knowledge of how the ocean system functions is still rudimentary, so that separating anthropogenic from natural change is difficult to achieve on current knowledge. Evidence from the geological record, however, indicates that changing the atmospheric CO_2 concentration changes global climate, involving changes in ocean circulation, productivity, sedimentation and chemical processes. The impact of global warming depends on positive and negative feedbacks in the ocean system, which as we have seen, are poorly constrained.

12.5 RELATIONSHIP BETWEEN CORAL REEFS AND CO_2

Because coral reefs grow by the secretion of $CaCO_3$ into skeletons of corals and other organisms, they draw some HCO_3^- from the environment, and because a reef contains a large content of calcareous biota, reefs act as sinks for part of the global HCO_3^- budget. Coral reefs cover 600 000 km^2, representing 0.17% of the global ocean surface, and 15% of shallow water to 30 m depth, and furthermore provide 0.05% of the global oceans' CO_2 fixation (Crossland *et al.*, 1991). As discussed in Chapter 5 coral reef organisms calcify by converting HCO_3^- into $CaCO_3$ by the equation:

$$2HCO_3^- + Ca^{2+} \Leftrightarrow CaCO_3 + H_2O + CO_2.$$

Therefore, rather than absorbing CO_2 from the sea, they release it instead. In the long term, the conversion of one HCO_3^- from seawater for the release of one CO_2 molecule means that part of the global CO_2 is locked up in carbonate, but over the short term the result is increased CO_2. Of course, in time, the released CO_2 would be expected to equilibriate with the atmosphere and be drawn back down to rivers as HCO_3^- by terrestrial weathering of silicates, as shown by the summary reaction:

$$2CO_2 + H_2O + CaSiO_3 \rightarrow Ca^{2+} + 2HCO_3^- + SiO_2$$

(this is a simplified version of the complex set of reactions shown in Section 4.3.2).

However, there is inevitably a lag time involved in this, during which global warming occurs. To make matters (potentially) worse, as seawater temperature increases, CO_2 solubility decreases, releasing CO_2 into the atmosphere.

Thus we are left with the continuing uncertainty of the real influence that coral reefs will have under a global warming scenario; it may be that the weathering rates, drawing down CO_2 back to the crust, will be important. Note also that it has been shown that sustained increase in ambient temperature can cause coral bleaching (loss of symbiotic algae from coral tissue) and concomitant reduction of growth, which presumably will reduce coral CO_2 output; also, increased CO_2 has been modelled to reduce the ability of corals to calcify because of the effects of falling pH within the oceans (Kleypas *et al.*, 1999).

Despite these problems, it is expected that coral reefs will survive predicted sea-level rises; estimated future rates vary from extremes of 0.8–8.2 mm/a within which 3.3–4.4 mm/a are mid-range figures (Spencer, 1995). If reefs cannot tolerate such a rise they will 'drown', as recorded for tropical reef systems during the last deglacial sea-level rise (Pirazzoli, 1996). Earlier suggestions that modern reefs may drown under the predicted sea-level rise are now considered unlikely (Spencer, 1995), and the expectation is that reefs will therefore thrive in future scenarios.

12.6 CONCLUSION

Global warming is likely to have some effect on ocean processes (ocean circulation, productivity, sedimentation and chemical processes), but the true impact depends on poorly understood positive and negative feedback processes within the ocean. The ocean system can act to reduce or amplify climate change, as it is a major component of the global climate system and a major sink for anthropogenic CO_2. The major worry is that ocean circulation may rapidly 'flip' to a different state giving rapid and unpredictable climate change, as has been identified for Younger Dryas processes (Chapter 11). This highlights the importance of understanding how the ocean system functions, and the response of the oceans to past changes in external (and internal) forcing mechanisms. As an illustration of the need to understand these systems, the current actual and potential breakup of the Antarctic ice margins have the capacity to raise sea levels by *c*.6m globally, with devastating consequences for world economy if they do so.

12.7 SUMMARY

1. The greenhouse effect is a naturally occurring feature, keeping the Earth surface warm. Enhanced greenhouse effect is the suspected unnatural increase in greenhouse gases by human action. Heat transference to the ocean, with its global circulation system, means that the potential for global warming of the oceans is high.
2. Warming of the ocean leads to a rise in SST. Because the interchange between ocean and atmosphere is such that small changes in ocean temperatures lead to large changes in atmospheric temperature (water has high heat capacity), the ocean has the potential as a major control on the atmosphere's behaviour, in the medium to long term (thousands to millions of years).

3. Both negative and positive feedback processes appear to operate in the ocean systems but the evidence for their control is poorly understood. Resolution of these will be critical in future analysis of ocean change.
4. Current evidence for changes in the ocean system is minimal, because records are not long enough to identify the changes. The two best examples, the Mediterranean circulation, and the Greenland sea ice, do however, provide sufficient warning that change can take place quickly enough for it to be taken seriously.
5. Discussion of predicted responses of coral reefs to changes in temperature and CO_2 provide depressing theories, but it is expected that the coral reef ecosystem will survive.

CONCLUSION: AN EARTH SCIENCE PERSPECTIVE OF THE OCEANS 13

The contents of this book attempt to emphasise the most important aspects of ocean processes, and their perspective through geological time. Key features and concepts are illustrated in Figure 13.1. The holistic approach of addressing ocean history in relation to its modern setting may provide a means of more clearly seeing the future. The oceans have had a magnificent past, a troublesome present, but hopefully great times ahead. Of the latter, there is no certainty. Perhaps the best lesson from studying palaeoceanography is that change in past times has been far more catastrophic than anything that is happening today. The 'gloom and doom' scenario has not reached crisis point yet, and it may or may not do so, but is it worth taking the chance? To misquote the phrase that has driven the Earth sciences ('The present is the key to the past'): 'The present knowledge of ocean processes is the key to their future.'

The final point harks back to a maxim applied at the beginning of Chapter 1 and in Figure 13.1, that the oceans continuously recycle through the crust, so that every 10 million years the volume recycled equals the entire ocean store. A report on later research (*Geoscientist*, 1999, **9**(11), 6–7) suggests that this may be wrong, and that the water carried down subduction zones is not all being returned to the oceans. Instead, it is being used to hydrate mantle minerals in the mantle, especially olivine, which is easily altered by water. Thus, under that argument, the oceans have been steadily draining into the mantle over geological time, so that in about 1000 million years the Earth's surface could be dry. If this is correct, and that sea levels have been gradually *falling* (superimposed on the observed fluctuations) over geological time, models of sea-level change applied to the geological past will need to be readdressed. The likely ultimate results of such water loss are that plate tectonics will stop, and without water, the planet's heat balance will go out of control, and of course, life will not be possible. Such an idea draws attention once again to the long-term inconstancy of Earth systems, and that humans live not at the *end* of a long period of geological history, but somewhere along the way.

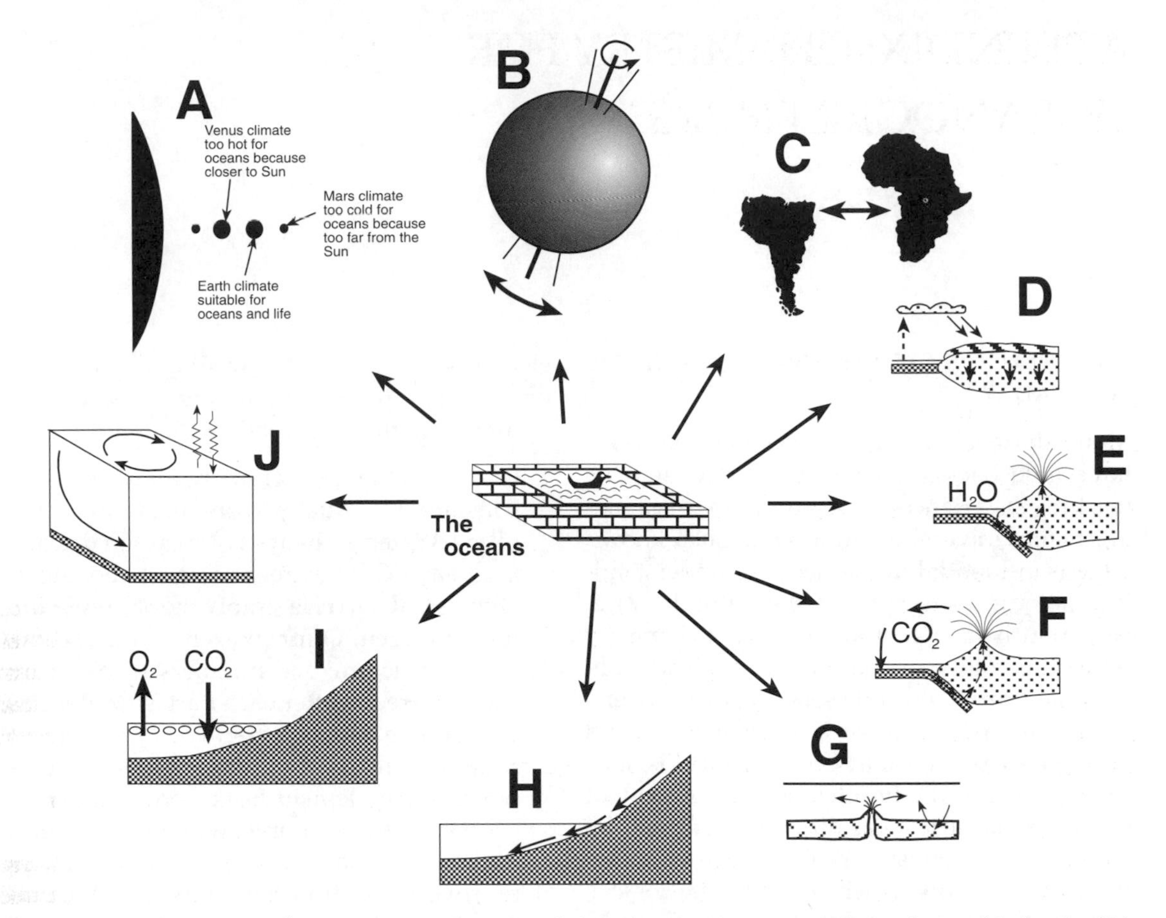

Figure 13.1 Summary of major interactions involving the oceans, treated throughout this book. (A) Earth's location in the Solar System allows abundant surface liquid water through the miracle of hydrogen bonding. (B) Earth's systems are influenced on a cyclic basis by orbital variations, generating orbitally forced sedimentary deposits in the rock record. (C) Plate tectonics affects the shape of ocean basins, the configuration of ocean circulation, and geographic controls on evolution of species on land and in oceans. (D) Climate change influences ocean processes; glaciations are influenced by location of continental masses near poles, and affect sea level. (E) The action of subduction is lubricated by water, which lowers the melting point of rocks, so that the presence of liquid water on the Earth's surface actively assists the process of plate tectonics, which itself has such a great influence on oceanographic processes. (F) The long-term carbon cycle is mediated partly through oceans, because organic and inorganic carbon deposited on ocean floors is a temporary repository of carbon. The influence of shifts in the carbon budget on global climate change over geological timescales is recognised and applied in palaeoceanographic modelling. (G) Ocean ridge processes and active submarine weathering modulate the ocean composition. It is estimated that over a 10 million year period, a volume of water equal to the global oceans is cycled through the crust. Currently in a probable steady state, evidence from the geological sedimentary record indicates that ocean composition was not always so stable. (H) Oceans are major sedimentary repositories of weathered and eroded dissolved and particulate matter, and play a major role in biogeochemical cycles. (I) Much productivity takes place in oceans, mostly by unicellular green algae, influencing the interchange of critical gases between the oceanic and atmospheric reservoirs. (J) Surface and deep ocean circulation distribute heat and dissolved matter around the globe, and therefore the oceans act as the principal thermal regulator, generally over longer time scales (the atmosphere regulates heat on short scales).

APPENDIX: CHEMISTRY FOR OCEANOGRAPHERS

A.1 DO YOU HAVE CHEMISTRY PHOBIA? IF SO, READ ON

Many students of Earth science in recent years have little background in chemistry, and the teaching experience of ourselves and colleagues at several universities shows that such knowledge is influential in the depth of understanding, amongst students, of the natural world. Importantly, a lack of knowledge of chemistry limits the ability to analyse Earth science concepts and problems. It is sobering to discover a student who does not know what an ion is, and it is not surprising that chemistry instils such dread. Most of the problem is the first step: of grasping the basic ideas of controls of chemical substances. Chemistry is all around you, and the principles on which it works are logical, and well within the understanding of everybody. Chemistry also involves physics, and it is possible to explain chemical processes in terms of physical interactions between substances. In this section, are explained in non-technical language, the important chemical features and processes which affect the oceans which apply in appropriate places throughout the book. Although it should be obvious to you that the controls on atoms and molecules are more complex than portrayed here, there is sufficient background for you to fully understand chemical aspects in this book.

A.2 JARGON-BUSTING: THE COMPONENTS OF CHEMISTRY

For most purposes, two types of material can be recognised in the world: compounds and elements, explained below. We also need to know what molecules, atoms, ions, atomic number, atomic mass, and isotopes are.

1. *Compounds* are materials made of simpler substances, usually combined in fixed ratios. Water is always H_2O; carbon dioxide is always CO_2, pure calcium carbonate is always $CaCO_3$ (H is simply the abbreviation for hydrogen, O for oxygen, C for carbon, Ca for calcium; the numbers of each are shown directly after each part, so water has 2 hydrogen and 1 oxygen).
2. *Elements* are the building blocks of compounds. An element has a certain range of properties (for example, whether it is a gas, liquid or solid at normal temperatures; how reactive it is with other substances). Compounds can be broken down to elements, but elements cannot be further broken down, without destroying their characteristics which define them. H, O, C and Ca are all elements. Hydrogen is the simplest element, and makes up virtually all the mass of the universe; all the other 100+ elements are present in tiny quantities in the cosmos, and their relative concentration on the Earth's surface is unusual.
3. *Molecules and atoms.* There is a lower limit to the amount of each compound and element that can exist in any sample. If you were able to divide a gram of water into smaller and smaller portions, you would eventually end up with an amount so small that further division would break it down into its elements; that small amount is a single *molecule*, and the hydrogen and oxygen particles would

then be the smallest possible amounts of the two elements; these are *atoms*. Atoms in turn are made of subatomic particles, of which there are three main types: *protons* and *neutrons* in a centrally positioned nucleus, and *electrons* in a series of concentric orbits (see Figure A.1 for the example of water, and Figure A.2 for the example of salt); however, the orbits are not limited to single planes, and in that respect are unlike planets orbiting a Sun, so they are called *electron shells*. Each proton carries a positive charge and each electron an equal negative charge; atoms contain equal numbers of the two and are electrically neutral. Neutrons have no charge. Protons and neutrons have equal mass (which we can refer to as one unit of mass), while electrons are so light that they have no effective mass; all the mass of an atom is in its nucleus.

4. *Ions.* Although atoms are electrically neutral, atoms of most elements have some degree of instability, and try to either lose a certain number of electrons, or gain them, to achieve stability. The reason for this is that each electron shell is capable of carrying a certain maximum number of electrons, and only when it has the maximum is the shell stable. However, only in a few elements does the outermost shell contain its stable number of electrons (the reason for this is explained in Section A.5, below). Electron shells achieve stability by sharing, receiving or donating electrons in complex interactions with surrounding atoms of other elements in a compound, so that the whole mass of atoms in the compound is electrically neutral. That must be true because minerals are electrically neutral (you do not get a shock if you pick up a piece of calcite – $CaCO_3$). When compounds (and therefore molecules) are in solid state, all these electrical charges hold the atoms and groups of atoms tightly held together, but when dissolved in water, they can move freely in the solution. An excellent example is salt (sodium chloride – NaCl), which happens to be one of the most important oceanic materials. Na wants to lose one electron to make it stable, and chlorine wants to gain one. In solid state, stability is achieved by the atoms being in close proximity, but when dissolved in water, they separate into Na^+ and Cl^- (Cl borrows an electron from Na) (Figure A.2). In two other important oceanic substances, $CaCO_3$ divides into Ca^{2+} and CO_3^{2-}, and $CaSO_4$ into Ca^{2+} and SO_4^{2-}. In these cases, Ca wants to lose 2 electrons while CO_3 and SO_4 want to gain 2. Some elements have alternative requirements; an important one in the oceans is iron (Fe). Fe can exist in two electrical states: Fe^{3+} and Fe^{2+}, under different conditions, which have critical results in the oceans. Negatively charged ions are called *cations*; positively charged ions are *anions*. Chemical substances which can be broken up into ions are often called ionic substances.

5. *Atomic number.* Despite what you might think, there is nothing magical about the differences between the various chemical elements. The only thing which defines an element is the number of protons. So, simply, hydrogen has only one proton, and it is the only element with one proton; helium is the only element with two protons, carbon has 6, oxygen has 8, calcium has 20; the number of protons is the atomic number, and the atomic number gives an element its unique properties. If you look in a chemistry textbook at the Periodic Table of the Elements, they are all laid out in ascending order of atomic number. Knowledge of atomic number helps us understand why ions exist. Hydrogen has only one proton, so it follows that it must only have one electron. We have seen that atoms have concentric shells of electrons. The first shell, closest to the nucleus, is only stable when it has 2 electrons; the second and third shells need 8 each. Hydrogen only has the first shell, and its one electron makes it unstable; it achieves stability by losing that electron, and becomes positively charged, thus

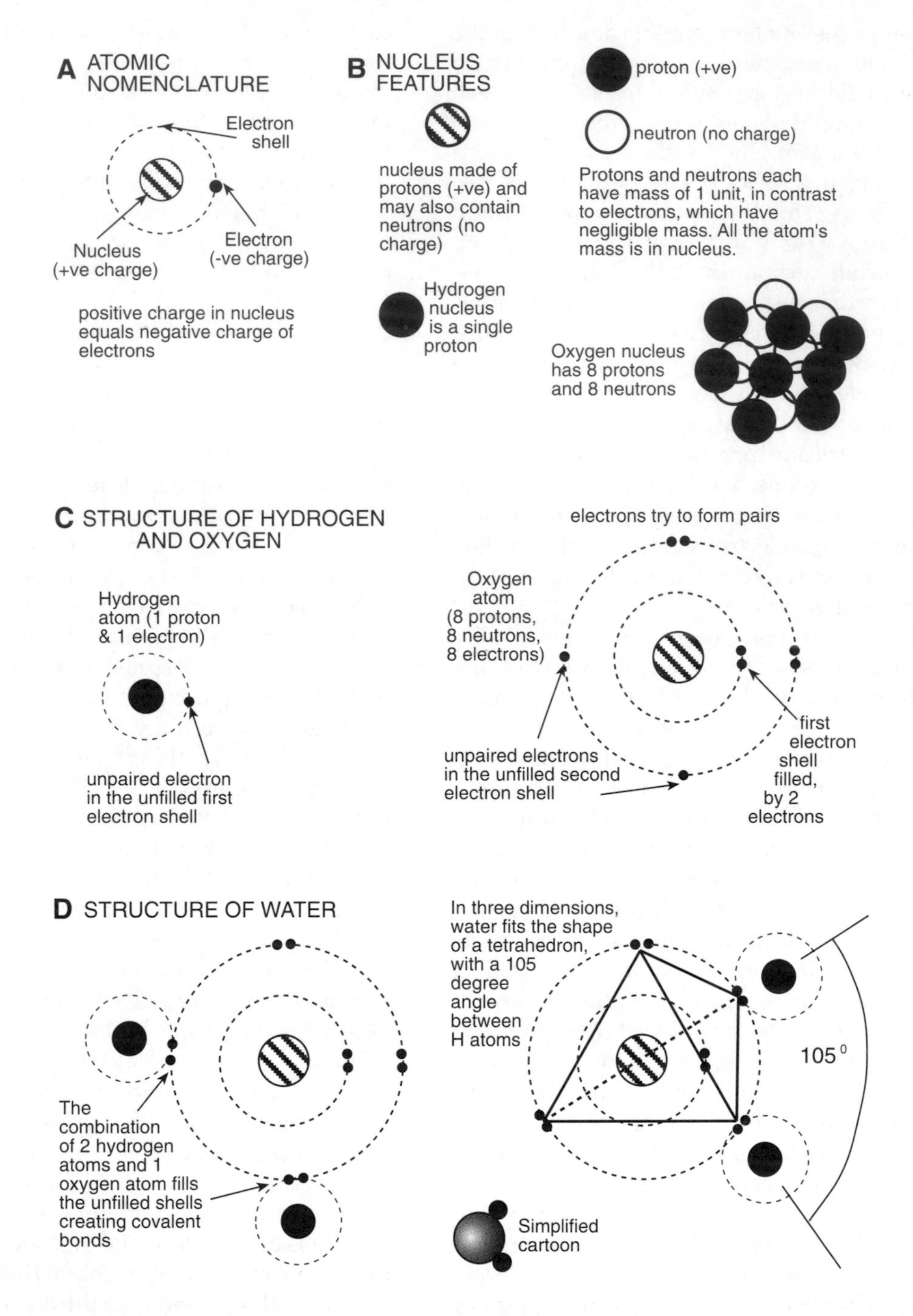

Figure A.1 (A, B) Principal features of atoms and their nuclei. (C) Structures of hydrogen and oxygen. (D) The asymmetric structure of water. These diagrams are not to scale; early physicists discovered that if an atom was the size of St Paul's Cathedral in London, then the nucleus would be the size of a pea suspended in the centre of the cathedral!

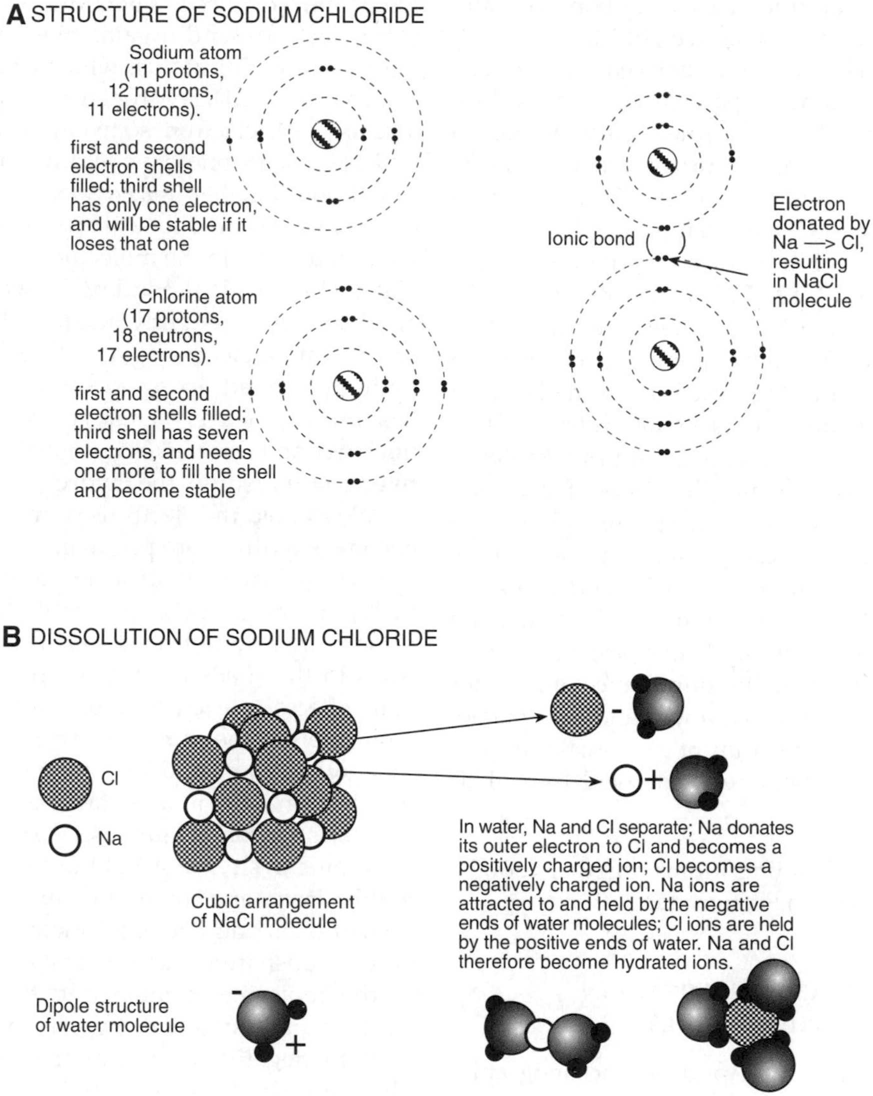

A STRUCTURE OF SODIUM CHLORIDE

Sodium atom (11 protons, 12 neutrons, 11 electrons).

first and second electron shells filled; third shell has only one electron, and will be stable if it loses that one

Chlorine atom (17 protons, 18 neutrons, 17 electrons).

first and second electron shells filled; third shell has seven electrons, and needs one more to fill the shell and become stable

Ionic bond

Electron donated by Na —> Cl, resulting in NaCl molecule

B DISSOLUTION OF SODIUM CHLORIDE

Cl

Na

Cubic arrangement of NaCl molecule

In water, Na and Cl separate; Na donates its outer electron to Cl and becomes a positively charged ion; Cl becomes a negatively charged ion. Na ions are attracted to and held by the negative ends of water molecules; Cl ions are held by the positive ends of water. Na and Cl therefore become hydrated ions.

Dipole structure of water molecule

Figure A.2 Molecular diagram showing what happens to salt when it dissolves.

converting to an ion. Sodium, with 11 electrons (2+8+1), is lacking seven needed to fill the third shell, it becomes stable by *losing* its one electron in that shell (2+8). That electron goes to chlorine; chlorine has 17 electrons (2+8+7), so it uses one from sodium to fill the third shell, making 18 (2+8+8).

6. *Atomic mass*. Often this is referred to as atomic weight. Protons and neutrons each have one unit of mass, but electrons have negligible mass. So the atomic mass is the number of protons + neutrons. Hydrogen normally has no neutrons, so its atomic mass and number are the same: 1. Helium

normally has two protons and two neutrons: atomic mass 4. Carbon normally has 6 of each: atomic weight 12.

7. *Isotope*. This word is derived from Greek (*iso*=the same; *tope*=place), and is best explained by example. Carbon has 6 protons, and as we noted above, *normally* has 6 neutrons, but the number of neutrons is not fixed. Carbon can have 6, 7 or even 8 neutrons, giving three alternative atomic masses of 12, 13 or 14. These are written ^{12}C, ^{13}C and ^{14}C. Because they are all carbon, they have the same chemical properties, so they are all in the *same place* in the Periodic Table. ^{12}C and ^{13}C are *stable isotopes*, while ^{14}C is an *unstable isotope*, and breaks down into other elements; it is used for carbon dating. Similarly, hydrogen usually has no neutrons but it can have either one, or two, making atomic masses of 2 and 3 respectively; these isotopes are called deuterium and tritium, although they are still chemically hydrogen. The importance of isotopes in oceanography cannot be underestimated; there is a huge amount of critically important data about oceans derived from their study.

The forces holding atoms together (bonds) are considered in Section A.3.

A.3 IMPORTANT CHEMICALS IN OCEANOGRAPHIC STUDIES

The following description of oceanographically critical chemicals is a general introduction. It is not exhaustive, but will give you a useful background.

A.3.1 WATER AND ITS HYDROGEN BONDING

Atoms react with other atoms in order to become more stable, forming compounds. For water, this process has unusual consequences. A single atom of oxygen has two electron shells; the first is filled by two electrons, the second contains only six, but needs 8 to be filled and stabilised. Therefore, it requires 2 extra electrons, and it gains these by reacting with an atom, or atoms, which can lose electrons (Figure A.1). Hydrogen has a tendency to lose its single electron, so two hydrogen atoms will react with one oxygen atom and chemically bond together in a covalent bond (where electrons are shared to satisfy the need for electrical neutrality in the molecule). In the case of water, this covalent bonding between oxygen and hydrogen is slightly unequal. The oxygen atom is more electronegative (it is better than hydrogen at attracting electrons, because it has more positively charged protons in its nucleus), and this makes it slightly more negative; consequently, the hydrogen atoms are unable to hold the electrons near to them, and become slightly more positive.

This imbalance of attraction leads to asymmetry in the shape of water molecules. For reasons that do not need to concern us, electrons in the shells of oxygen are arranged in pairs. Because each pair of electrons in the molecule has the same negative charge, pairs repel other pairs, and the shape which allows the electron pairs to be as far apart as possible is a tetrahedron (Figure A.1). The hydrogen ions come to sit at any 2 of the 4 corners of the tetrahedron, and therefore always lie near each other at one side of the molecule. The two pairs of unshared electrons at the oxygen side of the molecule, combined with the higher electronegativity of the oxygen atom, produce a slight negative charge at the oxygen end. Therefore, because oxygen is several times larger than hydrogen, the whole water molecule behaves as a sphere with two small lumps near each other on one side. The molecule therefore develops an overall polarity of charge, called a dipole. So, in contrast to most other liquids (which are usually a collection of freely moving molecules), in water the dipoles interact and weakly bond together, where the negatively charged end of a water molecule electrostatically attracts the positive end of another water molecule. This weak attractive

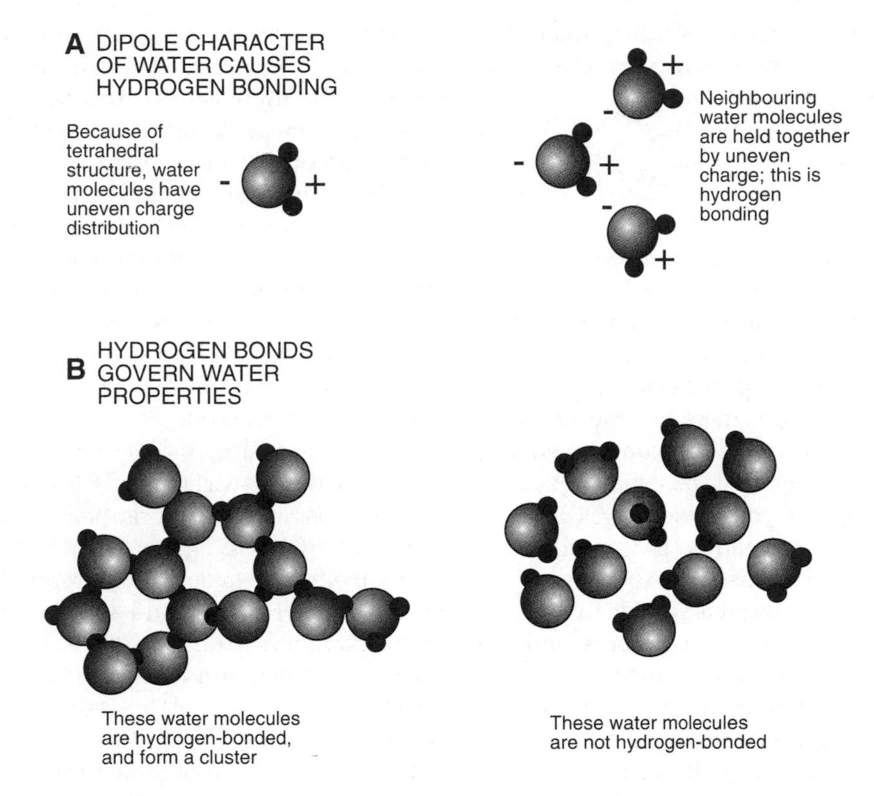

Figure A.3 (A) Dipole nature of water molecules and hydrogen bonds. (B) The ordered arrangement of clusters of hydrogen-bonded water molecules in comparison with unclustered water molecules which are not hydrogen-bonded.

force is termed hydrogen bonding (Figure A.3); without it water would be a gas, not a liquid, on the Earth's surface, with the consequences outlined in Section 1.6.

A.3.2 DISSOLUTION OF MATERIALS BY WATER

Water is a very effective solvent, and the dissolution capability of water can be illustrated by examining what happens to NaCl in contact with water. In the solid state, the Na and Cl are held together by ionic bonds, which involves the *transfer* of electrons, rather than the *sharing* of electrons characteristic of covalent bonds. Na donates an electron to Cl, by which process both atoms achieve greater stability (Figure A.2). Because a Na atom has lost an electron, it is positively charged (Na^+, a cation), whereas each Cl atom, having gained an electron, is negatively charged (Cl^-, an anion). In the solid state, the two are attracted to each other electrostatically, in an ionic bond, which provides the strength to make salt a solid. When immersed in water, the negatively charged end of the H_2O molecule takes Na ions from the surface of the solid, while Cl ions are pulled away by the positively charged ends of other water molecules (Figure A.2). The water molecules surround the ions, which are now too far apart to exert any attractive force on each other, so the ions may be regarded as dissolved. Because of this surrounding effect, water can dissolve large amounts of material.

The process continues until all the salt has dissolved, or the water has become saturated, when it can hold no more ions.

A.3.3 FERROUS (Fe^{2+}) AND FERRIC (Fe^{3+}) IRON

Two key environmental parameters affecting the oceans are the acid–alkaline (the familiar pH) and the reduction–oxidation (the less familiar Eh) states of water, both affecting seabed and subsurface processes. The presence of different forms of iron reflect the Eh of an environment. Under oxidised conditions, iron exists as Fe^{3+}, the ferric state, usually as ferric hydroxide ($Fe(OH)_3$) or oxide (Fe_2O_3); this is insoluble and is precipitated in the sediment. In anaerobic conditions, Fe^{2+} (ferrous iron) dominates, and because it is soluble in water, it is taken up into cements in sediments, and can be identified by chemical staining, especially when the iron is incorporated in calcite cement (which is common in sedimentary materials). Thus the oxidation state of iron can help interpretations of sediments and can show ancient oxidation states of the sea floor; it has proved invaluable in understanding the Precambrian banded iron formations, for example. An important feature is the redox (reduction–oxidation) boundary, beneath which the environment is anaerobic, and above it is aerobic. The redox boundary might be located beneath the surface of sediment, but in particularly anoxic settings, such as the land-locked Black Sea, it may be high in the water, and underneath it, the water has little life. Redox boundaries may be diffuse, allowing subdivision into anaerobic, dysaerobic and aerobic with increasing oxygen levels.

A.3.4 STABLE ISOTOPES

The major stable isotopes used in sedimentary studies are carbon (^{12}C and ^{13}C), oxygen (^{16}O and ^{18}O), sulphur (^{32}S and ^{34}S) and strontium (^{86}Sr and ^{87}Sr). These are studied as ratios between the lighter and heavier versions, usually related to some standard ratio, and in literature the standard notations are: for carbon, $\delta^{13}C$; for oxygen, $\delta^{18}O$; and for sulphur, $\delta^{34}S$. The δ means a difference in the amount of that isotope from a standard, so a value of +5‰ $\delta^{13}C$ means enrichment of ^{13}C over the standard by 5 parts per thousand, and a value of −5‰ $\delta^{13}C$ means depletion of ^{13}C under the standard by 5 parts per thousand. If depleted, that means the ^{12}C is relatively enriched compared with ^{13}C. For strontium, the direct $^{87}Sr/^{86}Sr$ ratio is quoted.

Carbon. Of carbon, ^{12}C = 98.89%, ^{13}C = 1.11%; ^{14}C is created in the upper atmosphere by the breakdown of stable ^{14}N under the influence of cosmic rays. ^{13}C is heavier than ^{12}C, therefore $^{12}CO_2$ is lighter than $^{13}CO_2$ and is more likely to be taken up by organisms, especially leaves in trees, and marine photosynthetic plankton. Aquatic CO_2 is taken up by shelly organisms and inorganic $CaCO_3$ precipitates, and ends up in limestones. Measurements of ^{12}C and ^{13}C in limestones therefore can give evidence of organic productivity. For example, high productivity is often suspected when the limestones have high levels of ^{13}C; this is because photosynthesis by marine algae draw $^{12}CO_2$ out of the water in preference to $^{13}CO_2$. If this happens when there is a phytoplankton bloom, then the algae will die in great numbers and take the $^{12}CO_2$ to the seabed, burying it in deep water shales. The seawater left is consequently enriched in $^{13}CO_2$, and organisms which secrete $CaCO_3$ in shallow water pick up this $^{13}CO_2$ signal (mostly studied in limestones).

Oxygen. Of oxygen, ^{16}O = 99.63%, ^{18}O = 0.1995%; tiny amounts of a third isotope, ^{17}O = 0.0375%, are too low to be useful in sediments and are not used. ^{16}O is lighter than ^{18}O, and seawater contains both $H_2{}^{18}O$ and $H_2{}^{16}O$. When seawater evaporates, the resulting rain is enriched in $H_2{}^{16}O$, so that non-marine $CaCO_3$ cements, and ice sheets, are often likewise enriched. However, the $^{18}O/^{16}O$ ratio is also affected by temperature. This provides the main tool for determining palaeotemperatures in the geological past, and works very well for Mesozoic and Cainozoic rocks. Good discussions of controls are given by Lowe and Walker (1997) and Tucker and Wright (1990).

Sulphur. Of sulphur, ^{32}S = 95.02%, ^{34}S = 4.21% (^{33}S = 0.75% and ^{36}S = 0.02% are not used). The most important control on sulphur isotopes is the reduction of SO_4 ions by anaerobic sulphate-reducing bacteria, especially the species *Desulfurovibrio desulfuricans*; this results in excretion of H_2S gas which is enriched in ^{32}S (and therefore leaves behind high $\delta^{34}S$ values). This passes to the shales in anaerobic settings and therefore betrays ancient seabed anoxia.

Strontium. Sr is derived from the erosion of igneous rocks, and is common in sedimentary systems, especially in carbonate rocks (because of the similarity of properties of Sr and Ca), but is also found in clastic rocks. Sr gives information about weathering and erosion trends because it arrives in sediments by these agents. The current $^{87}Sr/^{86}Sr$ of 0.709 06 ± 0.000 33 in the oceans is a combination of weathering of young volcanic rocks, older continental rocks and Phanerozoic marine carbonates, and the $^{87}Sr/^{86}Sr$ of rocks through geological time fluctuates as a result of changes in the amount of these rock types exposed to weathering. Despite the variations of $^{87}Sr/^{86}Sr$ in time, the $^{87}Sr/^{86}Sr$ ratio *at any one time* is globally the same. This indicates how well mixed the oceans were through time, as they are now, and illustrates that ocean mixing must have operated through geological time, despite shifts in continental positions which may have caused changes in ocean circulation patterns.

Although both ^{86}Sr and ^{87}Sr are stable, their geochemistry is complicated by the fact that ^{87}Sr is derived from the radioactive decay of ^{87}Rb (rubidium) in igneous rocks (Rb–Sr is a key radiometric dating system). The $^{87}Sr/^{86}Sr$ ratio has changed over time and cannot therefore be compared with a standard, so is quoted as a ratio. In measurements for sedimentary purposes, both the age of the rock and the Rb content must be known, so that the ^{87}Sr content can be estimated for the time when the rock formed; for a Devonian rock, for example, the original $^{87}Sr/^{86}Sr$ would be less than now, because between then and now, some ^{87}Rb will have been converted to ^{87}Sr. The Earth's primordial $^{87}Sr/^{86}Sr$ was *c.*0.699 and now it is 0.7044 ± 0.002 in the mantle. Although these ratios seem close, the differences are significant. Sr is a powerful indicator of Earth processes; e.g. evidence that the mantle began to differentiate at *c.*4000 Ma is the low $^{87}Sr/^{86}Sr$ ratios (0.700–0.702) in granitic gneisses dated at 3700 Ma. $^{87}Sr/^{86}Sr$ ratios in sediments therefore reflect the source material, but are influenced by the differences in rates of weathering of different minerals. Most Sr entering the oceans (*c.*60%) comes from weathering of previously deposited marine carbonates, which hold a lot of Sr. Large-scale temporal fluctuations in the $^{87}Sr/^{86}Sr$ curve show prominent reductions in Permian and much of Mesozoic times, interpreted as due to an increase in ^{86}Sr from volcanic sources (compared with carbonate sources, which were more common in the earlier Palaeozoic because there was more carbonate then). The Mesozoic was a period of rapid seafloor spreading as Pangaea broke up. Both Sr and S show similar trends through geological time, and, although they are controlled by different processes, seem to indicate global controls.

A.3.5 RARE EARTH ELEMENTS (REE)

The REE constitute a group of 15 elements with similar properties occurring in minute proportions in all rock types. Like Sr, they are derived from weathering and erosion. The array of elements is divided into a light and a heavygroup (LREE and HREE, respectively) which behave differently in erosion. They are usually plotted as a group, and therefore it is possible to identify trends in vertical sequences in relation to basin evolution. Also two REE (cerium (Ce) and europium (Eu)) have more than one oxidation state (like ferrous and ferric iron), so they may produce troughs in the REE curve, which indicate basin anoxia. However, many REE also have an important use as immobile elements, because only Ce and Eu are affected by oxidation state changes in the aquatic environments.

A.3.6 IMMOBILE ELEMENTS

These are elements which are unaffected by the chemical conditions of their environments, and are transported from a source rock to depositional site without modification. They are therefore valuable in determining the source of the sediment (called *provenance studies*). Common immobile elements are thorium (Th), hafnium (Ha) and tantalum (Ta), and some of the REE.

A.3.7 URANIUM AND THORIUM

Comparisons between the concentrations of these two elements provides a powerful tool for the recognition of anoxia in sediments. Thorium (Th) is immobile, but uranium (U) is affected by oxidation state; U is soluble in oxygenated waters, and insoluble in anoxic water. The ratio of U/Th has been used to provide a scale of anoxia.

REFERENCES

Aldridge, R.J., Jeppsson, L. and Dorning, K.J. (1993) Early Silurian oceanic episodes and events. *Journal of the Geological Society*, London, **150**, 501–513.

Alexander, R.B., Smith, R.A. and Schwarz, G.E. (2000) Effect of stream channel size on the delivery of nitrogen in the Gulf of Mexico. *Nature*, **403**, 758–761.

Alvarez, W. and Asaro, F. (1990) An extraterrestrial impact. *Scientific American*, Oct, 44–52.

Anderson, D.M. (1994) Red tides. *Scientific American*, **271**, 62–68.

Anderson, R. (1996) Seasonal sedimentation: a framework for reconstructing climatic and environmental change. In: *Palaeoclimatology and Palaeoceanography from Laminated Sediments*. (ed. Kemp, A.E.S.) *Geological Society Special Publication* No.116. The Geological Society, London, pp.1–15.

Armstrong, H.A. (1995) High resolution biostratigraphy (conodonts and graptolites) of the Upper Ordovician and Lower Silurian – evaluation of the Late Ordovician mass extinction. *Modern Geology*, **20**(1), 41–68.

Armstrong, H.A. (1996) Biotic recovery after mass extinction: the role of climate and ocean-state in the post-glacial (Late Ordovician-Early Silurian) recovery of the conodonts. In: *Biotic Recovery from Mass Extinction Events* (ed. Hart, M.B.) *Geological Society Special Publication*. The Geological Society, London, pp.105–117.

Arthur, M.A. (1986) Rhythmic bedding in Upper Cretaceous pelagic carbonate sequences: varying sedimentary response to climatic forcing. *Geology*, **14**, 153–156.

Bauer, J.E. and Druffel, E.R.M. (1998) Ocean margins as a significant source of organic matter to the deep open ocean. *Nature*, **392**, 482–484.

Bearman, G. (ed.) (1989a) *The Ocean Basins: Their Structure and Evolution*. Open University and Pergamon Press, Oxford, 171 pp.

Bearman, G. (ed.) (1989b) *Seawater: Its Composition, Properties and Behaviour*. Open University and Pergamon Press, Oxford, 165pp.

Benitez-Nelson, C. and Buesseler, K.O. (1999) Variability of inorganic and organic phosphorus turnover rates in the coastal ocean. *Nature*, **398**, 502–505.

Benton, M. (1995) Diversification and extinction in the history of life. *Science*, **268**, 53.

Berner, R.A. (1991) A model for atmospheric CO_2 over Phanerozoic time. *American Journal of Science*, **291**, 339–376.

Berner, R.A. and Canfield, D.E. (1989) A new model for atmospheric oxygen over Phanerozoic time. *American Journal of Science*, **289**, 333–361.

Berner, R.A., Lasaga, A.C. and Garrels, R.M. (1983) The carbonate-silicate geochemical cycle and its effect on atmospheric carbon dioxide over the past 100 million years. *American Journal of Science*, **283**, 641–683.

Berry, W.B.N., Wilde, P. and Quinby-Hunt, M.S. (1990) Late Ordovician graptolite mass mortality and subsequent Early Silurian re-radiation. In: *Extinctions in Earth History*. (ed. Kauffman, E.G. and Walliser, O.H.). Springer, Berlin, pp.115–123.

Berry, W.B.N. (1996) Recovery of post-Late Ordovician extinction graptolites: a western North American perspective. In: *Biotic Recovery from Mass Extinction Events*. (ed. Hart, M.B.). *Geological Society Special Publication*. The Geological Society, London, pp.119–126.

Berry, W.B.N., Quinby-Hunt, M.S. and Wilde, P. (1996) Post-pacificus event graptolite recovery: faunal-environmental interrelationships at Dob's Linn, southern Scotland. *The James Hall Symposium: Second International Symposium of the Silurian System, Abstracts*, Rochester, New York, USA, pp.31.

Béthoux, J.-P. and Pierre, C. (1999) Mediterranean functioning and sapropel formation: respective influences of climate and hydrological changes in the Atlantic and the Mediterranean. *Marine Geology*, **153**, 29–39.

Bianchi, G.G. and McCave, I.N. (1999) Holocene periodicity in North Atlantic climate and deep-ocean flow south of Iceland. *Nature*, **397**, 515–517.

Bottjer, D.J., Arthur, M.A., Dean, W.E., Hattin, D.E. and Savrda, C.E. (1986) Rhythmic bedding produced in Cretaceous pelagic carbonate environments: sensitive recorders of climatic cycles. *Paleoceanography*, **1**(4), 467–481.

Bowring, S.A. *et al.* (1998) U/Pb zircon geochronology and tempo of the end-Permian mass extinction. *Science*, **280**, 1039–1045.

Bralower, T.J. *et al.* (1994) Timing and palaeoceanography of oceanic dysoxia/anoxia in the Late Barremian to Early Aptian (Early Cretaceous). *Palaios*, **9**, 335–369.

Bralower, T.J. and Thierstein, H.R. (1984) Low productivity and slow deep-water circulation in mid-Cretaceous oceans. *Geology*, **12**, 614–618.

Brasier, M.D. (1992) Nutrient-enriched waters and the early skeletal fossil record. *Journal of the Geological Society, London*, **149**(4), 621–630.

Brasier, M.D. (1995a) Fossil indicators of nutrient levels 1: Eutrophication and climate change. In: *Marine Palaeoenvironmental Analysis from Fossils*. (ed. Bosence, D.W.J. and Allison, P.A.). *Geological Society Special Publication* No. 83. The Geological Society, London, pp.113–132.

Brasier, M.D. (1995b) Fossil indicators of nutrient levels 2: Evolution and extinction in relation to oligotrophy. In: *Marine Palaeoenvironmental Analysis from Fossils*. (ed. Bosence, D.W.J. and Allison, P.A.). *Geological Society Special Publication* No. 83. The Geological Society, London, pp.133–150.

Brasier, M.D. and McIlroy, D. (1998) *Neonereites uniserialis* from c.600 Ma year old rocks in western Scotland and the emergence of animals. *Journal of the Geological Society, London*, **155**(1), 5–12.

Brenchley, P.J. and Newall, G. (1984) Late Ordovician environmental changes and their effects on faunas. In: *Aspects of the Ordovician System*. (ed. Bruton, D.L.). Palaeontological Contributions from the University of Oslo, Oslo, pp.65–79.

Brenchley, P.J., Marshall, J.D., Carden, G.A.F., Robertson, D.B.R., Long, D.G.F., Meidla, T., Hints, L. and Anderson, T.F. (1994) Bathymetric and isotopic evidence for a short-lived Late Ordovician glaciation in a greenhouse period. *Geology*, **22**, 295–298.

Brenchley, P.J., Carden, G.A.F. and Marshall, J.D. (1995) Environmental changes associated with the 'First Strike' of the Late Ordovician mass extinction. *Modern Geology*, **20**(1), 69–82.

Breuer, D. and Spohn, T. (1995) Possible flush instability in mantle convection at the Archaean-Proterozoic transition. *Nature*, **378**, 608–610.

Briggs, D.E.G. and Crowther, P.R. (1990) *Palaeobiology: A Synthesis*. Blackwell Scientific, Oxford, 583pp.

Brocks, J., Logan, G., Buick, R. and Summons, R. (1999) Archaean molecular fossils and the early rise of eukaryotes. *Science*, **285**, 1033–1036.

Broecker, W.S. (1994) Massive iceberg discharges as triggers for global climate change. *Nature*, **372**, 421–424.

Broecker, W.S. (1997) Future directions of palaeoclimate research. *Quaternary Science Reviews*, **16**, 821–825.

Broecker, W.S. (1998) Palaeocean circulation during the last deglaciation: a bipolar seesaw? *Paleoceanography*, **13**(2), 119–121.

Broecker, W.S. (1999) What if the conveyor were to shut down? Reflections on a possible outcome of the great global experiment. *GSA Today*, **9**(1), 1–7.

Broecker, W.S. and Denton, G.H. (1990a) The role of ocean–atmosphere reorganisations in glacial cycles. *Quaternary Science Reviews*, **9**, 305–341.

Broecker, W.S. and Denton, G.H. (1990b) What drives glacial cycles? *Scientific American*, January, 43–50.

Broecker, W.S., Peteet, D. and Rind, D. (1985) Does the ocean–atmosphere have more than one stable mode of operation? *Nature*, **315**, 21–25.

Brunton, F.R., Copper, P. and Dixon, O.A. (1997) Silurian reef-building episodes. *Proceedings of the 8th Coral Reef Symposium*, Panama, pp.1643–1650.

Brunton, F. *et al.* (1998) Silurian reef episodes, changing seascapes, and paleobiogeography. In: *Silurian Cycles: Linkages of Dynamic Stratigraphy with Atmospheric, Oceanic and Tectonic Changes* (eds Landing, E. and Johnson, M.). New York State Museum Bulletin 491, pp.265–282.

Buick, R., Thornett, J.R., McNaughton, N.J., Smith, J.B., Barley, M.E. and Savage, M. (1995) Record of emergent continental crust c.3.5 billion years ago in the Pilbara craton of Australia. *Nature*, **375**, 574–577.

Bull, D. and Kemp, A.E.S. (1996) Composition and origins of laminae in late Quaternary and Holocene sediments from the Santa Barbara Basin. In: *Palaeoclimatology and Palaeoceanography from Laminated Sediments*. (ed. Kemp, A.E.S.). *Geological Society Special Publication* No. 116. The Geological Society, London, pp.143–156.

Burke, W.H. *et al.* (1982) Variations of seawater 87Sr/86Sr through Phanerozoic time. *Geology*, **10**, 516–519.

Busch, R.M. and Rollins, H. (1984) Correlation of Carboniferous strata using hierarchy of transgressive–regressive units. *Geology*, **12**, 471–474.

Busch, R.M. and West, R.R. (1987) Hierarchical genetic stratigraphy: a framework for palaeoceanography. *Paleoceanography*, **2**, 141–164.

Butcher, S.S., Charlson, R.J., Orians, G.H. and Wolfe, G.V. (ed.) (1992). *Global Biogeochemical Cycles*. Academic Press, London, 379 pp.

Butterfield, N.J. and Chandler, F.W. (1992) Palaeoenvironmental distribution of Proterozoic microfossils, with an example from the Agu Bay Formation, Baffin Island. *Palaeontology*, **35**(4), 943–957.

Cairns-Smith, A.G. (1985) *Seven Clues to the Origin of Life*. Cambridge University Press, Cambridge.

Canfield, D.E. and Teske, A. (1996) Late Proterozoic rise in atmospheric oxygen concentration inferred from phylogenetic and sulphur-isotope studies. *Nature*, **382**, 127–132.

Carlowicz, M. (1996) Warming by methane? *Eos*, **77**, 321–322.

Castanier, S., Le Métayer-Levrel, G. and Perthuisot, J.-P. (1999) Ca-carbonates and limestone genesis – the micro biogeologist point of view. *Sedimentary Geology*, **126**, 9–23.

Chester, R. (1990) *Marine Geochemistry*. Chapman and Hall, London, 698 pp.

Clarke, R.B. (1993) *Marine Pollution*. Clarendon Press, Oxford.

Clauzon, G., Suc, J.-P., Gautier, F., Berger, A. and Loutre, M.-F. (1996) Alternate interpretation of the Messinian salinity crisis: controversy resolved? *Geology*, **24**(4), 363–366.

Clift, P.D., Carter, A. and Hurford, A.J. (1998) The erosional and uplift history of NE Atlantic passive margins: constraints on a passing plume. *Journal of the Geological Society, London*, **155**, 787–800.

Coale, K.H. *et al.* (1996) A massive phytoplankton bloom induced by an ecosystem-scale iron fertilization experiment in the equatorial Pacific Ocean. *Nature*, **383**, 495–501.

Compston, W. and Pidgeon, R.T. (1986) Jack Hills, evidence of more very old zircons in Western Australia. *Nature*, **321**, 766–769.

Cook, P.J. (1992) Phosphogenesis around the Proterozoic-Phanerozoic transition. *Journal of the Geological Society, London*, **149**(4), 615–620.

Cook, R.M., Sinclair, A. and Stefansson, G. (1997) Potential collapse of North Sea cod stocks. *Nature*, **385**, 521–522.

Cope, J.C.W. (1994) A latest Cretaceous hotspot and the southeasterly tilt of Britain. *Journal of the Geological Society, London*, **151**, 905–908.

Crossland, C.J., Hatcher, B.G. and Smith, S.V. (1991) Role of coral reefs in global ocean production. *Coral Reefs*, **10**, 55–64.

Croudace, I.W. and Cundy, A.B. (1995) Heavy metal and hydrocarbon pollution in Recent sediments from Southampton Water, southern England: a geochemical and isotopic study. *Environmental Science and Technology*, **29**, 1288–1296.

Crowley, T.J. and Baum, S.K. (1991) Toward reconciliation of Late Ordovician (c.440 Ma) glaciation with very high CO_2 levels. *Journal of Geophysical Research*, **96**(D12), 22597–22610.

Crowley, T.J. and Baum, S.K. (1995) Reconciling Late Ordovician (440 Ma) glaciation with very high (14×) CO_2 levels. *Journal of Geophysical Research*, **100**(D1), 1093–1101.

Crowley, T.J. and North, G.R. (1991) *Paleoclimatology*. Oxford University Press, Oxford, 349 pp.

Cundy, A.B. and Croudace, I.W. (1995) Sedimentary and geochemical variations in a salt marsh/mud flat environment from the mesotidal Hamble Estuary, southern England. *Marine Chemistry*, **51**, 115–132.

Cundy, A.B., Croudace, I.W., Thomson, J. and Lewis, J.T., (1997) Reliability of salt marshes as 'geochemical recorders' of pollution input: a case study from contrasting estuaries in southern England. *Environmental Science Technology*, **31**, 1093–1101.

Dam, G., Larsen, M. and Sonderholm, M. (1998) Sedimentary response to mantle plumes: implications from Palaeocene onshore successions, west and east Greenland. *Geology*, **26**(3), 207–210.

Dasgupta, H.C., Sambasiva Rao, V.V. and Krishna, C. (1999) Chemical environments of deposition of ancient iron- and manganese-rich sediments and cherts. *Sedimentary Geology*, **125**, 83–98.

de Ronde, C.E.J. and Ebbesen, T.W. (1996) 3.2 b.y. of organic compound formation near sea-floor hot springs. *Geology*, **24**(9), 791–794.

de Wit, M.J. and Hynes, A. (1995) The onset of interaction between the hydrosphere and oceanic crust, and the origin of the first continental lithosphere. In: *Early Precambrian Processes*. (ed. Coward, M.P. and Ries, A.C.). *Geological Society of London Special Publication* No. 95, pp.1–9.

Deming, D. (1999) On the possible influence of extraterrestrial volatiles on Earth's climate and the origin of the oceans. *Palaeogeography, Palaeoclimatology, Palaeoecology*, **146**, 33–51.

Dickson, B. (1999) All change in the Arctic. *Nature*, **397**, 389–391.

Dickson, B., Meincke, J., Vassie, I., Jungclaus, J. and Osterhus, S. (1999) Possible predictability in overflow from the Denmark Strait. *Nature*, **397**, 243–246.

DiTullio, G.R. *et al.* (2000) Rapid and early export of *Phaeocystis antarctica* blooms in the Ross Sea, Antarctica. *Nature*, **404**, 595–600.

Droser, M.L. (1991) Ichnofabric of the Palaeozoic *Skolithos* ichnofacies and the nature and distribution of Skolithos piperock. *Palaios*, **6**(3), 316–325.

Edwards, D.C. (1990) Cement reefs. Unpublished PhD Thesis, University of Wales, Cardiff, 201 pp.

Ekdale, A.A. and Stinnesbeck, W. (1998) Trace fossils in Cretaceous-Tertiary (KT) boundary beds in northeastern Mexico: implications for sedimentation during the KT boundary event. *Palaios*, **13**(6), 593–602.

Elrick, M. and Hinnov, L.A. (1996) Millennial-scale climate origins for stratification in Cambrian and Devonian deep-water rhythmites. *Palaeogeography, Palaeoclimatology, Palaeoecology*, **123**, 353–372.

Emery, D. and Myers, K.J. (eds) (1996) *Sequence Stratigraphy*. Blackwell Science, Oxford, 297 pp.

Eriksson, K.A. (1995) Crustal growth, surface processes, and atmospheric evolution on the early Earth. In: *Early Precambrian Processes*. (ed. Coward, M.P. and Ries, A.C.). *The Geological Society, Special Publication* No. 95, London, pp.11–25.

Erwin, D.H. (1994) The Permo-Triassic extinction. *Nature*, **367**, 231–236.

Eyles, N. and Young, G.M. (1994) Geodynamic controls on glaciation in Earth history. In: *Earth's Glacial Record; International Geological Correlation Project 260*. (ed. Deynoux, M. *et al.*). Cambridge University Press, Cambridge, pp.1–28.

Fairbanks, R. (1989) A 17,000 year glacio-eustatic sea-level record: influence of glacial melting rates on the Younger Dryas event and deep ocean circulation. *Nature*, **342**, 637–642.

Faure, K., de Wit, M.J. and Willis, J.P. (1995) Late Permian global coal hiatus linked to 13C-depleted CO_2 flux into the atmosphere during the final consolidation of Pangaea. *Geology*, **23**(6), 507–510.

Fiennes, R. (1993) *Mind Over Matter*. Mandarin.

Francis, P. and Dise, N. (1997) *Atmosphere, Earth and Life. S269 Earth and Life*. The Open University, Milton Keynes, UK, 195 pp.

Freiwald, A., Wilson, J.B. and Henrich, R. (1999) Grounding Pleistocene icebergs shape recent deep-water coral reefs. *Sedimentary Geology*, **125**, 1–8.

Frost, B.W. (1996) Phytoplankton bloom on iron rations. *Nature*, **383**, 475–476.

Gerlach, S.A. (1994) Oxygen conditions improve when the salinity in the Baltic Sea decreases. *Marine Pollution Bulletin*, **28**, 413–416.

Gibbons, W.A. (1981) *The Weald. Rocks and Fossils Geological Guides*. Unwin Paperbacks, London.

Gill, R. (1996) *Chemical Fundamentals of Geology*. Chapman and Hall, London, 289 pp.

Glennie, K.W. (ed.) (1990a) *Introduction to the Petroleum Geology of the North Sea*. Blackwell Scientific, Oxford, 402 pp.

Glennie, K.W. (1990b) Lower Permian – Rotliegend. In: *Introduction to the Petroleum Geology of the North Sea*. (ed. Glennie, K.W.). Blackwell Scientific, Oxford, pp.120–152.

Glynn, P.W. and Croz, L.D. (1990) Experimental evidence for high-temperature stress as the cause of El Niño co-incident coral mortality. *Coral Reefs*, **8**, 181–191.

Gorsline, D.S., Nava-Sanchez, E. and Murillo de Nava, J. (1996) A survey of occurrences of Holocene laminated sediments in Californian Borderland Basins: products of a variety of depositional processes. In: *Palaeoclimatology and Palaeoceanography from Laminated Sediments*. (ed. Kemp, A.E.S.). *Geological Society Special Publication* No. 116. The Geological Society, London, pp.93–110.

Graham, J.B., Dudley, R., Aguilar, N. and Gans, C. (1995) Implications of the late Palaeozoic oxygen pulse for physiology and evolution. *Nature*, **375**, 117–120.

Grotzinger, J.P. (1994) Trends in Precambrian carbonate sediments and their implication for understanding evolution. In: *Early Life on Earth*. (ed. Bengtson, S.). Columbia University Press, New York, pp.245–258.

Grotzinger, J.P., Bowring, S.A., Saylor, B.Z. and Kaufman, A.J. (1995) Biostratigraphic and geochronologic constraints on early animal evolution. *Science*, **270**, 598–604.

Grotzinger, J.P. and Kasting, J.F. (1993) New constraints on Precambrian ocean composition. *Journal of Geology*, **101**, 235–243.

Grotzinger, J.P. and Knoll, A.H. (1995) Anomalous carbonate precipitates: is the Precambrian the key to the Permian? *Palaios*, **10**, 578–596.

Guo, Z. *et al.* (1998) Climate extremes in loess of China coupled with the strength of deep-water formation in the North Atlantic. *Global and Planetary Change*, **18**, 113–128.

Haczewski, G. (1996) Oligocene laminated limestones as a high resolution correlator of palaeoseismicity, Polish Carpathians. In: *Palaeoclimatology and Palaeoceanography from Laminated Sediments*. (ed. Kemp, A.E.S.). *Geological Society Special Publication* No. 116. The Geological Society, London, pp.209–220.

Hagadorn, J.W. (1996) Laminated sediments of Santa Monica Basin, California continental borderland. In: *Palaeoclimatology and Palaeoceanography from Laminated Sediments*. (ed. Kemp, A.E.S.). *Geological Society Special Publication* No. 116. The Geological Society, London, pp.111–120.

Hallam, A. (1992) *Phanerozoic Sea-Level Changes*. Columbia University Press, New York, 266 pp.

Hallam, A. and Wignall, P.B. (1997) *Mass Extinctions and Their Aftermath*. Oxford University Press, Oxford, 320 pp.

Hallberg, R.O. (1992) Sediments: their interaction with biogeochemical cycles through formation and diagenesis. In: *Global Biogeochemical Cycles*. (ed. Butcher, S.S., Charlson, R.J., Orians, G.H. and Wolfe, G.V.). Academic Press, London, pp.155–174.

Hallock, P. and Schlager, W. (1986). Nutrient excess and the demise of coral reefs and carbonate platforms. *Palaios*, **1**, 389–398.

Han, T.-M. and Runnegar, B. (1992) Megascopic eukaryotic algae from the 2.1-billion-year-old Negaunee Iron Formation, Michigan. *Science*, **257**, 232–235.

Hardie, L. (1996) Secular variation in seawater chemistry: an explanation for the coupled secular variation in the mineralogies of marine limestones and potash evaporites over the past 600m.y. *Geology*, **24**(3), 279–283.

Harper, D.A.T. and Rong, J.-Y. (1995) Patterns of change in brachiopod faunas through the Ordovician-Silurian interface. *Modern Geology*, **20**(1), 83–100.

Harvey, L.D.D. and Huang, Z. (1995) Evaluation of the potential impact of methane clathrate destabilization on future global warming. *Journal of Geophysical Research-Atmospheres*, **100**, 2905–2926.

Haszeldine, S. and Smythe, D. (1997) Why was Sellafield rejected as a disposal site for radioactive waste? *Geoscientist*, **7**(7), 18–20.

Henrich, R. (1998) Dynamics of Atlantic water advection to the Norwegian-Greenland Sea – a time-slice record of carbonate distribution in the last 300 ky. *Marine Geology*, **145**, 95–131.

Herbert, T.D. *et al*. (1998) Depth and seasonality of alkenone production along the California margin inferred from a core top transect. *Paleoceanography*, **13**(3), 263–271.

Hoffman, A., Gruszczynski, M. and Malkowski, K. (1991) On the interrelationship between temporal trends in $\delta^{13}C$, $\delta^{18}O$ and $\delta^{34}S$ in the world ocean. *Journal of Geology*, **99**, 355–370.

Hoffman, P.F. (1999) The break-up of Rodinia, birth of Gondwana, true polar wander and the snowball Earth. *Journal of African Earth Sciences*, **28**(1), 9–27.

Hoffman, P.F., Kaufman, A.J., Halverson, G.P. and Schrag, D.P. (1998) A Neoproterozoic snowball Earth. *Science*, **281**, 1342–1346.

Holland, H.D. (1978) *The Chemistry of the Atmosphere and Oceans*. Wiley-Interscience, New York, 582 pp.

Holland, H.D. (1984) *The Chemical Evolution of the Atmosphere and Oceans. Princeton Series in Geochemistry*. Princeton University Press, Princeton, NJ, 582 pp.

Holland, H.D. and Kasting, J.F. (1992) The environment of the early Earth. In: *The Proterozoic Biosphere*. (ed. Schopf, J.W. and Klein, C.). Cambridge University Press, Cambridge, pp.21–24.

Hollister, C.D. and Nadis, S. (1998) Burial of radioactive waster under the seabed. *Scientific American*, January, 60–65.

Holmén, K. (1992) The global carbon cycle. In: *Global Biogeochemical Cycles*. (ed. Butcher, S.S., Charlson, R.J., Orians, G.H. and Wolfe, G.V.). Academic Press, London, pp.239–262.

Holmes, B. (1996) Imagine a future where most of the fish . . . *New Scientist*, **152**(2059), 32–36.

Horodyski, R.J. and Knauth, L.P. (1994) Life on land in the Precambrian. *Science*, **263**, 494–498.

House, M.R. (1995) Orbital forcing timescales: an introduction. In: *Orbital Forcing Timescales and Cyclostratigraphy*. (ed. House, M.R. and Gale, A.S.). *Geological Society of London Special Publication* No. 85, pp.1–18.

House, M.R. and Gale, A.S. (eds) (1995) *Orbital Forcing Timescales and Cyclostratigraphy. Geological Society of London Special Publication* No. 85, London, 204 pp.

Howe, J.A., Pudsey, C.J. and Cunningham, A.P. (1997) Pliocene-Holocene contourite deposition under the Antarctic Circumpolar Current, Western Falkland Trough, South Atlantic Ocean. *Marine Geology*, **138**, 27–50.

Hsü, K.J. *et al*. (1977) History of the Messinian salinity crisis. *Nature*, **267**, 399–403.

Hsü, K.J., Ryan, W.B.F. and Cita, M.B. (1973) Late Miocene desiccation of the Mediterranean. *Nature*, **242**, 240–244.

Hughen, K.A., Overpeck, J.Y., Peterson, L.C. and Anderson, R.F. (1996) The nature of varved sedimentation in the Cariaco Basin, Venezuela, and its palaeoclimatic significance. In: *Palaeoclimatology and Palaeoceanography from Laminated Sediments*. (ed. Kemp, A.E.S.). *Geological Society Special Publication* No. 116. The Geological Society, pp.171–183.

IPCC (1990) *Climate Change: The IPCC Scientific Assessment*. Cambridge University Press, Cambridge, 365 pp.

IPCC (1996) *Climate Change 1995. Impacts, Adaptations and Mitigation of Climate Change: Scientific-Technical Analyses*. Cambridge University Press, Cambridge, 878 pp.

Ittekkot, V. (1993) The abiotically-driven biological pump in the ocean and short-term fluctuations in atmospheric CO_2 contents. *Global and Planetary Change*, **8**, 17–25.

Jahnke, R.J. (1992) The phosphorus cycle. In: *Global Biogeochemical Cycles*. (ed. Butcher, S.S., Charlson, R.J., Orians, G.H. and Wolfe, G.V.). Academic Press, London, pp.301–315.

James, N.P. and Clarke, J.A.D. (eds) (1997) *Cool-Water Carbonates, 56*. SEPM Special Publication, Tulsa, Oklahoma.

Jefferies, D.F., Preston, A. and Steele, A.K. (1973) Distribution of caesium-137 in British coastal waters. *Marine Pollution Bulletin*, **4**, 118–122.

Jenkins, W.J. (1992). Tracers in oceanography. *Oceanus*, **35**(1), 47–56.

Jeppsson, L. (1987) Lithological and conodont distributional evidence for episodes of anomalous oceanic conditions during the Silurian. In: *Palaeobiology of Conodonts*. (ed. Aldridge, R.J.). Ellis Horwood Ltd., Chichester.

Jeppsson, L. (1990) An oceanic model for lithological and faunal changes tested on the Silurian record. *Journal of the Geological Society*, **147**, 663–674.

Jeppsson, L., Aldridge, R.J. and Dorning, K.J. (1995) Wenlock (Silurian) oceanic episodes and events. *Journal of the Geological Society, London*, **152**, 487–498.

Jin, J. and Copper, P. (1996) Species-level extinction and recovery of brachiopods across the Ordovician-Silurian boundary, Anticosti Island, eastern Canada. In: *The James Hall Symposium: Second International Symposium of the Silurian System, Abstracts*, Rochester, New York, USA, p.60.

Johnson, C.C. *et al.* (1996) Middle Cretaceous reef collapse linked to ocean heat transport. *Geology*, **24**(4), 376–380.

Johnson, M.E., Kaljo, D. and Rong, D.-Y. (1991) Silurian Eustasy. In: *The Murchison Symposium*. (ed. Bassett, M.G., Lane, P.D. and Edwards, D.). *Special Papers in Palaeontology 44*, pp.145–163.

Jones, E.J.W., Cande, S.C. and Spathopoulos, F. (1995) Evolution of a major oceanographic pathway: the equatorial Atlantic. In: *The Tectonics, Sedimentation and Palaeoceanography of the North Atlantic Region*. (ed. Scrutton, R.A., Stoker, M.S., Shimmield, G.B. and Tudhope, A.W.). *Geological Society Special Publication* No. 90, London, pp.199–213.

Jones, P.D., Wigley, T.M.L. and Wright, P.B. (1986) Global temperature variations between 1861 and 1984. *Nature*, **322**, 430–434.

Kaljo, D. (1996) Diachronous recovery patterns in Early Silurian corals, graptolites and acritarchs. In: *Biotic Recovery from Mass Extinction Events*. (ed. Hart, M.B.). *Geological Society Special Publication*. The Geological Society, London, pp.127–133.

Karhu, J. and Holland, H.D. (1996) Carbon isotopes and the rise of atmospheric oxygen. *Geology*, **24**(10), 867–870.

Kasting, J.F. and Chang, S. (1992) Formation of the Earth and the origin of life. In: *The Proterozoic Biosphere*. (ed. Schopf, J.W. and Klein, C.). Cambridge University Press, Cambridge, pp.9–12.

Kasting, J.F., Eggler, D.H. and Raeburn, S.P. (1993) Mantle redox evolution and the oxidation state of the Archaean atmosphere. *Journal of Geology*, **101**, 245–257.

Kawahata, H. and Eguchi, N. (1996) Biogenic sediments on the Eauripik Rise of the western equatorial Pacific during the late Pleistocene. *Geochemical Journal*, **30**, 201–215.

Kawahata, H., Suzuki, A. and Goto, K. (1997) Coral reef ecosystems as a source of atmospheric CO_2: evidence from PCO_2 measurements of surface waters. *Coral Reefs*, **16**, 261–266.

Kawahata, H., Suzuki, A. and Ohta, H. (1998) Sinking particles between the equatorial and subarctic regions (0°N-46°N) in the central Pacific. *Geochemical Journal*, **32**, 125–133.

Kemp, A.E.S. (ed.) (1996) *Palaeoclimatology and Palaeoceanography from Laminated Sediments*. *Geological Society Special Publication* No. 116. The Geological Society, London, 258 pp.

Kemp, A.E.S., Baldauf, J.F. and Pearce, R.B. (1996) Origins and palaeoceanographic significance of laminated diatom ooze from the Eastern Equatorial Pacific Ocean. In: *Palaeoclimatology and Palaeoceanography from Laminated Sediments*. (ed. Kemp, A.E.S.). *Geological Society Special Publica-*

tion No. 116. The Geological Society, London, pp.243–252.

Kemp, A.E.S., Pearce, R.B., Koizumi, I., Pike, J. and Rance, S.J. (1999) The role of mat-forming diatoms in the formation of Mediterranean sapropels. *Nature*, **398**, 57–61.

Kempe, S. and Degens, E.T. (1985) An early soda ocean? *Chemical Geology*, **53**, 95–108.

Kempe, S. and Kazmierczak, J. (1994) The role of alkalinity in the evolution of ocean chemistry, organisation of living systems, and biocalcification processes. *Bulletin de l'Institut océanographique, Monaco*, **13**, 61–117.

Kerr, A.C. (1998) Oceanic plateau formation: a cause of mass extinction and black shale deposition around the Cenomanian-Turonian boundary? *Journal of the Geological Society, London*, **155**, 619–626.

Kershaw, S. (1993) Sedimentation control on growth of stromatoporoid reefs in the Silurian of Gotland. *Journal of the Geological Society, London*, **150**, 197–205.

Kershaw, S., Zhang, T. and Lan, G. (1999) A microbialite crust at the Permian-Triassic boundary in south China, and its palaeoenvironmental significance. *Palaeogeography, Palaeoclimatology, Palaeoecology*, **146**, 1–18.

Klein, G.D. (1992) Climatic and tectonic sea-level gauge for Mid-continent Pennsylvanian cyclothems. *Geology*, **20**, 363–366.

Kleypas, J.A. *et al.* (1999) Geochemical consequences of increased atmospheric carbon dioxide on coral reefs. *Science*, **284**, 118–120.

Knoll, A.H. (1994) Neoproterozoic evolution and environmental change. In: *Early Life on Earth*. (ed. Bengtson, S.). Columbia University Press, New York, pp.439–449.

Knoll, A.H. and Simonson, B. (1981) Early Proterozoic microfossils and penecontemporaneuous quartz cementation in the Sokoman Iron Formation, Canada. *Science*, **211**, 478–480.

Knoll, A.H., Bambach, R.K., Canfield, D.E. and Grotzinger, J.P. (1996) Comparative Earth history and Late Permian mass extinction. *Science*, **273**, 452–457.

Knutson, T.R. and Manabe, S. (1994) Impact of increased CO_2 on simulated ENSO-like phenomena. *Geophysical Research Letters*, **21**, 2295–2298.

Konhauser, K.O. (1998) Diversity of bacterial iron mineralization. *Earth-Science Reviews*, **43**, 91–121.

Kouwenhoven, T.J., Seidenkrantz, M.-S. and van der Zwaan, G.J. (1999) Deep-water changes: the near synchronous disappearance of a group of benthic foraminifera from the Late Miocene Mediterranean. *Palaeogeography, Palaeoclimatology, Palaeoecology*, **152**, 259–281.

Kuypers, M.M.M., Pancost, R.D. and Damste, J.S.S. (1999) A large and abrupt fall in atmospheric CO_2 concentration during Cretaceous times. *Nature*, **399**, 342–345.

Kvenvolden, K.A. (1998) A primer on the geological occurrence of gas hydrate. In: *Gas Hydrates: Relevance to World Margin Stability and Climate Change* (eds Henriet, J.-P. and Miniert, J.). Geological Society, London, Special Publication No. 137, pp. 9–30.

Lahvis, G.P. *et al.* (1995) Decreased in vitro lymphocyte responses in free-ranging bottlenose dolphins (*Tursiops truncatas*) are associated with increased whole blood concentrations of polychlorinated biphenyls (PCBs) and p,p'-DDT; p,p'-DDE; and o,p'DDE. *Environmental Health Perspectives*, **103**(4), 67.

Lambeck, K. (1999) Shoreline displacements in southern-central Sweden and the evolution of the Baltic Sea since the last maximum glaciation. *Journal of the Geological Society, London*, **156**, 465–486.

Lambert, I.B. and Donnelly, T.H. (1991) Atmospheric oxygen levels in the Precambrian: a review of isotopic and geological evidence. *Palaeogeography, Palaeoclimatology, Palaeoecology (Global and Planetary Change Section)*, **97**, 83–91.

Lambert, I.B. and Donnelly, T.H. (1992) Global oxidation and a supercontinent in the Proterozoic: evidence from stable isotope trends. In: Early Organic Evolution; Implications for Mineral and Energy Resources. (ed. Schidlowski, M., Golubic, S., Kimberley, M.M., McKirdy, D.M. and Trudinger, P.A.). Springer-Verlag, Berlin, pp.404–418

Leary, P.N. and Rampino, M.R. (1990) A multicausal model of mass extinctions: increase in trace metals in the oceans. In: *Extinctions in Earth History*. (ed. Kauffman, E.G. and Walliser, O.H.). Springer, Berlin, pp.45–55.

Leeder, M. (1982) *Sedimentology: Process and Product*. Unwin-Hyman, London, 344 pp.

Lehman, S. (1993) Flickers within cycles. *New Scientist* (4th February), 404–405.

Leng, M. and Greenwood, P. (1999) Isotopes as climate archives: climate and diatoms. *NERC News*, Autumn, 8–9.

Lerche, I., Yu, Z., Torudbakken, B. and Thomsen, R.O. (1997) Ice loading effects in sedimentary basins with reference to the Barents Sea. *Marine and Petroleum Geology*, **14**(3), 277–338.

Li, Z.X., Li, X.H., Kinny, P.D. and Wang, J. (1999) The breakup of Rodinia: did it start with a mantle plume beneath south China? *Earth and Planetary Science Letters*, **173**, 171–181.

Libes, S. (1992) *An Introduction to Marine Biogeochemistry*. John Wiley and Sons, New York, 734 pp.

Liss, P. and Duce, R. (eds) (1997) *The Sea Surface and Global Change*. Cambridge University Press, Cambridge, 519 pp.

Little, C.T.S., Herrington, R.J., Maslennikov, V.V. and Zaykov, V.V. (1998) The fossil record of hydrothermal vent communities. In: *Modern Ocean Floor Processes and the Fossil Record*. (ed. Mills, R.A. and Harrison, K.). *Geological Society Special Publication* No. 148. The Geological Society, London, pp.259–270.

Livingston, H.D. and Bowen, V.T. (1977) Windscale effluent in the waters and sediments of the Minch. *Nature*, **269**, 586–588.

Loehle, C. (1993) Geologic methane as a source for post-glacial CO_2 increases – the hydrocarbon pump hypothesis. *Geophysal Research Letters*, **20**, 1415–1418.

Long, A.J. and Shennan, I. (1994) Sea-level changes in Washington and Oregon and the 'Earthquake Deformation Cycle'. *Journal of Coastal Research*, **10**(4), 825–838.

Lowe, J.J. and Walker, M.J.C. (1997) *Reconstructing Quaternary Environments*. Longman, London, 446 pp.

Ludwig, W., Amiotte-Suchet, P. and Probst, J.-L. (1996) River discharges of carbon to the world's oceans: determining local inputs of alkalinity and of dissolved and particulate organic carbon. *C.R. Academie Science Paris*, **323**(Serie IIa), 1007–1014.

Mackenzie, D. (1995) The cod that disappeared. *New Scientist* (16th September), 24–29.

MacKenzie, F.T., Lerman, A. and Ver, L.M.B. (1998) Role of the continental margin in the global carbon balance during the past three centuries. *Geology*, **26**(5), 423–426.

Manabe, S. and Stouffer, R.J. (1993) Century-scale effects of increased atmospheric CO_2 on the ocean-atmosphere system. *Nature*, **364**, 215–218.

Marincovich, L. and Gladenkov, A. (1999) Evidence for an early opening of the Bering Strait. *Nature*, **397**, 149–151.

Markun, C.D., Randazzo, A.F. and Simonson, B.M. (1988) Early silica cementation and subsequent diagensis in arenites from four early Proterozoic iron formations of North America; discussion and reply. *Journal of Sedimentary Petrology*, **58**, 544–549.

Martin, J.H. (1992). Iron as a limiting factor. In: *Primary Productivity and Biogeochemical Cycles in the Sea*. (ed. Falkowski, P.G. and Woodhead, A.D.). Plenum, New York, pp.123–137.

Martin, R.E. (1995) Cyclic and secular variation in microfossil biomineralisation: clues to the biogeochemical evolution of Phanerozoic oceans. *Global and Planetary Change*, **11**, 1–23.

Martin, R.E. (1996) Secular increase in nutrient levels through the Phanerozoic: implications for productivity, biomass, and diversity of the marine biosphere. *Palaios*, **11**, 209–219.

Maslin, M. (1998) Equatorial western Atlantic Ocean circulation changes linked to the Heinrich events: deep-sea sediment evidence from the Amazon Fan. In: *Geological Evolution of Ocean Basins: Results from the Ocean Drilling Programme*. (ed. Cramp, A., MacLeod, C.J., Lee, S.V. and Jones, E.J.W.). *Geological Society Special Publication* No. 131. The Geological Society, London, pp.111–127.

May, R.M. (ed.) (1976) *Theoretical Ecology; Principles and Applications*. Blackwell Scientific, Oxford, 317 pp.

McArthur, A.G. and Tunnicliffe, V. (1998) Relics and antiquity revisited in the modern vent fauna. In: *Modern Ocean Floor Processes and the Fossil Record*. (ed. Mills, R.A. and Harrison, K.). *Geological Society Special Publication* No.148. The Geological Society, London, pp.271–291.

McCabe, A.M. and Clark, P.U. (1998) Ice-sheet variability around the North Atlantic Ocean during the last deglaciation. *Nature*, **392**, 373–377.

McCall, J. (1996) The early history of the Earth. *Geoscientist*, **6**(1), 10–14.

McKerrow, W.S., Scotese, C.R. and Brasier, M.D. (1992) Early Cambrian continental reconstructions. *Journal of the Geological Society, London*, **149**(4), 599–606.

McMahon, N.A. and Turner, J. (1998) The documentation of a latest Jurassic–earliest Cretaceous uplift throughout southern England and adjacent offshore areas. In: *Development, Evolution and Petroleum Geology of the Wessex Basin*. (ed. Underhill, J.R.). *Geological Society Special Publication* No. 133. The Geological Society, London, pp.215–240.

Melchin, M.J. (1996) Graptolite diversity, survivorship and sea level change through the late Ashgill, Llandovery and Wenlock in Arctic Canada. *The James Hall Symposium: Second International Symposium of the Silurian System*, Rochester, New York, USA, pp.75.

Melchin, M.J. and Mitchell, C.E. (1991) Late Ordovician extinction in the Graptoloidea. In: *Advances in Ordovician Geology* (eds Barnes, C. and Williams, S.H.). Paper 90–9, Geological Survey of Canada, pp. 143–156.

Mil, H.-S. and Grossman, E.L. (1994) Late Pennsylvanian seasonality reflected in the ^{18}O and elemental composition of a brachiopod shell. *Geology*, **22**, 661–664.

Miller, S.L. (1992) The prebiotic synthesis of organic compounds as a step toward the origin of life. In: *Major Events in the History of Life*. (ed. Schopf, J.W.). Jones and Bartlett, Boston, pp.1–28.

Mojzsis, S.J. *et al.* (1996) Evidence for life on Earth before 3800 million years ago. *Nature*, **384**, 55–59.

Monster, J., Appel, P.W.U., Thode, H.G., Schidlowski, M., Carmichael, C.M. and Bridgwater, D. (1979) Sulphur isotope studies in early Archaean sediments from Isua, West Greenland: implications for the antiquity of bacterial sulphate reduction. *Geochimica et Cosmochimica acta*, **43**, 405–413.

Moore, G.T., Hayashida, D.N. and Ross, C.A. (1993) Late Early Silurian (Wenlockian) general circulation model-generated upwelling, graptolitic black shales, and organic-rich source rocks – an accident of plate tectonics? *Geology*, **21**, 17–20.

Morse, J.W. and MacKenzie, F.T. (1998) Hadean ocean carbonate geochemistry. *Aquatic Geochemistry*, **4**, 301–319.

Morse, J.W. and Wang, Q. (1997) Influences of temperature and Mg:Ca ratio on $CaCO_3$ precipitates from seawater. *Geology*, **25**(1), 85–87.

Motluk, A. (1995) Deadlier than the harpoon? *New Scientist* (1st July), 12–13.

Murray, J.W. (1992) The oceans. In: *Global Biogeochemical Cycles*. (ed. Butcher, S.S., Charlson, R.J., Orians, G.H. and Wolfe, G.V.). Academic Press, London, pp.175–212.

Murray, J.W. (1995) Microfossil indicators of ocean water masses, circulation and climate. In: *Marine Palaeoenvironmental Analysis from Fossils*. (ed. Bosence, D.W.J. and Allison, P.A.). *Geological Society Special Publication* No. 83. The Geological Society, London, pp.245–264.

Nadelhoffer, K. *et al.* (1999) Nitrogen deposition makes a minor contribution to carnon sequestration in temperate forests. *Nature*, **398**, 145–148.

Nadirov, R.S., Bagirov, E., Tagiyev, M. and Lerche, I. (1997) Flexural plate subsidence, sedimentation rates, and structural development of the superdeep South Caspian Sea. *Marine and Petroleum Geology*, **14**(4), 383–400.

Nelson, A.R., Shennan, I. and Long, A.J. (1996) Identifying coseismic subsidence in tidal-wetland stratigraphic sequences at the Cascadia subduction zone of western North America. *Journal of Geophysical Research*, **101**(B3), 6115–6135.

New, A. (1998) Oceans and climate. *NERC News*, Spring, 16–17.

Nicholas, C.J. (1996) The Sr isotope evolution of the oceans during the 'Cambrian Explosion'. *Journal of The Geological Society, London*, **153**, 243–254.

Nisbet, E. (1995) Archaean ecology: a review of evidence for the early development of bacterial biomes, and speculations on the development of a global-scale biosphere. In: *Early Precambrian Processes*. (ed. Coward, M.P. and Ries, A.C.). *Geological Society Special Publication* No. 95, The Geological Society, London, pp.27–51.

Nisbet, E.G. (1987) *The Young Earth; An Introduction to Archaean Geology*. Allen & Unwin, Boston, 402 pp.

Ohmoto, H. (1996) Evidence in pre-2.2.Ga palaeosols for the early evolution of atmospheric oxygen and terrestrial biota. *Geology*, **24**, 1135–1138.

Owen, A. and Robertson, D.B.R. (1995) Ecological changes during the end-Ordovician extinction. *Modern Geology*, **20**(1), 21–39.

Özsoy, E. and Ünlüata, Ü. (1997) Oceanography of the Black Sea: a review of some recent results. *Earth Science Reviews*, **42**, 231–272.

Pedley, M. (1996) Miocene reef distributions and their associations in the central Mediterranean region: an overview. *SEPM Concepts in Sedimentology and Palaeontology*, **5**, 73–87.

Pinet, P.R. (1992) *Oceanography: An Introduction to the Planet Oceanus*. West Publishing Company, St Paul, MN, 571 pp.

Pirazzoli, P.A. (1996) *Sea-level Changes; The Last 20,000 Years. Coastal Morphology and Research*. John Wiley and Sons, Chichester, 211 pp.

Pirazzoli, P.A., Mastronuzzi, G., Saliège, J.F. and Sansò, P. (1997) Late Holocene emergence in Calabria, Italy. *Marine Geology*, **141**, 61–70.

Poulsen, C.J., Seidov, D., Barron, E.J. and Peterson, W.H. (1998) The impact of palaeogeographic evolution on the surface oceanic circulation and the marine environment within the mid-Cretaceous Tethys. *Paleoceanography*, **13**(5), 546–559.

Price, G.D., Valdes, P.J. and Sellwood, B.W. (1997) Quantitative palaeoclimate GCM validation: Late Jurassic and mid-Cretaceous case studies. *Journal of the Geological Society, London*, **154**, 769–772.

Price, G.D., Sellwood, B.W., Corfield, R.M., Clarke, L. and Cartlidge, J.E. (1998) Isotopic evidence for palaeotemperatures and depth stratification of Middle Cretaceous planktonic foraminifera from the Pacific Ocean. *Geological Magazine*, **135**(2), 183–191.

Price, J.F. (1992) Overflows: the source of new abyssal ocean waters. *Oceanus*, **35**, 28–34.

Pudsey, C.J. and Howe, J.A. (1998) Quaternary history of the Antarctic Circumpolar Current evidence from the Scotia Sea. *Marine Geology*, **148**, 83–112.

Rabalais, N.N. *et al.* (1996) Nutrient changes in the Mississippi river and system responses on the adjacent continental-shelf. *Estuaries*, **19**, 386–407.

Rabalais, N.N. *et al.* (1998) Consequences of the 1993 Mississippi river flood in the Gulf of Mexico. *Regulated Rivers – Research and Management*, **14**, 161–177.

Rahm, L. (1987) Oxygen consumption in the Baltic proper. *Limnology and Oceanography*, **32**, 973–978.

Railsback, L.B. (1993) Original mineralogy of Carboniferous worm tubes: evidence for changing marine chemistry and biomineralization. *Geology*, **21**, 703–706.

Ramsbottom, W.H.C. (1973) Transgressions and regressions in the Dinantian: a new synthesis of British Dinantian stratigraphy. *Proceedings of the Yorkshire Geological Society*, **39**, 567–607.

Ramsbottom, W.H.C. (1977) Major cycles of transgression and regression (Mesothems) in the Namurian. *Proceedings of the Yorkshire Geological Society*, **41**, 261–291.

Read, W.A. (1991) The Millstone Grit (Namurian) of the southern Pennines viewed in the light of eustatically controlled sequence stratigraphy. *Geological Journal*, **26**, 157–165.

Reasoner, M. and Jodry, M. (2000) Rapid response of alpine timberline vegetation to the Younger Dryas climate oscillation in the Colorado Rocky Mountains, USA. *Geology*, **28**, 51–54.

Riding, R. (1982) Cyanophyte calcification and changes in ocean chemistry. *Nature*, **299**, 814–815.

Riding, R. (1992) The algal breath of life. *Nature*, **359**, 13–14.

Riding, R. and Awramik, S. (2000) *Microbial Sediments*. Springer-Verlag, Berlin.

Riding, R., Braga, J.C., Martin, J.M. and Sánchez-Almazo, I.M. (1998) Mediterranean Messinian salinity crisis: constraints from a coeval marginal basin, Sorbas, SE Spain. *Marine Geology*, **146**,

Roberts, N. (1998) *The Holocene*. Blackwell, Oxford.

Roether, W. *et al.* (1996) Recent changes in eastern Mediterranean deep waters. *Science*, **271**, 333–335.

Rohling, E.J. *et al.* (1998) Abrupt cold spells in the northwest Mediterranean. *Paleoceanography*, **13**, 316–322.

Rosell-Mele, A. and Koc, N. (1997) Paleoclimate significance of the stratigraphic occurrence of photosynthetis biomarker pigments in the Nordic seas. *Geology*, **25**(1), 49–52.

Rothwell, R.G., Thomson, J. and Kähler, G. (1998) Low-sea-level emplacement of a very large Late Pleistocene 'megaturbidite' in the western Mediterranean Sea. *Nature*, **392**, 377–379.

Rouchy, J.M. and Saint-Martin, J.P. (1992) Late Miocene events in the Mediterranean as recorded by carbonate–evaporite relations. *Geology*, **20**, 629–632.

Royal Society (1993) Understanding the North Sea system. *Meteorological Magazine*, **122**, 162–166.

Runnegar, B. (1982) Oxygen requirements, biology and phylogenetic significance of the late Precambrian worm *Dickinsonia*, and the evolution of the burrowing habit. *Alcheringa*, **6**, 223–239.

Russell, M.J. and Hall, A.J. (1997) The emergence of life from iron monosulphide bubbles at a submarine hydrothermal redox and pH front. *Journal of the Geological Society, London*, **154**, 377–402.

Russell, M.J., Hall, A.J. and Turner, D. (1990) *In vitro* growth of iron sulphide chimneys: Possible culture chambers for origin-of-life experiments. *Terra Nova*, **1**, 238–241.

Rust, D and Kershaw, S. (2000) Holocene tectonic uplift patterns in northeastern Sicily: evidence from marine notches in coastal outcrops. *Marine Geology* (in press).

Rye, R., Kuo, P.H. and Holland, H.D. (1995) Atmospheric carbon dioxide concentrations before 2.2 billion years ago. *Nature*, **378**, 603–605.

Safina, C. (1995) The world's imperiled fish. *Scientific American*, November, 30–37.

Sandberg, P.A. (1983) An oscillating trend in Phanerozoic non-skeletal carbonate mineralogy. *Nature*, **305**, 19–22.

Schidlowski, M. (1988) A 3800-million-year isotopic record of life from carbon in sedimentary rocks. *Nature*, **333**, 313–318.

Scholle, P.A., Bebout, D.G. and Moore, C.H. (eds) (1983) *Carbonate Depositional Environments*. American Association of Petroleum Geologists, Tulsa, OK, 708 pp.

Schopf, J.W. (1992a) Evolution of the Proterozoic biosphere: benchmarks, tempo and mode. In: *The Proterozoic Biosphere*. (ed. Schopf, J.W. and Klein, C.). Cambridge University Press, Cambridge, pp.583–599.

Schopf, J.W. (ed.) (1992b) *Major Events in the History of Life*. Jones and Bartlett, Boston, 190 pp.

Schopf, J.W. (1994) The oldest known records of life: Early Archaean stromatolites, microfossils and organic matter. In: *Early Life on Earth*. (ed. Bengtson, S.). Columbia University Press, New York, pp.193–206.

Schultz, H., von Rad, U. and Erlenkeuser, H. (1998) Correlation between Arabian Sea and Greenland climate oscillations of the past 110,000 years. *Nature*, **393**, 54–57.

Schultz, H., von Rad, U. and von Stackelberg, U. (1996) Laminated sediments from the oxygen-minimum zone of the northeastern Arabian Sea. In: *Palaeoclimatology and Palaeoceanography from Laminated Sediments*. (ed. Kemp, A.E.S.). *Geological Society Special Publication* No.116. The Geological Society, London, pp.185–207.

Segar, D. (1998) *Introduction to Ocean Sciences* (with Infotrac) 1st edn. Brooks/Cole Publishing, Thomson Learning, CA, USA.

Seilacher, A. (1992) Vendobionta and Psammocorallia: lost constructions of Precambrian evolution. *Journal of the Geological Society, London*, **149**(4), 607–614.

Sheehan, P.M. and Coorough, P.J. (1990) Brachiopod zoogeography across the Ordovician-Silurian extinction event. In: *Palaeozoic Palaeogeography and Biogeography*. (ed. McKerrow, W.S. and Scotese, C.R.). *Geological Society of London Memoir*. The Geological Society, London, pp.181–187.

Siegenthaler, U. and Sarmiento, J.L. (1993) Atmospheric carbon dioxide and the ocean. *Nature*, **365**, 119–125.

Simkiss, K. (1989) Biomineralisation in the context of geological time. *Transactions of the Royal Society of Edinburgh: Earth Sciences*, **80**, 193–199.

Simó, R. and Pedrós-Alló, C. (1999) Role of vertical mixing in controlling the oceanic production of dimethyl sulphide. *Nature*, **402**, 396–399.

Simonson, B.M. (1985) Sedimentological constraints on the origins of Precambrian iron-formations. *Geological Society of America Bulletin*, **96**, 244–252.

Simonson, B.M. (1987) Early silica cementation and subsequent diagenesis in arenites from four early Proterozoic iron formations of North America. *Journal of Sedimentary Petrology*, **57**(3), 494–511.

Simonson, B.M. (1996) Was the deposition of large Precambrian Iron Formations linked to major marine transgressions? *The Journal of Geology*, **104**, 665–676.

Simonson, B.M. and Lanier, W.P. (1987) Early silica cementation and microfossil preservation in cavities in iron-formation stromatolites, Early Proterozoic of Canada. In: *Precambrian Iron-Formations*. (ed. Appel, P.W.U. and LaBerge, G.L.). Theophrastus Publications, Athens, Greece, pp.187–213.

Skelton, P., Spicer, R. and Rees, A. (1997) *Evolving Life and the Earth. S269 Earth and Life*. The Open University, Milton Keynes, UK, 199 pp.

Skelton, P.W. (ed.) (1993) *Evolution: A Biological and Palaeontological Approach*. Addison-Wesley, New York, 1064 pp.

Skoog, S.Y., Venn, C. and Simpson, E.L. (1994) Distribution of *Diatpatra cuprea* across modern tidal flats: implications for *Skolithos*. *Palaios*, **9**, 188–201.

Smith, J.E. *et al.* (1997) Rapid climate change in the North Atlantic during the Younger Dryas recorded by deep-sea corals. *Nature*, **386**, 818–820.

Soria, J.M. Fernández, J. and Viseras, C. (1999) Late Miocene stratigraphy and palaeogeographic evolution of the intramoutane Guadix Basin (Central Betic Cordillera, Spain): implications for an Atlantic–Mediterranean connection. *Palaeogeography, Palaeoclimatology, Palaeoecology*, **151**, 255–266.

Spencer, T. (1995) Potentialities, uncertainties and complexities in the response of coral reefs to future sea-level rise. *Earth Surface Processes and Landforms*, **20**, 49–64.

Spray, J.G., Kelley, S.P. and Rowley, D.B. (1998) Evidence for a Late Triassic multiple impact event on Earth. *Nature*, **392**, 171–173.

Stanley, S.M. and Hardie, L.A. (1998) Secular oscillations in the carbonate mineralogy of reef-building and sediment-producing organisms driven by tectonically forced shifts in seawater chemistry. *Palaeogeography, Palaeoclimatology, Palaeoecology*, **144**, 3–19.

Stanley, S.M. and Hardie, L.A. (1999) Hypercalcification: palaeontology links plate tectonics and geochemistry to sedimentology. *GSA Today*, **9**(2), 1–7.

Stauffer, B. *et al.* (1998) Atmospheric CO_2 concentration and millennial-scale climate change during the last glacial period. *Nature*, **392**, 59–61.

Stearn, C.W., Scoffin, T.P. and Martindale, W. (1977) Calcium carbonate budget of a fringing reef on the west coast of Barbados. Part I – zonations and productivity. *Bulletin of Marine Science*, **27**, 479–510.

Steuber, T. (1996) Stable isotope sclerochronology of rudist bivalves: growth rates and Late Cretaceous seasonality. *Geology*, **24**(4), 315–318.

Stewart, I.S., Cundy, A., Kershaw, S. and Firth, C. (1997) Holocene coastal uplift in the Taormina area, northeastern Sicily: implications for the southern prolongation of the Calabrian seismogenic belt. *Journal of Geodynamics*, **24**, 37–50.

Stigebrandt, A. (1991) Computations of oxygen flux through the sea surface and the net production of organic matter with application to the Baltic and adjacent seas. *Limnology and Oceanography*, **36**, 444–454.

Strong, G.E. *et al.* (1994) The petrology and diagenesis of Permo-Triassic rocks of the Sellafield area, Cumbria. *Proceedings of the Yorkshire Geological Society*, **50**(1), 77–89.

Stuiver, M., Quay, P.D. and Ostlund, H.G. (1983) Abyssal water carbon-14 distribution and the age of the world oceans. *Science*, **219**, 849–851.

Sykes, T.J.S., Ramsay, A.T.S. and Kidd, R.B. (1998) Southern hemisphere Miocene bottom-water circulation: a palaeobathymetric analysis. In: *Geological Evolution of Ocean Basins: Results from the Ocean Drilling Programme.* (ed. Cramp, A., MacLeod, C.J., Lee, S.V. and Jones, E.J.W.). *Geological Society Special Publication* No. 131. The Geological Society, London, pp.43–54.

Tassell, J.V. (1994) Evidence for orbitally-driven sedimentary cycles in the Devonian Catskill Delta Complex. In: *Tectonic and Eustatic Controls on Sedimentary Cycles. Concepts in Sedimentology and Palaeontology*. (ed. Dennison, J.M. and Ettensohn, F.R.). SEPM, Tulsa, OK, pp.121–131.

Taylor, K.C. *et al.* (1993) The 'flickering switch' of late Pleistocene climate change. *Nature*, **361**, 432–436.

Taylor, P.D. and Allison, P.A. (1998) Bryozoan carbonates through time and space. *Geology*, **26**, 459–462.

Timmerman, A. *et al.* (1999) Increased El Niño frequency in a climate model forced by future greenhouse warming. *Nature*, **398**, 694–697.

Tobin, K.J. and Walker, K.R. (1997) Ordovician oxygen isotopes an palaeotemperatures. *Palaeogeography, Palaeoclimatology, Palaeoecology*, **129**, 269–290.

Tortell, P.D., Maldonado, M.T. and Price, N.M. (1996) The role of heterotrophic bacteria in iron-limited ocean ecosystems. *Nature*, **383**, 330–332.

Trompette, R. (1996) Temporal relationship between cratonization and glaciation: the Vendian–early Cambrian glaciation in Western Gondwana. *Palaeogeography, Palaeoclimatology, Palaeoecology*, **123**, 373–383.

Tucker, M.E. (1992) The Precambrian–Cambrian boundary: seawater chemistry, ocean circulation and nutrient supply in metazoan evolution, extinction and biomineralisation. *Journal of the Geological Society, London*, **149**(4), 655–668.

Tucker, M.E. and Wright, V.P. (1990) *Carbonate Sedimentology*. Blackwell Scientific, Oxford, 482 pp.

Valdes, P.J., Sellwood, B.W. and Price, G.D. (1995) Modelling Late Jurassic Milankovitch climate variations. In: *Orbital Forcing Timescales and Cyclostratigraphy*. (ed. House, M.R. and Gale, A.S.). *Geological Society of London Special Publication* No. 85. The Geological Society, London, pp.115–132.

van Andel, T.H. (1994) *New Views on an Old Planet; A History of Global Change*. Cambridge University Press, Cambridge, 439 pp.

Veizer, J. *et al.* (1997) Strontium isotope stratigraphy: potential resolution and event correlation. *Palaeogeography, Palaeoclimatology, Palaeoecology*, **132**, 65–77.

Villareal, T.A. *et al.* (1999) Upward transport of oceanic nitrate by migrating diatom mats. *Nature*, **397**, 423–426.

Wang, K., Chatterton, B.D.E., Attrep Jr, M. and Orth, C.J. (1992) Iridium abundance maxima at the latest Ordovician mass extinction horizon, Yangtze Basin, China: terrestrial or extraterrestrial? *Geology*, **20**, 39–42.

Wang, K. *et al.* (1993) The great latest Ordovician extinction on the South China Plate: chemostratigraphic studies of the Ordovician–Silurian boundary interval on the Yangtze Platform. *Palaeogeography, Palaeoclimatology, Palaeoecology*, **104**, 61–79.

Wang Xiaofeng and Chai, Z. (1990) Ordovician–Silurian boundary extinction and its relationship to iridium and carbon isotope anomalies. *Acta Geologica Sinica*, **3**(1), 81–92.

Waterhouse, H.K. (1995) High-resolution palynofacies investigation of Kimmeridgian sedimentary cycles. In: *Orbital Forcing Timescales and Cyclostratigraphy*. (ed. House, M.R. and Gale, A.S.). *Geological Society of London Special Publication* No. 85. The Geological Society, London, pp.75–114.

Waterhouse, H.K. (1999) Orbital forcing of palynofacies in the Jurassic of France and the United Kingdom, *Geology*, **27**, 511–514.

Wells, M.L., Vallis, G.K. and Silver, E.A. (1999) Tectonic processes in Papua New Guinea and past productivity in the eastern equatorial Pacific Ocean. *Nature*, **398**, 601–604.

Wells, N. (1997) *The Atmosphere and Ocean.* John Wiley and Sons, Chichester, 394 pp.

West, R.R., Archer, A.W. and Miller, K.B. (1997) The role of climate in stratigraphic patterns exhibited by late Palaeozoic rocks exposed in Kansas. *Palaeogeography, Palaeoclimatology, Palaeoecology,* **128**, 1–16.

Wignall, P. (1992) The day the world nearly died. *New Scientist* (25 January), 51–55.

Wignall, P. (1994) *Black Shales. Oxford Monographs on Geology and Geophysics.* Clarendon Press, Oxford, 126 pp.

Wignall, P. and Hallam, A. (1992) Anoxia as a cause of the Permian/Triassic mass extinction: facies evidence from northern Italy and the western United States. *Palaeogeography, Palaeoclimatology, Palaeoecology,* **93**, 21–46.

Wilkin, R.T., Arthur, M.A. and Dean, W.E. (1997) History of water-column anoxia in the Black Sea indicated by pyrite framboid size distribution. *Earth and Planetary Science Letters,* **148**, 517–525.

Wilson, M. (1997) Thermal evolution of the central Atlantic passive margins: continental break-up above a Mesozoic super-plume. *Journal of The Geological Society of London,* **154**, 491–495.

Windley, B.F. (1995) *The Evolving Continents.* John Wiley & Sons, New York, 526 pp.

Woese, C.R. and Wächtershäuser, G. (1990) Origin of life. In: *Palaeobiology: A Synthesis.* (ed. Briggs, D.E.G. and Crowther, P.R.). Blackwell Scientific, Oxford, pp.3–9.

Wold, C.N. (1995) Palaeobathymetric reconstruction on a gridded database: the northern North Atlantic and southern Greenland-Iceland-Norwegian sea. In: *The Tectonics, Sedimentation and Palaeoceanography of the North Atlantic Region.* (ed. Scrutton, R.A., Stoker, M.S., Shimmield, G.B. and Tudhope, A.W.). *Geological Society Special Publication* No. 90. The Geological Society, London, pp.271–302.

Wright, V.P. and Burchette, T.P. (eds) (1998) *Carbonate Ramps. Geological Society Special Publication* No. 149. The Geological Society, London, 465 pp.

Wunsch, C. (1992) Observing ocean circulation from space. *Oceanus,* **33**(2), 9–17.

Zhang, T., Kershaw, S., Wan, Y. and Lan, G. (2000) Geochemical and facies evidence for palaeoenvironmental change during the Late Ordovician Hirnantian glaciation in south Sichuan Province, China. *Global and Planetary Change,* **24**, 133–152

Zhuravlev, A.Y. and Wood, R.A. (1996) Anoxia as the cause of the mid-Early Cambrian (Botomian) extinction event. *Geology,* **24**(4), 311–314.

INDEX

Numbers in **bold** type refer to figures and those in *italics* to chapter sections